ADVANCES ON COMPUTER MATHEMATICS AND ITS APPLICATIONS

ADVANCES ON COMPUTER MATHEMATICS AND ITS APPLICATIONS

edited by

Elias A. Lipitakis

Department of Informatics
Athens University of Economics
and Business

World Scientific
Singapore • New Jersey • London • Hong Kong

Published by

World Scientific Publishing Co Pte Ltd
5 Toh Tuck Link, Singapore 596224
USA office: 27 Warren Street, Suite 401-402, Hackensack, NJ 07601
UK office: 57 Shelton Street, Covent Garden, London WC2H 9HE

British Library Cataloguing-in-Publication Data
A catalogue record for this book is available from the British Library.

ADVANCES ON COMPUTER MATHEMATICS AND ITS APPLICATIONS

ISBN-13 978-981-02-1292-6
ISBN-10 981-02-1292-5

Foreword

The branch of Mathematics and Informatics called *Computer Mathematics* is mainly intended for elaborating efficient computational methods for solving numerically a very wide range of scientific problems and effective ways of using computer systems for this purpose. It should be noted that problems encountered in Computer Mathematics result from the most diverse fields of Science, Engineering, Economics and other disciplines.

There is currently considerable research activity in Computer Mathematics with an increasing emphasis towards the provision of efficient computational methods for a wide range of interesting applications. The scientific community requires Computer Mathematics to enable it to take advantage of the increasing computing power offered by the multiprocessor and highly parallel computer systems. The changes taking place in computer technology are partially reflected in the scope of this book, offering a cross section of ongoing research and applications in this area of scientific computing. Computer Mathematics will play undoubtedly an important role in the predominant contribution of the three *key-sciences*, i.e., Mathematics – Informatics – Economy, in the diffusion of scientific knowledge and in the evolution and progress of the human race in the 21st century.

The main purpose of this volume is to present characteristic representative samples of the current trends in various directions of research being done in the area of Computer Mathematics and its applications. The reader will easily find out from the works and the given references of the book, where the scientific efforts of many researchers all over the world may lead to. The topics covered here encompass different aspects of the subject.

This volume contains selected articles presented in the HERMIS '92 Conference by distinguished researchers. For the two days, September 25–26, 1992, at the Athens University of Economics (AUE), Athens, over 200 scientists attended the first Hellenic Conference on Mathematics and Informatics (HERMIS '92). The Conference HERMIS (HEllenic Research on Mathematics and InformaticS) was jointly organized by the Hellenic Mathematical Society (HMS) and the Department of Informatics of AUE. The main theme within the HERMIS '92 Conference was Computer Mathematics and its applications, with special emphasis on Computational Mathematics. Several other related topics were also included in the programme that made the Conference of interest to a wide audience of Mathematicians, Computer Scientists, Economists and many others who wished to become acquainted with the underlying basic ideas and the potential applications of its subjects.

This volume also includes selected articles of the recent HMS Bulletin special issues on Computer Mathematics. All the selected articles, each of which being devoted to a specialized, but to some extend independent topic, could be considered that form a scientific research *anthology* of representative samples in various directions of the research being done in Computer Mathematics.

Elias A. Lipitakis
Athens, Hellas
November 1992

Contents

Log barrier methods for semi-infinite programming calculations[1]

M.J.D. Powell

Abstract: We consider the application of several log barrier methods to some test problems that are derived from nonlinearly constrained optimization calculations in only 2 or 4 variables, where each constraint is replaced by several supporting tangent planes. Thus we investigate the efficiency of the methods for up to 10000 linear constraints, our conclusions being drawn from numerical results. The algorithms include the classical log barrier method, Karmarkar's original algorithm and Polyak's shifted log barrier method that employs estimates of Lagrange multipliers. We recommend the use of line searches along directions that are generated by the Newton–Raphson formula. Our results expose some weaknesses of the methods. Therefore they may be helpful to the development of new algorithms that are particularly suitable for semi-infinite programming calculations.

Department of Applied Mathematics and Theoretical Physics,
University of Cambridge,
Silver Street,
Cambridge CB3 9EW,
England.

[1]Presented at the Hellenic Research Conference on Mathematics and Informatics (Athens, September, 1992)

1. Introduction

We consider linear programming problems of the form

$$\left.\begin{array}{lll} \text{minimize} & \underline{c}^T\underline{x}, & \underline{x}\in\mathcal{R}^n, \\ \text{subject to} & \underline{a}_j^T\underline{x}\geq b_j, & j=1,2,\ldots,m, \end{array}\right\} \qquad (1.1)$$

where m is much larger than n. Let $\mathcal{S}$ be the set of vectors $\underline{x}$ that satisfy the constraints. We study some cases where the general shape of $\mathcal{S}$ is that it is bounded by approximations to smooth curved surfaces, so one has to zoom in on fine detail in order to observe the true piecewise linear character of the boundary. Thus edges that join adjacent vertices of $\mathcal{S}$ are relatively short, which causes the simplex method for linear programming to require a large number of iterations.

Interior point methods are designed to be suitable for such cases, because the main idea is that one generates vectors $\underline{x}$ that stay away from the boundary of $\mathcal{S}$ until they approach the required solution. Thus one expects the piecewise linear structure of the boundary to cause hardly any deterioration in efficiency for most of the calculation. It can happen, unfortunately, that this hope is not achieved. Indeed, the algorithm of Karmarkar (1984) employs about $m/8$ iterations when $n=2$ and the constraints of expression (1.1) are equally spaced tangents to the unit circle (Powell, 1991). We investigate numerically whether similar inefficiencies occur for some other semi-infinite linear programming problems. Our experiments include some tests with a "small-step" log barrier method that has the property that the number of iterations is bounded above by a multiple of $m^{1/2}$ (Gonzaga, 1992).

Therefore brief descriptions of log barrier methods and Karmarkar's algorithm are given in Sections 2 and 3 respectively. Then four test problems and some numerical results are presented in Section 4. We observe a phenomenon that was discussed by Wright (1992), namely that it is usual for some "large-step" log barrier methods to pick trial changes to the variables that overshoot the boundary of the feasible region. A useful remedy is to employ a line search when generating a new vector of variables, and it is shown in Section 5 that this enhancement requires hardly any extra computation on each iteration. Section 6 considers the interesting variant of the classical log barrier method that is recommended by Polyak (1992a). Finally, our results are reviewed in Section 7, and there it is noted that the TOLMIN software (Powell, 1989) requires fewer iterations for the test problems of Section 4 than the given calculations that apply log barrier algorithms.

The efficiency of interior point methods for semi-infinite programming is also addressed by Todd (1991). He studies the interesting question whether a method requires more iterations if m is increased to $2m$ by including each constraint of expression (1.1) twice.

2. Log barrier methods

We assume that the feasible region $\mathcal{S}$ of the linear programming problem (1.1) is bounded and has a nonempty interior, and that a point, $\underline{x}_1$ say, is given such that all the constraint residuals $\{\underline{a}_j^T \underline{x}_1 - b_j : j=1,2,\ldots,m\}$ are positive. Then a typical log barrier method for calculating an approximate solution of the problem is as follows.

Step 0: Set the parameter σ to a suitable positive real number and set $k=1$.

Step 1: Calculate $\underline{x}_{k+1}$ by applying an algorithm for unconstrained minimization to the function

$$\Phi_\sigma(\underline{x}) = \sigma\, \underline{c}^T \underline{x} - \sum_{j=1}^{m} \log\,(\underline{a}_j^T \underline{x} - b_j), \quad \underline{x} \in \mathcal{S}, \tag{2.1}$$

where $\underline{x}_k$ is the starting point of this calculation.

Step 2: Accept $\underline{x}_{k+1}$ as a sufficiently accurate approximation to the solution of the problem (1.1) if σ satisfies the condition

$$\sigma \geq m/\epsilon, \tag{2.2}$$

where ϵ is a prescribed positive tolerance on the difference between $\underline{c}^T \underline{x}_{k+1}$ and the optimal value of the objective function $\underline{c}^T \underline{x}$.

Step 3: Otherwise increase σ, replace k by $k+1$, and go back to Step 1.

Next we specify the method of this type that is applied in the numerical experiments of Section 4. We include an important characteristic of log barrier methods, namely that usually the unconstrained algorithm of Step 1 makes only a few changes to the variables when generating $\underline{x}_{k+1}$ from $\underline{x}_k$, even if σ is very large. Therefore a suitable criterion for ending the unconstrained calculation is required, and Step 3 increases σ gradually. Further, the initial choice of σ is guided by the remark that it is sensible to make $\sigma\,\|\underline{c}\|$ comparable to the norms of the gradients at $\underline{x} = \underline{x}_1$ of the terms under the summation sign of expression (2.1), the vector norm being Euclidean. Therefore we impose the constraints

$$\sigma\,\|\underline{c}\| \geq \|\underline{a}_j\| \,/\, (\underline{a}_j^T \underline{x}_1 - b_j), \quad j = 1,2,\ldots,m, \tag{2.3}$$

on the initial σ, and we take up the remaining freedom by making the initial gradient norm $\|\underline{\nabla}\Phi_\sigma(\underline{x}_1)\|$ as small as possible.

The boundedness of $\mathcal{S}$ implies that the vectors $\{\underline{a}_j : j=1,2,\ldots,m\}$ span $\mathcal{R}^n$. Thus it is straightforward to deduce analytically that the second derivative matrix of the function (2.1) is positive definite at every interior point of $\mathcal{S}$. Further, n

is small in our experiments. Therefore it is inexpensive to calculate the Newton–Raphson step of the unconstrained calculation of Step 1. Consequently, each change to the variables during Step 1 has the form

$$\underline{x} \leftarrow \underline{x} - \alpha \left[\nabla^2 \Phi_\sigma(\underline{x})\right]^{-1} \underline{\nabla}\Phi_\sigma(\underline{x}), \tag{2.4}$$

where α is a positive steplength. We require the new $\underline{x}$ to be an interior point of $\mathcal{S}$. Therefore we define $\bar{\alpha}$ to be the positive value of α such that the vector on the right hand side of expression (2.4) is on the boundary of $\mathcal{S}$, and we use the steplength

$$\alpha = \min\left[1,\, 0.9\,\bar{\alpha}\right] \tag{2.5}$$

in the log barrier calculations of Section 4.

Some analysis in Gonzaga (1992) and Powell (1992a) shows that it is suitable to terminate the minimization calculation of Step 1 when the difference between $\Phi_\sigma(\underline{x})$ and the least value of Φ_σ is less than a constant from the interval $[0.1, 0.3]$. Now the quadratic Taylor series approximation to $\Phi_\sigma(\underline{x})$ that gives the Newton–Raphson step of the previous paragraph also provides an estimate of this difference. Therefore Step 1 adjusts the variables recursively by applying formulae (2.4) and (2.5) until the condition

$$[\underline{\nabla}\Phi_\sigma(\underline{x})]^T \left[\nabla^2 \Phi_\sigma(\underline{x})\right]^{-1} \underline{\nabla}\Phi_\sigma(\underline{x}) \le \tfrac{1}{4} \tag{2.6}$$

is satisfied, and then we let $\underline{x}_{k+1}$ be the current value of $\underline{x}$. Although the choice (2.5) does not guarantee that inequality (2.6) will hold eventually, the convergence of the unconstrained calculations was found to be adequate in all the experiments of Section 4.

It is also proved by Gonzaga (1992) and Powell (1992a) that the test (2.2) is appropriate for accepting $\underline{x}_{k+1}$ as a solution to the linear programming problem. Their procedure, however, includes an extra cycle through Step 3, in order to compensate for the fact that $\underline{x}_{k+1}$ is usually not the exact minimizer of the function (2.1). We ignore this complication, however, and justify condition (2.2) under the assumption that the unconstrained minimization of Step 1 is exact.

In this case $\underline{\nabla}\Phi_\sigma(\underline{x}_{k+1})$ is zero, so the definition (2.1) implies the equation

$$\sigma\,\underline{c} = \sum_{j=1}^{m} \frac{\underline{a}_j}{\underline{a}_j^T \underline{x}_{k+1} - b_j}. \tag{2.7}$$

Let $\underline{x}^*$ be a solution of the problem (1.1). Then equation (2.7) and the constraints of the linear programming problem provide the bound

$$\underline{c}^T(\underline{x}_{k+1} - \underline{x}^*) = \frac{1}{\sigma} \sum_{j=1}^{m} \frac{\underline{a}_j^T \underline{x}_{k+1} - \underline{a}_j^T \underline{x}^*}{\underline{a}_j^T \underline{x}_{k+1} - b_j} \le \frac{1}{\sigma} \sum_{j=1}^{m} \frac{\underline{a}_j^T \underline{x}_{k+1} - b_j}{\underline{a}_j^T \underline{x}_{k+1} - b_j} = \frac{m}{\sigma}. \tag{2.8}$$

Thus condition (2.2) gives the property $\underline{c}^T(\underline{x}_{k+1}-\underline{x}^*) \le \epsilon$ as required. Further, $\underline{x}_{k+1}$ is feasible because it is an interior point of $\mathcal{S}$.

The method of increasing σ in Step 3 is simply to multiply it by a constant, Θ say. In "small-step" log barrier methods the difference $\Theta-1$ is of magnitude $m^{-1/2}$, and we employ the value

$$\Theta = (m^{1/2}+0.5)\,/\,(m^{1/2}+0.1), \tag{2.9}$$

while, in the "large-step" calculations, each choice of Θ is from the interval $[1.5, 10]$. One of the main reasons for the experiments of Section 4 is to investigate the remark of Gonzaga (1992) that, in practice, it is more efficient to prefer "large" rather than "small" changes to σ. In theory, however, it is known that the factor (2.9) gives the property that the number of iterations of the algorithm of this section is of magnitude $m^{1/2}$. On the other hand, the complexity results that have been proved already for "large-step" methods are less favourable.

3. The Karmarkar algorithm

In the original version of Karmarkar's algorithm the only constraints are equality conditions and the requirement that all the variables shall be nonnegative. The problem (1.1) can be expressed in this form by first changing the variables if necessary so that n of the constraints become simple bounds, and then the remaining constraints are treated by the introduction of $n-m$ slack variables. Thus the number of variables increases from n to m. Fortunately, however, there is an equivalent version of Karmarkar's algorithm that allows general homogeneous inequality constraints. Therefore it is sufficient to work with only $n+1$ variables. Specifically, we define the vectors

$$\hat{\underline{c}} = \begin{pmatrix} \underline{c} \\ \cdots \\ \gamma \end{pmatrix} \quad \text{and} \quad \hat{\underline{a}}_j = \begin{pmatrix} \underline{a}_j \\ \cdots \\ -b_j \end{pmatrix}, \quad j=1,2,\ldots,m, \tag{3.1}$$

where γ is a parameter, in order to write the calculation (1.1) in the form

$$\left.\begin{array}{lll} \text{minimize} & \hat{\underline{c}}^T\underline{y}, & \underline{y}\in\mathcal{R}^{n+1}, \\ \text{subject to} & y_{n+1} = 1, & \\ \text{and to} & \hat{\underline{a}}_j^T\underline{y} \ge 0, & j=1,2,\ldots,m. \end{array}\right\} \tag{3.2}$$

The condition $y_{n+1}=1$ causes the optimal variables to be independent of the value of γ. Karmarkar's algorithm, however, requires the optimal value of the objective function to be zero, which fixes γ. This γ is prescribed in each of the numerical experiments, although it is unusual to know the final value of $\hat{\underline{c}}^T\underline{y}$ in advance.

Let $\hat{\mathcal{S}}$ be the set of vectors $\underline{y}$ in $\mathcal{R}^{n+1}$ such that the left hand sides $\{\hat{\underline{a}}_j^T\underline{y} : j = 1,2,\ldots,m\}$ of the inequality constraints are all positive. Karmarkar's algorithm employs the function

$$V(\underline{y}) = m\log(\hat{\underline{c}}^T\underline{y}) - \sum_{j=1}^{m}\log(\hat{\underline{a}}_j^T\underline{y}), \quad \underline{y}\in\hat{\mathcal{S}}. \tag{3.3}$$

Further, it requires an initial vector of variables, $\underline{y}_1$ say, that is in $\hat{\mathcal{S}}$ and whose last component is one. Further, the termination of the algorithm depends on a prescribed parameter, Δ say. Further, the constraints must have the properties that the vectors $\{\hat{\underline{a}}_j : j = 1,2,\ldots,m\}$ span $\mathcal{R}^{n+1}$ and the feasible region of the problem (3.2) is bounded. Then our implementation of the algorithm is as follows.

Step 0: Set $k=1$.

Step 1: Go to Step 4 if $\underline{y}_k$ satisfies the condition

$$V(\underline{y}_k) \le -\Delta\, m. \tag{3.4}$$

Step 2: Otherwise calculate the search direction

$$\underline{d}_k = -\left[\sum_{j=1}^{m}\frac{\hat{\underline{a}}_j\hat{\underline{a}}_j^T}{(\hat{\underline{a}}_j^T\underline{y}_k)^2}\right]^{-1}\underline{\nabla}V(\underline{y}_k). \tag{3.5}$$

Step 3: Let $\underline{y}_{k+1}$ be the vector $\underline{y}_k+\alpha_k\underline{d}_k$, where α_k is a positive steplength that is calculated by a line search to minimize $V(\underline{y}_{k+1})$. Then replace k by $k+1$ and go back to Step 1.

Step 4: Let $\underline{y}^*$ be the multiple of $\underline{y}_k$ whose last component is one. This vector is provided as the calculated solution of the problem (3.2).

This algorithm has some beautiful features that depend strongly on the homogeneity of the function (3.3), the homogeneity condition being the identity $V(\lambda\underline{y}) = V(\underline{y})$, where $\underline{y}$ is any point in $\hat{\mathcal{S}}$ and where λ is any positive number. In particular, the scaling of $\underline{y}_k$ does not matter in Steps 0–3, which allows the equality constraint $y_{n+1} = 1$ to be ignored until Step 4. There the given construction of $\underline{y}^*$ is possible, because the boundedness of the feasible region implies that the last component of every vector in $\hat{\mathcal{S}}$ is positive. Further, every nonzero vector in the closure of $\hat{\mathcal{S}}$ also has a positive last component.

The matrix inside the square brackets of formula (3.5) is the second derivative matrix of the function $-\sum_{j=1}^{m}\log(\hat{\underline{a}}_j^T\underline{y})$ at $\underline{y} = \underline{y}_k$. It follows from the definition (2.1) that $\underline{d}_k$ is analogous to the search direction of expression (2.4), especially if one takes the view that σ might have the value $m/\hat{\underline{c}}^T\underline{y}_k$. It is both elegant and disadvantageous, however, that the function (3.3) does not include a parameter that is analogous to the choice of σ in log barrier methods.

A practical stopping condition is needed for the line search in Step 3. In the experiments of Section 4, we require the actual value of α_k to agree with the optimal steplength to 10 decimal places. This tolerance is also used to ensure that each $\underline{y}_{k+1}$ is in the open set $\hat{\mathcal{S}}$.

In order to explain the stopping condition (3.4), we note that this inequality, the definition (3.3) and the construction of Step 4 imply the bound

$$\log(\hat{\underline{c}}^T\underline{y}^*) \leq -\Delta + m^{-1}\sum_{j=1}^{m}\log(\hat{\underline{a}}_j^T\underline{y}^*). \tag{3.6}$$

Now we expect the sum on the right hand side to be negative because each solution to the problem (3.2) is on the boundary of at least one inequality constraint. In this case expression (3.6) provides the condition

$$\hat{\underline{c}}^T\underline{y}^* \leq e^{-\Delta}, \tag{3.7}$$

and we recall that the optimal value of the objective function is zero. Thus the accuracy of Karmarkar's algorithm is comparable to the accuracy of the log barrier method at the beginning of Section 2 if we set $\Delta = -\log\epsilon$.

The most remarkable property of Karmarkar's algorithm is that the reduction in the function (3.3) on each iteration satisfies the inequality

$$V(\underline{y}_{k+1}) \leq V(\underline{y}_k) - (1-\log 2) \approx V(\underline{y}_k) - 0.307 \tag{3.8}$$

(see Powell, 1992a, for instance). Therefore, if $V(\underline{y}_1)$ and Δ are of magnitude one, then the number of iterations that are needed to achieve the termination condition (3.4) is bounded above by a multiple of m. One purpose of the following experiments is to investigate whether this bound is pessimistic in practice.

4. Four test problems

The first problem is the one from Powell (1991) that is mentioned in the second paragraph of Section 1. It is the minimization of the second component of $\underline{x} \in \mathcal{R}^2$ subject to the constraints

$$\cos\left(\frac{2j\pi}{m}\right)x_1 + \sin\left(\frac{2j\pi}{m}\right)x_2 \geq -1, \quad j=1,2,\ldots,m, \tag{4.1}$$

where x_i denotes the i-th component of $\underline{x}$. It has the disadvantage, however, that the minimizer of the function (2.1) is on the x_2-axis for each value of σ, due to symmetry. Therefore the symmetry is broken in the second problem, by letting m be even, and by replacing the constraints by the conditions

$$\left.\begin{array}{ll} \cos\left(\frac{4j\pi}{m}\right)x_1+\sin\left(\frac{4j\pi}{m}\right)x_2\geq -1, & j=1,2,\ldots,\frac{1}{2}m, \\ x_1\geq -1, & j=\frac{1}{2}m+1,\frac{1}{2}m+2,\ldots,m, \end{array}\right\} \qquad (4.2)$$

there being no change to the objective function. Thus the term $\log(x_1+1)$ occurs $\frac{1}{2}m+1$ times in the sum (2.1).

The third and fourth problems are derived from one-sided approximation to the function $\{f(\xi,\eta)=\xi^2\eta : (\xi,\eta)\in\mathcal{R}^2\}$ by a linear polynomial on a point set $\{(\xi_j,\eta_j) : j=1,2,\ldots,\frac{1}{2}m\}$, where m is a positive even integer. Both linear programming calculations have four variables, namely the three coefficients of the linear polynomial and the maximum residual of the fit. Indeed, each calculation is the minimization of the last component of $\underline{x}\in\mathcal{R}^4$ subject to the constraints

$$\left.\begin{array}{l} x_1+\xi_j x_2+\eta_j x_3+x_4\geq \xi_j^2\eta_j \\ x_1+\xi_j x_2+\eta_j x_3\leq \xi_j^2\eta_j \end{array}\right\}, \qquad j=1,2,\ldots,\tfrac{1}{2}m. \qquad (4.3)$$

In the third problem the points $\{(\xi_j,\eta_j) : j=1,2,\ldots,\frac{1}{2}m\}$ are spaced at equal intervals along the perimeter of the square $\{(\xi,\eta) : 0\leq\xi\leq 1, 0\leq\eta\leq 1\}$, while in the fourth problem they are equally spaced on the circle $\{(\xi,\eta) : (\xi-\frac{1}{2})^2+(\eta-\frac{1}{2})^2=\frac{1}{2}\}$, where one of the points is the origin in both cases. It is harder to identify the constraints that are active at the solution in the latter case, because circles are smoother than squares.

The remaining parameters of the calculations are as follows. The initial vector of variables, namely $\underline{x}_1$, is given the components $(0.8, 0.5)$ in the first and second problems, and in the other two cases it has the components $(-0.1, 0.0, 0.0, 1.2)$. The values of ϵ and Δ are set to 10^{-8} and 15 in conditions (2.2) and (3.4) respectively. The calculations that test Karmarkar's algorithm are constructed from the given problems by adding an extra variable in the way that is described in the opening paragraph of Section 3, each initial vector $\underline{y}_1$ being the $\underline{x}_1$ that has been mentioned augmented by a last component that is set to one. In every case m takes 5 different values from the interval $[20, 10000]$. Further, the parameter Θ of the log barrier experiments is given the "small-step" value (2.9) and 3 "large-step" values.

Table 1 shows the amount of work of these calculations. Here "number of steps" denotes the number of iterations and the number of applications of formula (2.4) for Karmarkar's algorithm and the log barrier methods respectively. The "number of values of σ" for the log barrier methods depends only on the initial value that is specified in the second paragraph of Section 2, on Θ and on the ratio m/ϵ. The term "small-step" in the second column of the table indicates that Θ has the value (2.9).

	Method	$m=20$	$m=100$	$m=500$	$m=2000$	$m=10000$
First problem	Karmarkar	5/0	14/0	47/0	215/0	1047/0
	Small-step	31/223	52/520	102/1234	201/2607	436/6209
	$\Theta=1.5$	53/47	58/51	74/55	97/58	116/62
	$\Theta=2.0$	48/28	54/30	66/32	72/34	82/37
	$\Theta=5.0$	29/13	30/14	36/15	38/15	43/16
Second problem	Karmarkar	3/0	13/0	45/0	208/0	1021/0
	Small-step	31/227	48/520	93/1234	172/2607	370/6209
	$\Theta=1.5$	52/48	53/51	80/55	97/58	121/62
	$\Theta=2.0$	45/28	52/30	59/32	67/34	78/37
	$\Theta=5.0$	33/13	35/14	45/15	41/15	47/16
Third problem	Karmarkar	13/0	12/0	10/0	18/0	65/0
	Small-step	65/221	90/482	131/1061	198/2107	385/4692
	$\Theta=1.5$	55/46	70/47	70/47	83/47	96/47
	$\Theta=2.0$	102/28	99/28	102/28	100/28	97/28
	$\Theta=5.0$	30/12	33/13	40/13	45/13	46/13
Fourth problem	Karmarkar	12/0	12/0	20/0	87/0	328/0
	Small-step	56/200	82/468	124/1028	207/2041	411/4546
	$\Theta=1.5$	48/42	65/46	81/46	94/46	113/46
	$\Theta=2.0$	99/25	101/27	101/27	97/27	100/27
	$\Theta=5.0$	30/11	36/12	40/12	43/12	48/12

Table 1:

Number of steps / Number of values of σ in the methods of Sections 3 and 2

We see that Karmarkar's algorithm requires fewest iterations when m is small. The analysis in (Powell, 1991) of the first problem, however, proves that the number of iterations tends to be of magnitude m as $m \to \infty$ for many choices of $\underline{x}_1$. This behaviour seems to occur in all of the Karmarkar calculations of Table 1, but it is not very obvious in the case of the third problem. Therefore the "small-step" method requires fewer iterations for sufficiently large m, its $\mathcal{O}(m^{1/2})$ complexity being confirmed in the table. On the other hand, because this method employs more values of σ than steps, formula (2.9) is too conservative. Indeed, the table shows that the "large-step" values of Θ are much more efficient.

It is clear that $\Theta = 5.0$ is the most successful choice of multiplying factor in the given experiments with the log barrier method. Therefore it is remarkable that $\Theta = 2.0$ is less efficient than $\Theta = 1.5$ for the third and fourth problems. A possible explanation of this observation can be obtained by asking whether $\underline{x}_k$ is a good starting point for the unconstrained minimization of the function (2.1) by the modified Newton–Raphson procedure that applies formulae (2.4) and (2.5). We take the view that the suitability of Newton–Raphson is doubtful if we have $\bar{\alpha} < 1$ in equation (2.5). Therefore we counted the number of occurrences of this inequality in the calculations of some of the rows of Table 1, summing these numbers over the 5 values of m. The total counts for the $\Theta = 1.5$, $\Theta = 2.0$ and $\Theta = 5.0$ rows of the third problem are 38, 139 and 82 respectively, while the corresponding counts for the fourth problem are 44, 131 and 78. Therefore $\Theta = 1.5$ gains over $\Theta = 2.0$ because it gives better starting points for Step 1 of the algorithm of Section 2, while $\Theta = 5.0$ has the advantage that there are fewer unconstrained subproblems. We find in the next section, however, that the Newton–Raphson step from $\underline{x}_k$ in Step 1 can benefit greatly from a careful choice of α in formula (2.4). Thus the numerical results for $\Theta = 2.0$ will become uniformly better than the $\Theta = 1.5$ results.

5. Line searches

Let $\underline{x}(\sigma)$ be the vector of variables that minimizes the function (2.1). The algorithm of Section 2 generates points $\{\underline{x}_{k+1} : k = 1, 2, 3, \ldots\}$ that are close to the "central path" $\{\underline{x}(\sigma) : 0 \leq \sigma < \infty\}$, because, as $\sigma \to \infty$, the central path tends to a limit which is a solution of the linear programming problem (1.1) (Gonzaga, 1992). Therefore it is relevant to ask whether the initial Newton–Raphson step

$$\underline{x} \leftarrow \underline{x}_k - \left[\nabla^2 \Phi_\sigma(\underline{x}_k)\right]^{-1} \underline{\nabla} \Phi_\sigma(\underline{x}_k) \tag{5.1}$$

of the calculation of Step 1 is good for following the central path in the case when $\underline{x}_k$ is on this path. Specifically, we let $\underline{x}_k$ be the point $\underline{x}(\hat{\sigma})$, where $\hat{\sigma}$ is less than σ.

The assumption that $\underline{x}_k$ is the minimizer of $\{\Phi_{\hat{\sigma}}(\underline{x}) : \underline{x} \in \mathcal{S}\}$ provides the equation $\underline{\nabla}\Phi_{\hat{\sigma}}(\underline{x}_k)=0$, which we write in the form

$$\sum_{j=1}^{m} \frac{\underline{a}_j}{\underline{a}_j^T\underline{x}_k-b_k} = \hat{\sigma}\,\underline{c}. \tag{5.2}$$

It follows that expression (5.1) reduces to the formula

$$\underline{x} \leftarrow \underline{x}_k - \Big[\sum_{j=1}^{m} \frac{\underline{a}_j\underline{a}_j^T}{(\underline{a}_j^T\underline{x}_k-b_j)^2}\Big]^{-1}(\sigma-\hat{\sigma})\,\underline{c}. \tag{5.3}$$

Thus the length of the step from $\underline{x}_k$ to $\underline{x}$ is proportional to $\sigma-\hat{\sigma}$ but the direction of the step is independent of σ. Hence, by considering the limit $\sigma \rightarrow \hat{\sigma}$, we deduce from the second order convergence of the Newton–Raphson procedure that this direction is tangential to the central path at $\underline{x}_k$. Moreover, because we are assuming that the feasible region is bounded, formula (5.3) causes $\underline{x}$ to be infeasible if σ is greater than a particular value, $\check{\sigma}$ say.

The following argument suggests the estimate $\check{\sigma} \approx 2\hat{\sigma}$ when $\hat{\sigma}$ is large. Let $\underline{x}_\infty$ be the limit of the central path and let $\mathcal{J} \subset \{1,2,\ldots,m\}$ contain the indices of the constraints that are satisfied as equations at $\underline{x}_\infty$. Then the identity (5.2) provides the approximation

$$\underline{c} = \sum_{j=1}^{m} [\,\hat{\sigma}(\underline{a}_j^T\underline{x}_k-b_j)\,]^{-1}\underline{a}_j \approx \sum_{j\in\mathcal{J}} [\,\hat{\sigma}(\underline{a}_j^T\underline{x}_k-b_j\,]^{-1}\underline{a}_j, \tag{5.4}$$

when $\hat{\sigma}$ is so large that the factors $\{\,[\,\hat{\sigma}(\underline{a}_j^T\underline{x}_k-b_j)\,]^{-1} : j\notin\mathcal{J}\}$ are negligible. Now the zero residuals $\{\underline{a}_j^T\underline{x}_\infty-b_j : j\in\mathcal{J}\}$ imply that the point $\underline{\hat{x}}=(1-\theta)\underline{x}_k+\theta\underline{x}_\infty$ satisfies the equations

$$\underline{a}_j^T\underline{\hat{x}}-b_j = (1-\theta)\,(\underline{a}_j^T\underline{x}_k-b_j), \quad j\in\mathcal{J}, \tag{5.5}$$

where θ is a parameter from the interval $[0,1]$. Therefore expression (5.4) gives the property

$$\frac{\hat{\sigma}\,\underline{c}}{1-\theta} \approx \sum_{j\in\mathcal{J}} \frac{\underline{a}_j}{(1-\theta)\,(\underline{a}_j^T\underline{x}_k-b_j)} = \sum_{j\in\mathcal{J}} \frac{\underline{a}_j}{\underline{a}_j^T\underline{\hat{x}}-b_j} \approx \sum_{j=1}^{m} \frac{\underline{a}_j}{\underline{a}_j^T\underline{\hat{x}}-b_j}, \tag{5.6}$$

the last assertion being analogous to the approximation (5.4). This relation shows that, if σ has the value $\hat{\sigma}/(1-\theta)$, then the gradient of the function (2.1) is small at $\underline{x}=\underline{\hat{x}}$. Therefore it is reasonable to make the estimate

$$\underline{x}(\sigma) \approx \underline{\hat{x}} = \underline{x}_k + \theta\,(\underline{x}_\infty-\underline{x}_k) = \underline{x}_k + (1-\hat{\sigma}/\sigma)\,(\underline{x}_\infty-\underline{x}_k) \tag{5.7}$$

for every $\sigma>\hat{\sigma}$. Moreover, we know that the difference between expression (5.3) and $\underline{x}(\sigma)$ is of magnitude $(\sigma-\hat{\sigma})^2$ when σ is close to $\hat{\sigma}$. These two statements are compatible only if we have the relation

$$-\Big[\sum_{j=1}^{m} \frac{\underline{a}_j\underline{a}_j^T}{(\underline{a}_j^T\underline{x}_k-b_j)^2}\Big]^{-1}\underline{c} \approx \lim_{\sigma\rightarrow\hat{\sigma}} \frac{\underline{x}_\infty-\underline{x}_k}{\sigma} = \frac{\underline{x}_\infty-\underline{x}_k}{\hat{\sigma}}. \tag{5.8}$$

Therefore we can take the view that the Newton–Raphson step from $\underline{x}_k$ has a tendency to generate a vector that is near to the point $\underline{x}_k+(\sigma/\hat{\sigma}-1)(\underline{x}_\infty-\underline{x}_k)$. Thus overshooting the boundary of the feasible region is usual when the parameter $\Theta=\sigma/\hat{\sigma}$ satisfies $\Theta>2$. Further, we expect the full Newton–Raphson step to be unacceptable in the case $\Theta=2$, because the function (2.1) is unbounded at $\underline{x}_\infty$.

We recall from Table 1, however, that one can gain better efficiency by choosing $\Theta>2$ in our experiments. Therefore, because the above analysis suggests that often the direction of the Newton–Raphson step is very suitable, we retain formula (2.4), but we seek a way of generating the steplength that is less crude than equation (2.5). The obvious answer is a line search that picks a value of α that is close to the one that minimizes the function of one variable

$$\phi(\alpha)=\Phi_\sigma(\underline{x}-\alpha\,\underline{d}),\quad \alpha\geq 0, \tag{5.9}$$

where $\underline{d}$ is the Newton–Raphson direction

$$\underline{d}=\left[\nabla^2\Phi_\sigma(\underline{x})\right]^{-1}\underline{\nabla}\Phi_\sigma(\underline{x}). \tag{5.10}$$

It follows from the definition (2.1) that we seek an approximation to the solution of the equation

$$\phi'(\alpha)=-\sigma\,\underline{c}^T\underline{d}+\sum_{j=1}^{m}\frac{\underline{a}_j^T\underline{d}}{\underline{a}_j^T\underline{x}-b_j-\alpha\,\underline{a}_j^T\underline{d}}=0. \tag{5.11}$$

This task is straightforward because the derivative $\{\phi'(\alpha):0\leq\alpha\leq\bar{\alpha}\}$ increases strictly monotonically from a negative value to $+\infty$, where $\bar{\alpha}$ is introduced just before equation (2.5), and where $\phi'(0)<0$ and the strict monotonicity are consequences of the positive definiteness of $\nabla^2\Phi_\sigma(\underline{x})$ for every $\underline{x}$ in $\mathcal{S}$.

In order to reduce the effort of calculating $\phi'(\alpha)$ for several values of α, we can generate and store the scalars $\sigma\,\underline{c}^T\underline{d}$, $\{\underline{a}_j^T\underline{d}:j=1,2,\ldots,m\}$ and $\{\underline{a}_j^T\underline{x}-b_j:j=1,2,\ldots,m\}$ before beginning the line search. Then each application of formula (5.11) requires about $4m$ computer operations. On the other hand, the number of operations to form the matrix $\nabla^2\Phi_\sigma(\underline{x})$ is bounded below by 5 times the total number of nonzero components in the vectors $\{\underline{a}_j:j=1,2,\ldots,m\}$, where this task includes the computation of the scalars $\{\underline{a}_j^T\underline{x}-b_j:j=1,2,\ldots,m\}$ that have just been mentioned. Thus the inclusion of an efficient line search in the Newton–Raphson procedure should not increase the total work of an application of formula (2.4) by more than about 30% when n is small. Further, the relative cost tends to be less when there are more variables, because the number of elements of the matrix $\nabla^2\Phi_\sigma(\underline{x})$ is n^2.

We applied the log barrier method of Section 2 to the four test problems of Section 4, after replacing formula (2.5) by a highly accurate line search. The numerical results for four values of Θ are given in Table 2. We see uniform

	Method	$m=20$	$m=100$	$m=500$	$m=2000$	$m=10000$
First problem	$\Theta=1.5$	49/47	50/51	55/55	58/58	59/62
	$\Theta=2.0$	32/28	31/30	33/32	35/34	36/37
	$\Theta=5.0$	17/13	15/14	16/15	16/15	16/16
	$\Theta=10.0$	13/9	11/10	11/10	12/11	12/12
Second problem	$\Theta=1.5$	48/48	52/51	57/55	69/58	77/62
	$\Theta=2.0$	30/28	36/30	41/32	45/34	54/37
	$\Theta=5.0$	15/13	20/14	27/15	29/15	35/16
	$\Theta=10.0$	11/9	17/10	22/10	41/11	84/12
Third problem	$\Theta=1.5$	47/46	48/47	52/47	54/47	63/47
	$\Theta=2.0$	29/28	30/28	35/28	38/28	41/28
	$\Theta=5.0$	22/12	25/13	27/13	30/13	33/13
	$\Theta=10.0$	21/9	22/9	24/9	25/9	24/9
Fourth problem	$\Theta=1.5$	45/42	52/46	55/46	62/46	64/46
	$\Theta=2.0$	28/25	35/27	40/27	43/27	47/27
	$\Theta=5.0$	22/11	28/12	34/12	37/12	38/12
	$\Theta=10.0$	21/8	26/9	31/9	34/9	34/9

Table 2:

Number of steps / Number of values of σ in log barrier methods with line searches

improvements in the "number of steps" over the corresponding entries in Table 1. Further, the line searches cause the choice $\Theta=2.0$ to be more successful than $\Theta=1.5$ in every case, and we still have the property that increasing Θ from 2.0 to 5.0 reduces the amount of computation uniformly. Therefore the value $\Theta = 10.0$ was tried too, and Table 2 shows that further improvements occur except in the second problem for large m. In view of the tables, we recommend the use of line searches, and we note also that they ensure the termination of each unconstrained calculation of a log barrier method.

6. The shifted log barrier function

The success of log barrier methods for linear programming is surprising, because several optimization experts reached the conclusion in the 1970s that the augmented Lagrangian algorithm is more suitable than barrier functions for converting constrained problems to unconstrained ones. That algorithm employs a parameter that is analogous to σ in expression (2.1), except that the parameter has the advantage that it remains finite. Further, estimates of the Lagrange multipliers of the constraints are included, which are adjusted by an outer iteration. Further, the inner iterative procedure is an unconstrained minimization calculation that is analogous to Step 1 in the first paragraph of Section 2. It is possible that such techniques will provide improvements to log barrier algorithms for linear programming. Therefore the purpose of this section is to consider a method, proposed by Polyak (1992a), that combines several useful features of augmented Lagrangian and log barrier calculations.

We are going to apply the following interesting version of Polyak's procedure to the test problems of Section 4. The notation and wording of the specification are chosen to assist comparisons with the algorithm at the beginning of Section 2. We continue to assume that a feasible starting point $\underline{x}_1$ is available.

Step 0: Set the parameter σ to a suitable positive real number, set $\underline{\lambda}$ to the vector in $\mathcal{R}^m$ whose components are all one and set $k=1$.

Step 1: Calculate $\underline{x}_{k+1}$ by applying an algorithm for unconstrained minimization to the function

$$\Phi_\sigma(\underline{x}) = \sigma \underline{c}^T \underline{x} - \sum_{j=1}^{m} \lambda_j \log [1+\sigma(\underline{a}_j^T \underline{x} - b_j)], \quad \underline{x} \in \mathcal{S}, \tag{6.1}$$

where $\underline{x}_k$ is the starting point of this calculation, and where $\underline{x}$ is in $\mathcal{S}$ if and only if all the arguments of all the logarithms are nonnegative.

Step 2: Accept $\underline{x}_{k+1}$ as a sufficiently accurate approximation to the solution of the problem (1.1) if it satisfies the conditions

$$\sum_{j=1}^{m} \lambda_j \frac{\underline{a}_j^T \underline{x}_{k+1} - b_j}{1+\sigma(\underline{a}_j^T \underline{x}_{k+1} - b_j)} \le \epsilon \text{ and } \underline{a}_j^T \underline{x}_{k+1} \ge b_j - \epsilon, \quad j=1,2,\ldots,m, \tag{6.2}$$

where ϵ is a prescribed positive tolerance.

Step 3: Otherwise update the components of $\underline{\lambda}$ by the formula

$$\lambda_j \leftarrow \lambda_j / [1+\sigma(\underline{a}_j^T \underline{x}_{k+1} - b_j)], \quad j=1,2,\ldots,m, \tag{6.3}$$

replace k by $k+1$ and go back to Step 1.

We see that σ remains fixed throughout the calculation. Therefore we require the changes to $\underline{\lambda}$ to ensure that the sequence $\{\underline{x}_k : k=1,2,3,\ldots\}$ provides an acceptable vector of variables. It is proved by Powell (1992b) that this sequence converges to a solution of the given linear programming problem if all the unconstrained minimizations of Step 1 are exact. In practice, however, we need a stopping condition for the iterations of the Step 1 subproblem. Again each change to the variables during Step 1 has the form $\underline{x} \leftarrow \underline{x} - \alpha \underline{d}$, where α is determined by a highly accurate line search and where $\underline{d}$ is the Newton–Raphson direction (5.10), but the test (2.6) guarantees high accuracy only when σ is large, so it is unsuitable for algorithms that adjust $\underline{\lambda}$ instead of σ. Therefore each unconstrained calculation is terminated when the following conditions hold. Firstly, we require the inequalities

$$|\underline{a}_j^T \underline{d}| \le \max\left[0.01|\underline{a}_j^T \underline{x} - b_j|,\ \min(0.1/\sigma,\ \epsilon)\right], \quad j=1,2,\ldots,m, \tag{6.4}$$

to be satisfied, because otherwise we expect another Newton–Raphson step to make a substantial change to the current constraint residuals. Secondly, we require $\|\underline{\nabla}\Phi_\sigma(\underline{x})\|$ to be less than 0.1σ, as condition (6.4) is often inadequate when one of the numbers $\{1+\sigma(\underline{a}_j^T \underline{x} - b_j) : j=1,2,\ldots,m\}$ is very small, because then it is usual for the length of $\underline{d}$ to be much too short due to the high curvature of Φ_σ. On the other hand, the justification of Steps 2 and 3 assumes that $\underline{x}_{k+1}$ is the exact minimizer of the function (6.1). Indeed, if the gradient $\underline{\nabla}\Phi_\sigma(\underline{x}_{k+1})$ is zero, then we have the identity

$$\underline{c} = \sum_{j=1}^{m} \frac{\lambda_j\, \underline{a}_j}{1+\sigma(\underline{a}_j^T \underline{x}_{k+1} - b_j)}, \tag{6.5}$$

instead of expression (2.7). Thus, using a straightforward modification of the derivation of inequality (2.8), it follows that the first of the conditions (6.2) provides the bound $\underline{c}^T(\underline{x}_{k+1} - \underline{x}^*) \le \epsilon$, where $\underline{x}^*$ is any optimal vector of variables. Further, equation (6.5) suggests that formula (6.3) is appropriate for generating new estimates of Lagrange multipliers.

Stronger motivation for the updating formula (6.3) can be derived from the dual function

$$\phi(\underline{\lambda}) = \min_{\underline{x} \in \mathcal{S}} \Phi_\sigma(\underline{x}), \quad \underline{\lambda} \in \mathcal{R}_+^m, \tag{6.6}$$

where $\mathcal{R}_+^m$ is the set of vectors in $\mathcal{R}^m$ whose components are positive. By letting $\underline{x}$ be the optimal vector $\underline{x}^*$ in the definition (6.1), we deduce the bound $\phi(\underline{\lambda}) \le \sigma\,\underline{c}^T \underline{x}^*$ from the feasibility of $\underline{x}^*$, where $\underline{\lambda} \in \mathcal{R}_+^m$. Further, if $\underline{\lambda}^*$ is a true vector of Lagrange multipliers at $\underline{x}^*$ and if $\{\underline{\lambda}_k : k=1,2,3,\ldots\}$ is a sequence in $\mathcal{R}_+^m$ that converges to $\underline{\lambda}^*$, then it can be proved that $\sigma\,\underline{c}^T \underline{x}^*$ is the limit of the sequence $\{\phi(\underline{\lambda}_k) : k=1,2,3,\ldots\}$. It follows that the aim of increasing $\phi(\underline{\lambda})$ provides a suitable guide for changes to $\underline{\lambda}$. Now it is straightforward to establish the formula

$$d\phi(\underline{\lambda})/d\lambda_j = -\log\left[1+\sigma(\underline{a}_j^T \underline{x}(\underline{\lambda}) - b_j)\right], \quad j=1,2,\ldots,m, \tag{6.7}$$

where $\underline{x}(\underline{\lambda})$ is the minimizer of the function (6.1). Further, it can be deduced that ϕ has the second derivatives

$$\nabla^2\phi = -A^T(A\,\Lambda\,A^T)^{-1}A, \tag{6.8}$$

where A is the $n \times m$ matrix that has the columns $\{[\underline{a}_j^T \underline{x}(\underline{\lambda}) - b_j + \sigma^{-1}]^{-1} \underline{a}_j : j = 1, 2, \ldots, m\}$ and where Λ is the $m \times m$ diagonal matrix whose diagonal elements are the positive numbers $\{\lambda_j : j = 1, 2, \ldots, m\}$. Therefore the rank of $\nabla^2 \phi$ is n. Hence, because equation (6.8) implies the identity

$$(\Lambda^{-1} + \nabla^2 \phi)\, \Lambda\, A^T = 0, \tag{6.9}$$

we infer from the strict positive definiteness of Λ^{-1} that the symmetric matrix $(\Lambda^{-1} + \nabla^2 \phi)$ has n zero and $m-n$ positive eigenvalues. It is straightforward to verify that this matrix is the Hessian of the function

$$\psi(\underline{\lambda}) = \sum_{j=1}^{m} (\lambda_j \log \lambda_j - \lambda_j) + \phi(\underline{\lambda}) + \underline{\lambda}^T \underline{v} + c, \quad \underline{\lambda} \in \mathcal{R}_+^m. \tag{6.10}$$

Hence ψ is convex for any $\underline{v} \in \mathcal{R}^m$ and $c \in \mathcal{R}$. We now suppose that $\underline{x}(\hat{\underline{\lambda}})$ has been calculated, that $\underline{\nabla} \phi(\hat{\underline{\lambda}})$ is nonzero, and that we wish to replace $\hat{\underline{\lambda}}$ by a vector $\underline{\lambda}$ that satisfies the condition $\phi(\underline{\lambda}) > \phi(\hat{\underline{\lambda}})$. We pick $\underline{v}$ and c so that $\psi(\hat{\underline{\lambda}})$ and $\underline{\nabla} \psi(\hat{\underline{\lambda}})$ are zero. Then the convexity of ψ implies $\psi \geq 0$, so we have the property

$$\phi(\underline{\lambda}) \geq -\sum_{j=1}^{m} (\lambda_j \log \lambda_j - \lambda_j + \lambda_j v_j) - c, \quad \underline{\lambda} \in \mathcal{R}_+^m, \tag{6.11}$$

which holds as an equation at $\underline{\lambda} = \hat{\underline{\lambda}}$. Therefore we can achieve $\phi(\underline{\lambda}) > \phi(\hat{\underline{\lambda}})$ by letting $\underline{\lambda}$ maximize the right hand side of expression (6.11), which gives the new Lagrange multiplier estimates

$$\lambda_j = e^{-v_j} = \exp[\log \hat{\lambda}_j + d\phi(\hat{\underline{\lambda}})/d\lambda_j], \quad j = 1, 2, \ldots, m, \tag{6.12}$$

where the last assertion depends on the choice of $\underline{v}$. Remembering the derivative (6.7), we find that the updating formulae (6.3) and (6.12) are identical.

We apply the algorithm of this section to our four test problems for two choices of σ. It is reasonable to let σ be of magnitude m, because, if $\underline{x}$ is not close to any constraint boundaries, and if all the components of $\underline{\lambda}$ are one, then typically the gradient of the sum of expression (6.1) is $\mathcal{O}(m)$, and we would like the vector $\underline{\nabla}(\sigma \underline{c}^T \underline{x}) = \sigma \underline{c}$ to be of this magnitude too. Moreover, we retain the choices of $\underline{x}_1$ and the tolerance parameter $\epsilon = 10^{-8}$ that are given in Section 4. It should now be straightforward to run the given algorithm, but, unfortunately, the updating formula (6.3) causes the Lagrange multipliers of the inactive constraints to underflow frequently, and then the trapping of possible errors by our implementation of Fortran is so expensive that our calculations are very slow. Another difficulty is that in theory $\nabla^2 \Phi$ becomes singular at the solutions of both the first and second problems, because all but one of the multipliers $\{\lambda_j : j = 1, 2, \ldots, m\}$ in expression (6.1) tend to zero, due to the nonuniqueness of optimal values of the

	Method	$m=20$	$m=100$	$m=500$	$m=2000$	$m=10000$
Modified first problem	$\sigma=m$	28/17	44/36	99/109	239/373	275/528
	$\sigma=2m$	21/11	49/22	128/55	354/129	1351/208
	$\Theta=2.0$	16/7	16/11	19/15	20/18	22/21
	$\Theta=5.0$	15/6	14/8	15/10	14/12	17/16
	$\Theta=10.0$	14/5	13/7	13/8	12/10	14/13
Modified second problem	$\sigma=m$	21/9	45/18	92/44	235/109	782/314
	$\sigma=2m$	17/7	37/13	93/23	264/56	1098/141
	$\Theta=2.0$	17/6	32/10	31/13	37/16	36/20
	$\Theta=5.0$	16/5	26/7	50/9	32/11	88/15
	$\Theta=10.0$	16/5	24/6	55/9	31/9	102/12
Third problem	$\sigma=m$	32/18	47/27	147/163	138/185	181/314
	$\sigma=2m$	27/11	39/18	87/82	81/82	94/131
	$\Theta=2.0$	22/7	25/8	25/11	31/10	34/12
	$\Theta=5.0$	21/6	22/6	20/7	27/6	31/6
	$\Theta=10.0$	20/5	20/5	19/6	26/5	30/5
Fourth problem	$\sigma=m$	34/17	66/46	176/190	1252/2352	954/1787
	$\sigma=2m$	30/12	49/25	111/91	665/1143	666/834
	$\Theta=2.0$	19/5	31/8	40/10	43/13	51/12
	$\Theta=5.0$	20/5	28/6	33/6	42/7	46/6
	$\Theta=10.0$	18/4	26/5	32/5	40/6	44/5

Table 3:

Number of steps / Number of values of $\underline{\lambda}$ in the shifted log barrier method variables. This difficulty was removed by changing j to $j+\frac{1}{2}$ in the sine and cosine functions of expressions (4.1) and (4.2), which gives the "modified" problems that are mentioned in the first column of Table 3.

The numbers of iterations of these calculations when $\sigma=m$ and $\sigma=2m$ are shown in the rows of Table 3 that specify the value of σ. We see that, particularly for large m, the work of applying the given version of the shifted log barrier method is usually much greater than the corresponding work that is indicated in Table 2. The entries in the $m=10000$ column of Table 3 for the first problem are instructive. Here the number of values of $\underline{\lambda}$ is reduced substantially when σ is doubled, which can be explained theoretically (Powell, 1992b), but the number of line searches that change the variables increases from 275 to 1351. The increase is due entirely to the first application of Step 1, which generates $\underline{x}_2$ from $\underline{x}_1$. Indeed, this calculation requires 7 and 1242 line searches in the $\sigma=m$ and $\sigma=2m$ cases respectively, the latter case being laborious because the Newton–Raphson steps crawl round the boundary of the feasible region. Therefore it is very helpful if the inequality $\sigma<2m$ is satisfied. On the other hand, we expect $\underline{\lambda}$ to be updated more than 208 times for such a choice. It follows that much better efficiency may be achieved by an algorithm that increases σ automatically between successive visits to Step 1. A possible procedure is given next, which provides the rows of Table 3 that mention Θ.

We require any increases in σ to preserve the property that $\underline{x}_k$ can be used as a starting point for the unconstrained minimization of the function (6.1). Therefore it is necessary to keep the numbers $\{1+\sigma(\underline{a}_j^T\underline{x}_k-b_j) : j=1,2,\ldots,m\}$ positive. Specifically, letting σ_{old} and σ_{new} be the old and new values of σ, we satisfy the conditions

$$1+\sigma_{\text{new}}(\underline{a}_j^T\underline{x}_k-b_j) \geq \tfrac{1}{2}[1+\sigma_{\text{old}}(\underline{a}_j^T\underline{x}_k-b_j)], \quad j=1,2,\ldots,m, \tag{6.13}$$

where the factor $\frac{1}{2}$ was chosen arbitrarily. Further, we are influenced by the formula $\sigma_{\text{new}}=\Theta\,\sigma_{\text{old}}$ that occurs in Section 2. In view of these remarks, we extend Step 3 of the algorithm of this section by including the rule that σ be increased to the greatest number in the interval $[\sigma_{\text{old}}, \Theta\,\sigma_{\text{old}}]$ that is allowed by the constraints (6.13), where, as shown in Table 3, Θ take the values 2.0, 5.0 and 10.0. In these experiments the initial values of σ are the same as the ones that occur in the log barrier calculations of Tables 1 and 2.

A comparison of Tables 2 and 3 shows that the best performance of the shifted log barrier method can be made similar to that of the ordinary log barrier method with line searches, if one measures performance by the number of changes to the variables. Moreover, we find that the number of values of $\underline{\lambda}$ in the Θ-rows of Table 3 is usually less than the number of values of σ in the corresponding entries of Table 2. Therefore the use of formula (6.3) does have the advantage of reducing the number of unconstrained subproblems that occur. On the other hand, because this reduction is not inherited by the total number of steps, we conclude that our implementation of the shifted log barrier method tends to have unconstrained subproblems that are more difficult than the corresponding Step 1 calculations of Section 2.

7. Conclusions

The minimization of a linear function of 2 or 4 variables subject to linear constraints should be easy, because the feasible region is convex and there is no need for any nonlinearities. Therefore the numbers of iterations that occur for large m in the Karmarkar and small-step rows of Table 1 and in the σ-rows of Table 3 are highly unsatisfactory. The reasons for the inefficiencies are known. Specifically, Karmarkar's algorithm likes to follow vertices of the feasible region when $n = 2$ (Powell, 1991), the factor (2.9) condemns the small-step log barrier method to make of magnitude $m^{1/2}$ changes to σ, and the algorithm in the second paragraph of Section 6 has to drive the Lagrange multiplier estimates of the inactive constraints to zero, which is slow work when the updating formula (6.3) is applied for a moderate fixed value of σ, if some inactive constraints have small residuals at the required solution. Therefore, as shown in the tables, substantial improvements over these poor performances can be achieved.

But even our relatively efficient methods deserve criticism. In particular, if the argument of the logarithm is very small in the ℓ-th term under the summation sign of expression (2.1) or (6.1), then the Newton–Raphson direction (5.10) is usually almost orthogonal to $\underline{a}_\ell$, because of the huge multiple of $\underline{a}_\ell \underline{a}_\ell^T$ that occurs in $\nabla^2 \Phi_\sigma(\underline{x})$. It follows that a search along this direction is often severely restricted by the boundary of the feasible region in semi-infinite programming problems that replace curved constraints by many tangent planes. Further, it does not seem sensible to employ logarithms, and then to make quadratic approximations to them in order to calculate search directions, when all functions are linear initially. Further, when $n = 2$ and $m = 10000$, the need to adjust m parameters to particular values in the shifted log barrier method is unattractive. This comment applies also to primal-dual methods, including the techniques for following the "central-path" that are mentioned by Gonzaga (1992) and the implementation of the shifted log barrier method that is recommended by Polyak (1992b). These algorithms, however, may be more successful than the ones we have studied.

The author employed the first problem of Section 4 when developing TOLMIN (Powell, 1989), which is a Fortran package that minimizes a general differentiable function subject to linear constraints. It is designed to choose search directions that avoid early encounters with constraint boundaries. Indeed, constraints are treated explicitly, and no search direction is allowed to move closer to the boundary of any constraint that has a "small" residual, where "smallness" is measured by a tolerance parameter that is quite large initially, and that is updated only when

Problem	$m=20$	$m=100$	$m=500$	$m=2000$	$m=10000$
First	4	5	4	5	6
Second	2	4	5	4	5
Third	4	8	10	10	14
Fourth	6	10	12	14	14

Table 4:

Number of steps in TOLMIN

a reduction is necessary to provide further progress. The work of calculating each search direction is sometimes substantial, because one has to identify the dominant terms in the set of "small" residual constraints. Thus the effort of this task tends to increase with m, and the reward is that often the number of changes to the variables is quite small for large numbers of constraints as shown in Table 4. We conclude from these results that our linear programming problems are not difficult to solve, if one applies an algorithm that is designed to take account of many constraints when each search direction is generated.

The given problems may be valuable for investigating many other methods for linear programming, partly because the computational effort of including many constraints is not expensive. Further, we have found that a gradual sequence of inequality constraints can expose some disadvantages of certain algorithms, which may help future research. Indeed, many experts were surprised by the $\mathcal{O}(m)$ complexity of Karmarkar's algorithm that is shown in Table 1. It seems, therefore, that some crucial features of our test problems do not occur in the vast majority of linear programming calculations that have guided the development of algorithms so far.

Acknowledgement: The author is very grateful to Clovis Gonzaga for several useful comments on the first draft of this paper.

References

C.C. Gonzaga (1992), "Path-following methods for linear programming", *SIAM Review*, Vol. 34, pp. 167–224.

N. Karmarkar (1984), "A new polynomial-time algorithm for linear programming", *Combinatorica*, Vol. 4, pp. 373–393.

R. Polyak (1992a), "Modified barrier functions (theory and methods)", *Math. Programming*, Vol. 54, pp. 177–222.

R. Polyak (1992b), "Modified barrier functions in linear programming", Research Report RC 17790 (78331), IBM T.J. Watson Research Center.

M.J.D. Powell (1989), "TOLMIN: a Fortran package for linearly constrained optimization calculations", Report DAMTP 1989/NA2, University of Cambridge.

M.J.D. Powell (1991), "On the number of iterations of Karmarkar's algorithm for linear programming", Report DAMTP 1991/NA23, University of Cambridge.

M.J.D. Powell (1992a), "The complexity of Karmarkar's algorithm for linear programming", in *Numerical Analysis 1991*, eds. D.F. Griffiths and G.A. Watson, Longman Scientific & Technical (Burnt Mill), pp. 142–163.

M.J.D. Powell (1992b), "Some convergence properties of the shifted log barrier method for linear programming", Report DAMTP 1992/NA7, University of Cambridge.

M.J. Todd (1991), "Interior-point algorithms for semi-infinite programming", Technical Report No. 978, Cornell University.

M.H. Wright (1992), "Interior methods for large scale nonlinear optimization problems", presented at the SIAM Conference on Optimization (Chicago).

A DUAL ALGORITHM WITH MULTIPLE CONSTRAINT ADDITIONS FOR STRICTLY CONVEX QUADRATIC PROGRAMMING

I.C. DEMETRIOU

University of Athens
Department of Economics
Division of Mathematics and Informatics
7 Pesmatzoglou Street, Athens, Greece

Abstract An algorithm for strictly convex quadratic programming is presented. At a dual feasible point that is the solution of the quadratic programming problem for a given active set, several violated constraints are added to the active set for the calculation of a step direction. The dual analogue of Goldfarb's sufficient conditions for simultaneous multiple constraint dropping in feasible point methods is applied on the expense of little additional work compared to that of the usual approach where one constraint at a time is treated. Since dual feasibility corresponding to the enlarged active set may be violated, we show how to drop constraints in order to regain it.

1980 Mathematics Subject Classification (1985 Revision): 49D40
Key words and phrases: strictly convex quadratic programming, matrix factorization, dual method, feasible point method

1. Introduction

The strictly convex quadratic programming problem is to calculate the vector of variables x in $\mathbb{R}^n$ that minimizes the objective function

$$F(x) \quad F(0) + x^T\underline{\nabla}F(0) + \tfrac{1}{2}x^TGx \tag{1.1}$$

subject to the linear constraints

$$a_k^Tx - b_k \geq 0, \; k=1,2,\ldots,m, \tag{1.2}$$

where G is a given symmetric positive definite matrix, and, where the n-vectors $\underline{\nabla}F(0)$ and $\{a_k\colon\ k=1,2,\ldots,m\}$, and the numbers $\{b_k\colon\ k=1,2,\ldots,m\}$ are also given. The superscript denotes transposition. For the sake of simplicity we shall ignore that equality constraints may occur in (1.2). The minimization of the function (1.1) subject to the constraints (1.2) has a unique solution vector when the constraints are consistent.

References to several methods for quadratic programming can be found in Goldfarb and Idnani [7] and Fletcher [3,4]. Active set methods have been popular in solving quadratic programming problems. In these methods a sequence of subsets of the constraint indices $\{1,2,...,m\}$ is generated such that the vectors $\{\mathbf{a}_i: i \in E\}$ are linearly independent, for each member E of this sequence. We shall use A to denote the $n \times |E|$ matrix whose columns are the vectors $\{\mathbf{a}_i: i \in E\}$. Further, the vector $\mathbf{x} = \mathbf{x}(E) \in \mathbb{R}^n$ is calculated to minimize (1.1) subject to the equality constraints

$$A^T\mathbf{x} \quad \mathbf{b}_{E'} \tag{1.3}$$

where $\mathbf{b}_E$ is the vector whose components are $\{b_i: i \in E\}$. It follows that unique Lagrange multipliers $\{\lambda_i(E): i \in E\}$ are defined by the condition that the gradient $\underline{\nabla}F(\mathbf{x}(E))$ of $F(\mathbf{x})$ at $\mathbf{x}(E)$ is in the column space of the matrix A, that is

$$\underline{\nabla}F(\mathbf{x}(E)) = A\boldsymbol{\lambda}(E). \tag{1.4}$$

We define $\lambda_i = 0$ for all integers in $[1,m]\setminus E$. Thus $\boldsymbol{\lambda} = \boldsymbol{\lambda}(E)$ is a vector in $\mathbb{R}^m$. The Kuhn-Tucker conditions (see, for example, Fletcher [3, p. 200]) characterize the solution, stating that $\mathbf{x}(E)$ is optimal if and only if it satisfies the constraints (1.2) and $\boldsymbol{\lambda}(E)$, determined by (1.4), satisfies the inequalities

$$\lambda_i(E) \geq 0, \ i = 1,2,...,m. \tag{1.5}$$

The feasible point or primal methods (see Fletcher [3]) choose the active sets so that each $\mathbf{x}(E)$ satisfies all the constraints (1.2), while the dual methods (see Goldfarb and Idnani [7]) satisfy the inequalities (1.5).

In particular, the dual approach of Goldfarb and Idnani [7] was found to be very satisfactory when compared to other quadratic programming methods (see Powell [8,9]).

It is usual for the most popular variants of the mentioned methods to treat one constraint at a time. For strictly convex quadratic programming there exist also primal methods that allow multiple constraint drops (Goldfarb [5,6]). Support to those by evidence is that they reduce the overall number of iterations.

In this paper we describe the more recent dual active set method of Demetriou [1] that allows several violated constraints simultaneously to be added to the active set for the calculation of a search direction.

In Section 2 we outline this method and in Section 3 we provide a quadratic programming algorithm and discuss its termination. In Section 4 we address Goldfarb's sufficient conditions for multiple constraint drops to our dual algorithm for multiple constraint additions.

2. The method of calculation

The following dual method is presented by Demetriou [1] for solving strictly convex quadratic programming problems. The procedure begins from $E=\emptyset$, so the corresponding $x(E)$ gives the unconstrained minimum of $F(.)$. Then it generates infeasible estimates of the solution and in each iteration the value of the objective function moves strictly upwards. At the beginning of an iteration of the algorithm, E, $x(E)$ and $\lambda(E)$ are available so that $\lambda(E)$ satisfies the conditions

$$\lambda_i(E) > 0, \ i \in E. \tag{2.1}$$

If all the constraints (1.2) hold at $x=x(E)$, then the procedure stops because the Kuhn-Tucker conditions are satisfied. Thus, $x(E)$ is the required solution. Otherwise the indices of some violated constraints are added to the active set so as the Lagrange multipliers associated to these constraints to be strictly positive. If indices of negative Lagrange multipliers correspond to the augmented active set, then the procedure starts dropping indices from the active set until the conditions (2.1) are satisfied. At this point, in exact arithmetic, $F(x(E))$ has a strictly larger value than that it had in the beginning of the previous iteration and the constraints (1.2) are again tested.

This approach allows the simultaneous addition of more than one violated constraint to the active set. An immediate question is how to add constraints so as to obtain a dual feasible direction, because the resultant dual search direction after adding more than one constraint may be infeasible. One of the algorithms in Goldfarb [5] provides sufficient conditions for this question when primal feasibility is preserved in the iterations. We give a dual implementation of this algorithm in Section 4.

Adding, however, more than one constraint to the active set may cause division by zeros when recovering dual feasibility whenever it is lost. Demetriou [1] considers a suitable subset of the active set for determining the constraint drops and establishes finite termination.

Before we formally present an algorithm for this calculation we need to introduce some notation. We implement our algorithm by employing the matrix factorizations of Goldfarb and Idnani [7]. In addition to A and G, the matrices J and R are available when the active set changes. The matrix J is n×n and satisfies both the equation

$$JJ^T = G^{-1} \tag{2.2}$$

and the condition that

$$J^TA = \begin{bmatrix} R \\ O \end{bmatrix}, \tag{2.3}$$

where R is the $|E|\times|E|$ upper triangular matrix whose rows are the first $|E|$ rows of J^TA, the symbol $|.|$ denoting set cardinality. Further, the matrix J is partitioned so as

$$J = [\; J_1 \;|\; J_2 \;] \quad \underset{\leftarrow|E|\rightarrow}{} \; \underset{\leftarrow n-|E|\rightarrow}{}$$

When the iterations start (ie. $E=\emptyset$), we make the choice

$$J = L^{-T},$$

where $G=LL^T$ is the Cholesky factorization of G. Thus the initial vector is formed by

$$\mathbf{x} = -J_1J_1^T\underline{\nabla}F(0) \tag{2.4}$$

For any subsequent E, the matrices J and R are generated by updating methods and $\mathbf{x}(E)$ is given by

$$\mathbf{x}(E) = \hat{\mathbf{x}} - J_2J_2^T\,\underline{\nabla}F(\hat{\mathbf{x}}) \quad \text{or} \quad \mathbf{x}(E) = \hat{\mathbf{x}}, \text{ if } |E|=n, \tag{2.5}$$

where

$$\hat{\mathbf{x}} = J_1R^{-T}b(E). \tag{2.6}$$

The vector of the Lagrange multipliers associated to the vector $\mathbf{x}(E)$, may be formed by updating or it may be calculated from the formula

$$\lambda(E) = R^{-1}J_1^T\,\underline{\nabla}F(\hat{\mathbf{x}}), \tag{2.7}$$

where $\hat{\mathbf{x}}$ occurs in (2.6). Techniques for updating J and R when E changes are described in [7].

3. A dual active set algorithm that allows multiple additions

We present a quadratic programming algorithm that follows the approach outlined in the previous section. In order to prevent zero divisors in Step 4 below, when several constraints are simultaneously added to the active set at $\mathbf{x}(E)$, Demetriou [1] keeps in E only those indices that are associated to non-zero Lagrange multipliers. Thus he defines

$$E = \{i: \lambda_i(E)\neq 0,\ 1\leq i\leq m\}. \tag{2.8}$$

Algorithm 1 (Quadratic programming)

Step 0 Set $\tilde{E}=\emptyset$ and $\lambda(\tilde{E})=0$. Compute J_1 and find the unconstrained minimum $\mathbf{x}=\mathbf{x}(\tilde{E})$ from (2.4).

Step 1 End the calculation if $\mathbf{x}(\tilde{E})$ satisfies all the constraints (1.2), because then it follows from dual feasibility that the Kuhn-Tucker conditions hold, so $\mathbf{x}(\tilde{E})$ is the required solution.

Step 2 Otherwise the vector $\mu=\lambda(\tilde{E})$ is recorded for use in Step 4. Then apply, for example, Algorithm 2 (which is described in Section 4) in order to form the index sets $\tilde{V}$ and E such that $E=\tilde{E}\cup\tilde{V}$ and $\lambda_i(E)>0$, $i\in\tilde{V}$, where $\lambda(E)$ may be calculated from (2.7).

Step 3 If $\lambda_i(E)>0$, for $i\in E$, then replace $\tilde{E}$ by E and go to Step 1.

Step 4 Otherwise, let $\tau=\min\{\lambda_i(E)/\mu_i: \lambda_i(E)<0,\ i\in E\}$ and let $\theta=1/(1-\tau)$. Note that in view of the definitions of τ, μ and E it follows that θ is the largest number that satisfies the restrictions $(1-\theta)\mu_i+\theta\lambda_i(E)\geq 0$, $i\in E$ and the inequalities $0\leq\theta<1$. Then replace μ by the vector $(1-\theta)\mu+\theta\lambda(E)$ which also has non-negative components. Then replace E by $E\backslash\{i:\mu_i=0\}$, update J and R, calculate $\lambda(E)$ and go back to Step 3. □

In the implementation of Step 0 we have $\tilde{E}=\emptyset$ and $\lambda(\tilde{E})=\mathbf{0}$. Algorithm 1 either iterates between the Steps 1, 2 and 3, or, between the Steps 3 and 4. At least one cycle of the Steps 1, 2 and 3, without reaching Step 4, is performed after exiting Step 0. Step 1 is always reached with a set $\tilde{E}$ such that $\lambda_i(\tilde{E})>0$, for $i\in\tilde{E}$, or $\tilde{E}=\emptyset$. In the implementation of Step 2, the vector $\lambda(E)$ is calculated immediately after the completion of Algorithm 2. At exit of Algorithm 2 some multipliers in $E\backslash\tilde{V}$ may not be positive. In this case Step 4 is entered with $E\neq\emptyset$ and Algorithm 1 cycles between Steps 3 and 4 until it eventually exits to Step 1 with $\{\lambda_i(\tilde{E})>0: i\in\tilde{E}\}$. The first visit to Step 4 picks an index k such that $k\notin\tilde{V}$. At least this k cannot be in $\tilde{E}$ on the next visit to Step 1 although it was in the previous $\tilde{E}$. Further, due to the definition (2.8) all operations of Step 4 are well defined, while this step reduces E by one. If, however, Step 4 is not reached, then the new $\tilde{E}$ is the old one augmented by $\tilde{V}$. Thus cycling cannot occur through Step 4 of Algorithm 1 with no change to $\tilde{E}$.

The following lemma [1] states that the function (1.1) never decreases between successive feasible vectors of Lagrange multipliers in Step 4, and strictly increases any time Step 1 is reached, even after at least one iteration of Steps 3 and 4. Therefore Step 1 is reached with the set $\{\lambda_i(\tilde{E}): i\in\tilde{E}\}$ that is different from any past such set. Since $\tilde{E}$ is a subset of $\{1,2,\ldots,m\}$ and since there is only a finite number of these subsets, it follows that Algorithm 1 is a finite procedure for calculating the solution of the strictly convex quadratic programming problem that is stated in Section 1.

Lemma 1 Let E_1 and E_2 be the sets $\tilde{E}$ that occur on any two consecutive visits to Step 1 in Algorithm 1, where $\tilde{E}$ is a non-empty subset of the constraint indices $\{1,2,\ldots,m\}$ such that the vectors $\{\mathbf{a}_i: i\in\tilde{E}\}$ are linearly independent. Then the strict inequality

$$F(\mathbf{x}(E_1)) < F(\mathbf{x}(E_2))$$

is obtained. □

4. Goldfarb's sufficient conditions for multiple additions

Corresponding to the strictly convex quadratic programming problem of Section 1, there exists (see Fletcher [3, p. 222]) a unique dual problem whose solution can be obtained by maximizing the dual function

$$L(\lambda) = F(\mathbf{x}) - \sum_{i=1}^{m} \lambda_i \left[\mathbf{a}_i^T\mathbf{x}-b_i\right] \tag{4.1}$$

subject to the nonnegativity constraints $\lambda_i \geq 0$, $i=1,2,\ldots,m$. The gradient of the function (4.1) is easy to calculate because it is given by the constraints of the strictly convex quadratic program of Section 1,

$$\frac{\partial L(\lambda)}{\partial \lambda_i} \quad -\left[\mathbf{a}_i^T\mathbf{x}(E)-b_i\right], \tag{4.2}$$

where $\mathbf{x}(E)$ denotes the vector that satisfies (1.4) for given λ.

If we solve (1.4) for $\mathbf{x}=\mathbf{x}(E)$ in terms of this λ and eliminate $\mathbf{x}$ from (4.1) then the second derivative matrix of $L(\lambda)$ is negative definite. Thus $L(.)$ is a strictly concave differentiable function. Therefore, if E^* is the final active set of the calculation and $\lambda^*=\lambda(E^*)$, then the optimality conditions for the dual problem are

$$\begin{aligned} u_i &= \frac{\partial L(\lambda^*)}{\partial \lambda_i} < 0\,, \quad \text{if } \lambda_i^*=0,\ i\in E^* \\ u_i &= \frac{\partial L(\lambda^*)}{\partial \lambda_i} \quad 0\,, \quad \text{if } \lambda_i^*\geq 0,\ i\notin E^*, \end{aligned} \tag{4.3}$$

where $\{u_i: i=1,2,\ldots,m\}$ are the Lagrange multipliers of the dual problem. In view of (4.2) and (4.3) it follows that the Lagrange multipliers of the dual problem are given by the constraints (1.2).

With this in mind, Goldfarb's conditions [5] that allow several constraints to be simultaneously dropped from the active set in feasible point methods, are immediately stated for adding constraints in our dual method.

Lemma 2 (Goldfarb's sufficient conditions for multiple additions) Let E_q denote the non-empty set $\{1,2,\ldots,q\}$ such that $q<m$ and let λ be any

vector that satisfies the equations $\{\lambda_i=0:\ i\notin E_q\}$. If k is an integer such that $q<k\le m$ and

$$a_i^T x(E_{j-1}) < a_i^T x(E_j),\ q\le i<j,\ q+1<j\le k$$

and if

$$a_j^T x(E_j) < b_j\ ,\quad q<j\le k$$

then

$$\lambda_j(E_k) > 0 \quad \text{for } q<j\le k.\ \square$$

The conditions of Lemma 2 are sufficient but not necessary for determining a feasible dual direction. However, they allow one to easily check whether adding a violated constraint into the active set will give a dual feasible direction. Moreover, some strategies suggested by Goldfarb [6] for primal methods may analogously be applied to our dual algorithm. The algorithm below employs one of those strategies that allows only a single constraint at a time to be added to the active set. It requires only an additional n-vector storage with respect to Algorithm 1.

Algorithm 2 (Multiple additions to the active set)

Step 0 Let $\tilde{E}$ be a subset of the constraint indices $\{1,2,...,m\}$ such that the corresponding constraint normals are linearly independent and $\lambda_i(\tilde{E})>0$, for $i\in\tilde{E}$. We assume that

$$V = \{j:\ a_j^T x(\tilde{E})-b_j<0,\ \text{for } j\notin\tilde{E}\}$$

is a non-empty set of indices of violated constraints at $x(E)$, so that the constraint normals $\{a_i:\ i\in V\}$ are linearly independent. Set $q=|\tilde{E}|$, $p=0$ and $E_q=\tilde{E}$. Without loss of generality we assume that $E_{q+p}=\{1,2,...,q\}$ and $V=\{q+1,q+2,..., q+|V|\}$.

Step 1 Set $p=p+1$. Let $q+p$ be the index of the most negative constraint value in V. Move this index in E_{q+p}, and update J and R. Calculate $x(E_{q+p})$ from (2.5) after replacing E by E_{q+p}. Determine the set

$$V=\{j:\ a_j^T x(E_{q+p})-b_j<0,\ a_j^T x(E_i)\le a_j^T x(E_{i+1}),\ j\notin E_{q+p},\ i=q,q+1,...,q+p-1\}\}.$$

Step 2 If $V=\emptyset$, set $E=E_{q+p}$ and $\tilde{V}=V$, and exit from this algorithm. Otherwise branch to Step 1. $\square$

No numerical experience, so far, is available on the performance of

Algorithm 1 for the general strictly convex quadratic problem. Work is under way for developing a Fortran program that implements the algorithms of this work. However, the method was applied to certain structured large scale calculations [2] and special purpose programs have been written (cf references within [2] and [1]). The derived numerical results [1] show that Algorithm 1 compares favorably with those when only one constraint at a time is treated.

References

1. Demetriou, I.C. *On a quadratic programming algorithm for l_2 discrete convex approximation.* Report INFOS/ΠΥΜ-90-32, Athens University of Economics, Greece, 1990, (Revision for Math. of Computation).

2. Demetriou I.C. and Powell M.J.D. *The minimum sum of squares change to univariate data that gives convexity.* IMA Journal of Numerical Analysis, 11(1991), pp 433-448.

3. Fletcher, R. **Practical methods of optimization**, John Wiley & Sons, Chichester, 1987.

4. Fletcher, R. *Recent developments in linear and quadratic programming*, in **The State of the Art in Numerical Analysis**, eds A. Iserles & M.J.D. Powell, Academic Press, 1986, pp 213-243.

5. Goldfarb, D. *Extensions of Newton's method and Simplex methods for solving quadratic programs*, in **Numerical Methods for Non-linear Optimization**, ed F.A. Lootsma, Academic Press, 1972, pp 239-254.

6. Goldfarb, D. *Strategies for constraint deletion in active set algorithms* in **Numerical Analysis 1985**, eds D.F. Griffiths and G.A. Watson, Pitman 1986, pp 66-81.

7. Goldfarb, D. and Idnani A. *A numerically stable dual method for solving strictly convex quadratic programs*, Math. Programming, 27(1983), pp 1-33.

8. Powell M.J.D. *On the quadratic programming algorithm Goldfarb and Idnani*, in Math. Programming, *Essays in Honor of George B. Dantzig, Part* EE, ed. R.W. Cottle, Math. Programming, Series B', 25(1985).

9. Powell, M.J.D. *ZQPCVX: A Fortran subroutine for convex quadratic programming*, Report DAMTP/1983/NA17, University of Cambridge, England.

TWO NUMERICAL METHODS FOR THE CONFORMAL MAPPING OF SIMPLY- AND MULTIPLY-CONNECTED DOMAINS

N. Papamichael
Department of Mathematics and Statistics
Brunel University, Uxbridge, Middlesex UB8 3PH, United Kingdom

Abstract This paper is a report of recent developments concerning the theory and application of : (i) The well-known Bergman kernel method for the conformal mapping of simply-connected domains. (ii) A closely related orthonormalization method for the conformal mapping of multiply-connected domains.

1 INTRODUCTION

This paper is a report of recent developments concerning the theory and application of two numerical methods for conformal mapping. The methods under consideration are respectively the well-known Bergman kernel method (BKM), for approximating the conformal map of a simply-connected domain onto the unit disc, and a closely related (but lesser known) orthonormalization method (ONM) for approximating one of the standard conformal maps for multiply-connected domains (viz. the conformal map of a multiply-connected domain onto a slit circular ring).

Let Ω be a finite N-connected ($N \geq 1$) domain, whose boundary $\partial\Omega$ consists of N closed piecewise analytic Jordan curves, and let $L^2(\Omega)$ denote the Hilbert space of all square integrable functions which are analytic in Ω , ie.

$$L^2(\Omega) := \{ g : g \text{ analytic in } \Omega , \| g \|^2 = (g , g) < \infty \} , \tag{1.1}$$

where

$$(g , h) := \iint_{\Omega} g(z)\, \overline{h(z)}\, dxdy . \tag{1.2}$$

Also, let f denote the conformal map of Ω which is to be approximated by the BKM, in the case $N = 1$, or by the ONM in the case $N > 1$. Then, both these methods approximate f by explicit expressions, in terms of an appropriately chosen set of functions $\{\mu_j\}$ from the space $L^2(\Omega)$. More specifically, in the case $N = 1$, the BKM leads to approximations of the form

$$f_n(z) = \sum_{j=1}^{n} a_j\, \mu_j(z) , \tag{1.3}$$

whilst in the case $N > 1$ the ONM approximations to f are, typically, of the form

$$f_n(z) = z \exp \{ \sum_{j=1}^{n} a_j \mu_j(z) \} \tag{1.4}$$

The specific objectives of the paper are as follows :

(i) To present in a concise form a report of the substantial work that has been done in recent years (notably at Brunel University), in connection with the application of the BKM and the ONM to the mapping of " *difficult* " domains, such as domains with nonsmooth boundaries. (As might be expected, the choice of the function set $\{\mu_j\}$ plays a very decisive role in such applications. Thus, one of our main objectives is to outline the available techniques for selecting this set.)

(ii) To report on the present state of the theory of the two numerical methods and to highlight any questions that remain unanswered.

(iii) To indicate how the application of the ONM can be extended to some of the other standard conformal maps for multiply-connected domains.

In our work we shall refer frequently to the following well-known result concerning the inner products (1.2) of the space $L^2(\Omega)$.

— *Complex form of Green's formula* : Let g , $h \in L^2(\Omega)$ be continuous on $\partial\Omega$ apart, possibly, from a finite number of branch point singularities of the form

$$(z - z_0)^\alpha, \quad z_0 \in \partial\Omega, \quad -\tfrac{1}{2} < \alpha < 0. \tag{1.5}$$

Then,

$$(g , h) = \frac{1}{2i} \int_{\partial\Omega} g(z) \overline{\mu(z)} \, dz , \tag{1.6}$$

where $\mu' = h$.

In the literature (1.6) is usually proved under the assumption that the functions g and h are continuous on $\partial\Omega$ (cf. [2, pp.59, 96], [3, p.118], [10, p.240]). It is, however, easy to deduce from these proofs that the result also holds under the somewhat weaker continuity assumptions stated above.

2 SIMPLY-CONNECTED DOMAINS

2.1 The Bergman kernel method (BKM)

Let Ω be a simply-connected domain in the complex plane **C**, whose boundary $\partial\Omega$ is a closed piecewise analytic Jordan curve, assume that $0 \in \Omega$ and let $L^2(\Omega)$ denote the Hilbert space (1.1). Also, let $w = f(z)$ denote the conformal mapping of Ω onto the unit disc $D := \{ w : |w| < 1 \}$ such that

$$f(0) = 0 \quad \text{and} \quad f'(0) > 0 . \tag{2.1}$$

Finally, let $K(\cdot, 0)$ denote the Bergman kernel function of Ω with respect to 0.

The BKM is a method for approximating the conformal map $f : \Omega \to D$. The method is based on the following two properties of the kernel function $K(\cdot, 0)$ (cf. [2], [3], [4], [10]):

— *Reproducing property*: The kernel $K(\cdot, 0)$ is uniquely determined by the reproducing property

$$(\eta, K(\cdot, 0)) = \eta(0), \quad \forall \eta \in L^2(\Omega). \tag{2.2}$$

— *Relation of* $K(\cdot, 0)$ *with* f : The kernel $K(\cdot, 0)$ is related to the mapping function f by means of

$$f'(z) = \sqrt{\frac{\pi}{K(0,0)}}\, K(z, 0), \text{ ie. } f(z) = \sqrt{\frac{\pi}{K(0,0)}} \int_{t=0}^{z} K(t, 0)\, dt. \tag{2.3}$$

These two properties show respectively that the determination of the Fourier coefficients of the kernel function $K(\cdot, 0)$ (with respect to any orthonormal system for $L^2(\Omega)$) does not require the explicit knowledge of the kernel, and that the derivative f' of the mapping function is a scalar multiple of $K(\cdot, 0)$.

Let $\{\eta_j\}_1^\infty$ be a complete set of functions for the space $L^2(\Omega)$. Then, (2.2) and (2.3) suggest the following numerical procedure for approximating the mapping function f ·

(i) Orthonormalize the functions $\{\eta_j\}_1^n$ by means of the Gram-Schmidt process to obtain the orthonormal set $\{\eta_j^*\}_1^n$.

(ii) Approximate the kernel function $K(\cdot, 0)$ by the Fourier sum

$$\begin{aligned} K_n(z, 0) &= \sum_{j=1}^{n} (K(\cdot, 0), \eta_j^*)\, \eta_j^*(z) \\ &= \sum_{j=1}^{n} \overline{\eta_j^*(0)}\, \eta_j^*(z). \end{aligned} \tag{2.4}$$

(That is, $K_n(\cdot, 0)$ is the least squares approximation of $K(\cdot, 0)$ in span ($\eta_1, \eta_2, \ldots, \eta_n$).)

(iii) Approximate f' and f respectively by

$$f_n'(z) = \sqrt{\frac{\pi}{K_n(0,0)}}\, K_n(z, 0) \quad \text{and} \quad f_n(z) = \sqrt{\frac{\pi}{K_n(0,0)}} \int_{t=0}^{z} K_n(t, 0)\, dt. \tag{2.5}$$

We shall refer to the approximations (2.4)-(2.5) as the " *nth BKM approximations to* $K(\cdot, 0)$, f' *and* f *corresponding to the basis set* $\{\eta_j\}$ "

2.2 Choice of the basis set

The first step of the BKM involves the use of the Gram-Schmidt process which, as is well-known, can be extremely unstable. For this reason, the success of the method depends strongly on the speed of convergence of the approximating Fourier series (2.4) and this, in turn, depends critically on the basis set $\{\eta_j\}$. Thus, the choice of the basis set plays a very decisive role in numerical computations.

The classical BKM is based on the use of the " *monomial set* "

$$\eta_j(z) = z^{j-1}; \quad j = 1, 2, \ldots . \tag{2.6}$$

This set is complete in the space $L^2(\Omega)$ and provides a computationally convenient basis. Unfortunately, however, the convergence of the resulting approximating polynomial series is often extremely slow. Thus, because of the instability, the classical BKM does not always lead to approximations of acceptable accuracy. As might be expected, the slow convergence associated with the use of the basis set (2.6) is due to the presence of singularities that the kernel function $K(\cdot, 0)$ (or, equivalently, the derivative f' of the mapping function) may have either on the boundary $\partial\Omega$ or close to $\partial\Omega$ in $\mathbf{C} \setminus \overline{\Omega}$.

The above numerical difficulty can be remedied by using, instead of (2.6), a basis set that reflects the main singular behaviour of $K(\cdot, 0)$ on $\partial\Omega$ and in $\mathbf{C} \setminus \overline{\Omega}$. This approach was first proposed by Levin et al [9], who recommended the use of " *augmented basis sets* " formed by introducing appropriate " *singular functions* " into the monomial set (2.6). The singular functions needed for this purpose are determined, as outlined below, by considering the corner singularities of f on $\partial\Omega$ and also the so-called " *pole-type* " singularities that the analytic extension of f may have in $\mathbf{C} \setminus \overline{\Omega}$.

2.2.1 Corner singularities

Any boundary singularities of f are corner singularities, similar to those that arise in the study of elliptic boundary value problems. The asymptotic form of these singularities can be deduced easily from the results of Lehman [8], concerning the development of the mapping function at an analytic corner. The details are as follows :

Assume that part of the boundary $\partial\Omega$ consists of two analytic arcs which meet at a point z_c and form there a corner of interior angle $\omega\pi$, where $0 < \omega < 2$. Then, depending on whether ω is rational or irrational, the results of [8] lead to the following asymptotic expansions (cf. [9], [11], [17]) :

(i) If $\omega = p/q$, with p and q relative prime, then as $z \to z_c$

$$f(z) - f(z_c) = \sum_{k,l,m} B_{klm}(z - z_c)^{k+l/\omega}(\log(z - z_c))^m , \tag{2.7}$$

where k, l and m run over all integers $k \geq 0$, $1 \leq l \leq p$, $0 \leq m \leq k/q$ and where $B_{010} \neq 0$. Also the terms in (2.7) are ordered so that the term corresponding to B_{klm} precedes the term corresponding to $B_{k_1 l_1 m_1}$ if either $k + l/\omega < k_1 + l_1/\omega$ or $k + l/\omega = k_1 + l_1/\omega$ and $m > m_1$.

(ii) If ω is irrational, then as $z \to z_c$

$$f(z) - f(z_c) = \sum_{k,l} B_{kl} (z - z_c)^{k+l/\omega}, \tag{2.8}$$

where now k and l run over all integers $k \geq 0$, $l \geq 1$ and where $B_{01} \neq 0$.

(iii) If the two arms of the corner are both straight lines or circular arcs, then the expansions (2.7) and (2.8) simplify to

$$f(z) - f(z_c) = \sum_{l=1}^{\infty} B_l (z - z_c)^{l/\omega}, \quad \text{as } z \to z_c, \tag{2.9}$$

where $B_1 \neq 0$.

It follows from the above that the dominant term in the asymptotic expansion of f is always $(z - z_c)^{1/\omega}$ Therefore, when $1/\omega$ is not an integer a branch point singularity always occurs at the corner z_c. Furthermore, because of the logarithmic terms in (2.7), a branch point singularity might also occur even when $1/\omega$ is an integer.

Since the kernel $K(\cdot, 0)$ is related to the derivative of f by means of the first equation in (2.3), the BKM basis set is constructed by introducing into the monomial set (2.6) the derivatives of the first few singular terms from the appropriate series (2.7), (2.8) or (2.9). That is the singular functions for dealing with corner singularities are of the form

$$\eta(z) = \frac{d}{dz} \{ (z - z_c)^r \}, \quad \text{with } r = k + l/\omega \text{ or } r = l/\omega, \tag{2.10a}$$

or

$$\eta(z) = \frac{d}{dz} \{ (z - z_c)^{k+l/\omega} (\log(z - z_c))^m \}. \tag{2.10b}$$

2.2.2 Pole-type singularities

Apart from corner singularities the mapping function f (or more precisely its analytic extension) may also have serious singularities off the boundary in $\mathbf{C} \setminus \bar{\Omega}$. For example, if $\partial\Omega$ consists of straight line segments and circular arcs, then f may have simple pole singularities at the mirror images of 0 in the straight lines and at the geometric inverses of 0 with respect to the circular arcs. This follows easily from the well-known Schwartz reflection principle (cf. [9], [11]). For more general boundaries, the singularities of f in $\mathbf{C} \setminus \bar{\Omega}$ can be determined as indicated in [16] and [17, § 5.1], by using the symmetry principle for analytic arcs (cf. [18, p.102]). Let γ be an analytic arc of $\partial\Omega$ with analytic parametric equation

$$z = \tau(s), \quad s_1 \leq s \leq s_2. \tag{2.11}$$

Then, the general procedure for determining the singularities of the analytic extension of f across γ involves examining the zeros of the function $\tau(\zeta)$ (of the complex variable $\zeta = s + it$) and of its derivative $\tau'(\zeta)$. This can be seen by observing that the analytic extension of f across γ is given by

$$\{ \overline{f(I(z))} \}^{-1} \quad \text{where} \quad I(z) := \tau\{ \overline{\tau^{[-1]}(z)} \} \tag{2.12}$$

Although simple poles are, probably, the most frequently occurring singularities, other types of singularities may also occur. For example, f may have a double pole at some point $z_0 \in \mathbf{C} \setminus \bar{\Omega}$ or branch point singularities of the form

$$(z - z_0)^{1/2} \quad \text{or} \quad (z - z_0)^{-1/2} \tag{2.13}$$

Following [16] and [17], we shall refer to all such singularities that f may have in $\mathbf{C} \setminus \bar{\Omega}$ as " *pole-type singularities* "

The procedure for dealing with pole-type singularities is exactly the same as that used in the case of singular corners. That is, the BKM basis set is formed by introducing into the monomial set (2.6) singular functions that reflect the dominant singularities of f' in $\mathbf{C} \setminus \bar{\Omega}$. For example, the singular function for treating a simple pole at a point $z_0 \in \mathbf{C} \setminus \bar{\Omega}$ is

$$\eta(z) = \frac{d}{dz} \{ z / (z - z_0) \} . \tag{2.14}$$

Similarly, the singular functions for treating branch point singularities the form (2.13) are

$$\eta(z) = \frac{d}{dz} \{ (z - z_0)^{1/2} \} \quad \text{and} \quad \eta(z) = \frac{d}{dz} \{ (z - z_0)^{-1/2} \} \tag{2.15}$$

If the domain Ω has M-fold ($M \geq 2$) rotational symmetry about the origin, then the number of basis functions in the BKM can be reduced considerably, in the sense that the monomial set (2.6) can be replaced by the set

$$z^{jM} ; \quad j = 0, 1, \dots . \tag{2.16}$$

Similarly, any M singular basis functions, that correspond to corner or pole singularities at M symmetric points, can be combined into a single singular basis function (cf. [9], [11]).

2.3 Computation of inner products

The main computation involved in determining the BKM approximations (2.5) occurs in step (i) of the process, ie. in the orthonormalization of the basis set $\{\eta_j\}_1^n$ and, in particular, in the evaluation of the inner products (η_k, η_l). Since all the basis functions (2.6), (2.10), (2.14) and (2.15) satisfy the boundary continuity requirement needed for the result (1.6), these inner products can be expressed as

$$(\eta_k, \eta_l) = \frac{1}{2i} \int_{\partial\Omega} \eta_k(z) \overline{\mu_l(z)} \, dz , \quad \mu_l' = \eta_l , \tag{2.17}$$

and can then be computed, as indicated in [9] and [11], by Gauss - Legendre quadrature. Naturally, when performing the quadrature, care must be taken to deal with the integrand singularities that occur when, due to the presence of a corner, the basis set contains singular functions of the form (2.10). Fortunately, the effects of such singularities can either be reduced considerably or, in many cases, removed completely by using a suitable parametrization of the boundary (cf. [9], [11]).

2.4 Stability - Convergence - Error estimates

By comparison with other methods for numerical conformal mapping, the stability and convergence theories of the BKM are in a good shape. There is, however, one notable gap in the convergence theory.

With reference to stability, we have a geometrical characterization of the degree of instability and computable indicators which measure the level of instability at each step of the numerical process. Full details of these can be found in [15, pp. 642-649] where, for example, it is shown that :

(i) Best stability occurs when $\partial\Omega$ is " *nearly circular* " . Conversely, the orthonormalization process is seriously unstable when Ω is a " *thin domain* "

(ii) For the purposes of stability, the origin 0 should be positioned so that its distance from $\partial\Omega$ is as small as possible.

(iii) The introduction of singular basis functions can, in some cases, lead to a deterioration of the stability of the process. This occurs when singular functions of the form (2.14) or (2.15) are used to reflect pole-type singularities at points which lie far from $\partial\Omega$ in $\mathbf{C} \setminus \bar{\Omega}$.

Regarding convergence, it follows at once from (2.4) that

$$\lim_{n\to\infty} \| K_n(\cdot, 0) - K(\cdot, 0) \| = 0 \tag{2.18}$$

and, in the space $L^2(\Omega)$, this norm convergence implies that $K_n(z, 0) \to K(z, 0)$ uniformly on each compact subset of Ω . From this it follows that $f_n(z) \to f(z)$ uniformly on each compact subset of Ω (cf. [4, p.32]). In addition, there are several results which, under various assumptions on the smoothness of $\partial\Omega$, establish the uniform convergence in $\bar{\Omega} = \Omega \cup \partial\Omega$ of the BKM approximations corresponding to the monomial set (2.6), and also provide estimates for the rates of convergence. For example, it is known that if f has a regular extension to $\bar{\Omega}$, then

$$\max_{z\in\bar{\Omega}} | f_n(z) - f(z) | = 0(\rho^{-n}), \tag{2.19}$$

for some $\rho > 1$ which depends on the distance of the nearest singularities of the analytic extension of f from $\partial\Omega$. On the other hand, if f is not regular on $\partial\Omega$, then it is known that

$$\max_{z\in\bar{\Omega}} | f_n(z) - f(z) | = 0(n^{-\gamma}), \tag{2.20}$$

for some $\gamma > 1$ which depends on the smoothness of $\partial\Omega$. (See e.g. [1], [3, p.125], [4, pp.27-29], [7], [15], [19], [20] and, in particular, the paper by Gaier [5] which, as far as we are aware, contains the most recent convergence results for BKM approximations corresponding to the use of the monomial basis (2.6)). Finally, the available theory can be used to explain the substantially improved rates of convergence which are achieved by using appropriate augmented basis sets, in cases where f is regular in $\bar{\Omega}$ and the singularities nearest to $\partial\Omega$ are simple poles (cf. [15, pp.651-652]).

The only serious gap in the convergence theory of the BKM concerns the use of augmented basis sets involving functions of the form (2.10), for treating corner singularities of f on $\partial\Omega$. As is illustrated in [9], [11], [15] and [17], the use of such sets leads to substantial improvements in accuracy, but so far we do not have any theoretical estimates for these improvements.

From the more practical point of view, the principle of maximum modulus leads to a reliable method for determining an estimate of the maximum error in $|f_n(z)|$. Since $|f(z)|=1$ for $z \in \partial\Omega$, this estimate is given, quite simply, by

$$E_n := \max_j |\, 1 - |f_n(z_j)|\, |, \tag{2.21}$$

where $\{z_j\}$ is a set of " *boundary test points* ".

2.5 Optimum number of basis functions

The BKM should be programmed so that it computes recursively a sequence of approximations $\{f_n\}$, where at each stage the number of basis functions used is increased by one. If this is done, then it is quite simple to include a termination criterion for determining the " *optimum number* " n_{opt} of basis functions which gives a " *best* " approximation $f_{n_{opt}}$ in some predefined sense. In [9], [11] and [15], the number n_{opt} is determined by using essentially the following procedure :

A minimum number $n_{\min}$ of basis functions to be used is defined and, for each $n \geq n_{\min}$, the error estimate E_n is computed. If at the ($n+1$)-th stage the inequality

$$E_{n+1} < E_n \tag{2.22}$$

is satisfied, then the approximation f_{n+2} is computed. When for a certain value of n, due to numerical instability, (2.22) no longer holds, then the process is terminated and n is taken to be the optimum number n_{opt} of basis functions.

2.6 Exterior conformal maps

Let Γ be a closed piecewise analytic Jordan curve and let Ω_E denote the exterior domain $\Omega_E := \text{Ext}\,\Gamma$. Also, let $w = \phi(z)$ denote the conformal map of Ω_E onto the exterior of the unit circle $D_E := \{ w : |w| > 1 \}$, normalized by the conditions $\phi(\infty) = \infty$ and $\phi'(\infty) > 0$. Then, the inversion $z \to z^{-1}$ transforms Γ onto a closed piecewise analytic Jordan curve $\hat{\Gamma}$ and maps conformally Ω_E onto the interior domain $\hat{\Omega} := \text{Int}\,\hat{\Gamma}$ Therefore, if $\hat{f}$ is the mapping function

$$\hat{f} : \hat{\Omega} \to D := \{ w : |w| < 1 \} \tag{2.23}$$

then

$$\phi(z) = \{ \hat{f}(z^{-1}) \}^{-1} \tag{2.24}$$

Hence, the n th BKM approximation to the exterior mapping function ϕ can be computed from

$$\phi_n(z) = \{ \hat{f}_n(z^{-1}) \}^{-1}, \tag{2.25}$$

where $\hat{f}_n$ denotes the n th approximation to $\hat{f}$, obtained by applying the BKM to the interior domain $\hat{\Omega}$.

It follows from the above that all the details of the BKM remain essentially unchanged. In fact, the only item that requires some attention is the determination of the augmented basis sets which, because of (2.24) and (2.25), are now constructed by considering the singularities of the mapping function $\hat{f}$ on $\partial\hat{\Omega} := \hat{\Gamma}$ and in $\mathbf{C} \setminus \overline{\hat{\Omega}}$ (cf. [12], [17, § 5]).

2.7 Numerical examples

The results of many numerical experiments illustrating the high accuracy that can be achieved by applying the BKM with an appropriate augmented basis set can be found in [9], [11], [12], [15] - [17]. Here we merely present the results of two examples taken from [15]. These were computed on a CDC 7600 computer, using programs written in single precision Fortran. (Single length working on the CDC 7600 is between 14 and 15 significant figures.)

In presenting the results we use the abbreviations BKM/MB and BKM/AB to denote respectively the BKM with monomial basis and with augmented basis. We also use the notations $f_n^{(M)}$ and $f_n^{(A)}$ to denote respectively the n th BKM/MB and BKM/AB approximations to f

(I) Let Ω be the ellipse

$$\Omega := \{ (x, y) : x^2/a^2 + y^2 < 1 ,\ a > 1 \} , \tag{2.26}$$

Then, the mapping function f is regular on $\partial\Omega$, but its analytic extension has simple poles at an infinite array of points on the imaginary axis (cf. [15, p.656]). The poles nearest to $\partial\Omega$ occur at the points

$$z = = \pm 2ia / (a^2 - 1)^{1/2} =: \pm ip \tag{2.27}$$

(cf. [15, p.656]). (Notice that the numerical mapping problem becomes more difficult as a increases, in the sense that the singular points (2.27) approach the boundary points $\pm 2i$ as a increases.)

Because of the two-fold rotational symmetry, the monomial basis set is taken to be

$$z^{2(j-1)} ;\quad j = 1, 2, \dots . \tag{2.28}$$

For the same reason, in the BKM/AB, the augmented basis set is formed by introducing into the set (2.28) the single singular function

$$\eta(z) := \frac{d}{dz} \{ z / (z^2 + p^2) \} , \tag{2.29}$$

which reflects the two simple pole singularities of f at the points (2.27).

The optimum BKM/MB and BKM/AB results corresponding to the three cases $a = 2$, $a = 4$ and $a = 8$ are as follows :

(i) $a = 2$

BKM/MB : $n_{opt} = 19$, $E_{19} = 2.6 \cdot 10^{-9}$

BKM/AB : $n_{opt} = 6$, $E_6 = 3.5 \cdot 10^{-13}$

(ii) $a = 4$

BKM/MB : $n_{opt} = 14$, $E_{14} = 1.2 \cdot 10^{-3}$

BKM/AB : $n_{opt} = 11$, $E_{11} = 7.5 \cdot 10^{-12}$

(iii) $a = 8$

BKM/MB : $n_{opt} = 12$, $E_{12} = 6.6 \cdot 10^{-2}$

BKM/AB : $n_{opt} = 16$, $E_{16} = 5.6 \cdot 10^{-8}$

As is explained in [15, pp. 656-660], the above results are in full accord with the stability and convergence theories of the method. For example, the convergence theory predicts that

$$\max_{z \in \bar{\Omega}} | f(z) - f_n^{(M)}(z) | = 0(\rho^{-n}), \quad \forall \rho, \ 1 < \rho < \rho^{(M)}, \tag{2.30}$$

and

$$\max_{z \in \bar{\Omega}} | f(z) - f_n^{(A)}(z) | = 0(\rho^{-n}), \quad \forall \rho, \ 1 < \rho < \rho^{(A)}, \tag{2.31}$$

where :

(i) $\rho^{(M)} = 3.0$ and $\rho^{(A)} = 243$ when $a = 2$.

(ii) $\rho^{(M)} = 1.66667$ and $\rho^{(A)} = 12.8601$ when $a = 4$.

(iii) $\rho^{(M)} = 1.28571$ and $\rho^{(A)} = 3.51336$ when $a = 8$.

(II) Let Ω be the L-shaped domain

$$\Omega := \{ (x, y) : -1 < x < 3, |y| < 1 \} \cup \{ (x, y) : |x| < 1, -1 < y < 3 \}. \tag{2.32}$$

In this case the mapping function f has a serious branch point singularity at the re-entrant corner $z_c = 1 + i$, in the sense that

$$f(z) - f(z_c) \sim (z - z_c)^{2/3}, \quad \text{as } z \to z_c, \tag{2.33}$$

(see § 2.2, Eq. (2.9)). In addition, f has simple pole singularities in $\mathbf{C} \setminus \bar{\Omega}$, of which those closest to $\partial\Omega$ occur at the points

$$p_1 = -2i \quad \text{and} \quad p_2 = -2. \tag{2.34}$$

If the monomial set (2.6) is used, then because of the singularity at z_c the convergence of the method is extremely slow. As a result, the BKM can only manage the following disappointing optimum results :

$$\text{BKM/MB}: \quad n_{opt} = 13 , \quad E_{13} = 2.4\cdot 10^{-1}$$

This is in accordance with the recent convergence theory of Gaier [5] which, for the L-shaped domain under consideration, predicts that

$$\max_{z\in\bar{\Omega}} | f(z) - f_n^{(M)}(z)| = O(n^{-\gamma}) , \tag{2.35}$$

for any $\gamma < 1/3$ but for no $\gamma > 1/3$.

By contrast, the BKM corresponding to the augmented basis formed by introducing into the set (2.6) the six functions

$$(z - z_c)^{2l/3-1} , \quad l = 1, 2, 4, 5, 7, 8 , \tag{2.36}$$

reflecting the corner singularity at z_c, and the two rational functions

$$\frac{d}{dz} \{ z / (z - p_j) \} , \quad j = 1, 2 , \tag{2.37}$$

that reflect the pole singularities at the points (2.34), leads to the following excellent results :

$$\text{BKM/AB}: \quad n_{opt} = 26 , \quad E_{26} = 2.5\cdot 10^{-3}$$

Unfortunately, as we remarked in § 2.4 , we do not yet have any theoretical estimates for this remarkable improvement in accuracy.

2.8 The numerical conformal mapping package BKMPACK

BKMPACK is a Fortran package due to Warby [21], for computing BKM approximations to the mapping function $f : \Omega \to D$. More precisely, the package is a collection of routines for performing the various tasks of the BKM outlined in Sections 2.1-2.6. In particular, BKMPACK contains a routine which, given a parametric description of $\partial\Omega$, attempts to determine the main corner and pole-type singularities of f and hence to construct an appropriate augmented basis set.

After a successful application, the package can provide the following information :

(i) The location and nature of the main singularities of f on $\partial\Omega$ and in $\mathbf{C}\setminus\bar{\Omega}$.

(ii) The list of the basis functions $\{\eta_j\}$ used.

(iii) The coefficients a_j , $j = 1, 2, \cdots , n_{opt}$, corresponding to the optimal approximation

$$f_{n_{opt}}(z) = \sum_{j=1}^{n_{opt}} a_j \, \mu_j(z) , \tag{2.38}$$

where μ_j are integrals of the basis functions η_j.

(iv) The values of certain indicators that predict the level of instability at the nth step, $n = 1, \dots , n_{opt}$, of the process.

(v) An estimate E_n of

$$\max_{z \in \bar{\Omega}} | f_n(z) - f(z) | , \tag{2.39}$$

for any n , $n = 1, \dots, n_{opt}$.

(vi) The values of $f_n(z)$ and $f'_n(z)$ at specific points $z \in \bar{\Omega}$, for any n , $n = 1, \dots, n_{opt}$.

3 MULTIPLY-CONNECTED DOMAINS

3.1 Canonical domains

Let Ω be a finite N-connected domain of **C** with $N \geq 2$ and with boundary

$$\partial\Omega = \bigcup_{j=1}^{N} \Gamma_j ,$$

where Γ_j , $j = 1, 2, \dots, N$ are closed piecewise analytic Jordan curves, and assume that Γ_1 is the outer component of $\partial\Omega$, ie. Γ_1 encloses all the other Γ_j. There are several standard conformal maps associated with such multiply-connected domains. The best known of these correspond respectively to the following five canonical domains (cf. [2, Ch.VI, §2], [3, Ch.V, § 5.1], [10, Ch.VII, § 1]).

(i) *The slit circular ring* $S_{1,N}$: This consists of a circular ring

$$R := \{ w : r_N < | w | < r_1 \} \tag{3.1}$$

slit along $N-2$ arcs of circles

$$| w | = r_j , \quad j = 2, \dots, N-1 , \tag{3.2a}$$

where

$$r_N < r_j < r_1 , \quad j = 2, \dots, N-1 . \tag{3.2b}$$

(ii) *The slit disc* $S^{(d)}$: This consists of a disc

$$D := \{ w : | w | < r_1 \} \tag{3.3}$$

slit along $N-1$ arcs of circles

$$| w | = r_j , \quad j = 2, \dots, N , \tag{3.4a}$$

where

$$r_j < r_1 , \quad j = 2, \dots, N \tag{3.4b}$$

(iii) *The circular slit domain* $S^{(c)}$ · This consists of the entire plane slit along N arcs of circles

$$| w | = r_j , \quad j = 1, \dots, N \tag{3.5}$$

(iv) *The radial slit domain* $S^{(r)}$: This consists of the entire plane slit along N segments of rays

$$\arg w = \theta_j , \quad j = 1, \dots, N \tag{3.6}$$

emanating from the origin.

(v) *The parallel slit domain* $S^{(p)}$: This consists of the entire plane slit along N parallel straight line segments, subtending an angle θ with the positive real axis.

The objectives of this chapter are as follows : (a) To present a report on an orthonormalization method (ONM) which has been studied recently in [13] and [6], in connection with the conformal mapping associated with the first canonical domain, ie. with a slit circular ring of the form (3.1)-(3.2). (b) To indicate the modifications needed for extending the application of this method to the conformal mappings associated with the other four canonical domains.

In all our subsequent work the approximations will be in terms of functions taken from the space

$$L_s^2(\Omega) := \{ g : g \in L^2(\Omega) \text{ and } g \text{ has a single-valued indefinite integral } \} , \tag{3.7}$$

which is in fact a closed subspace of $L^2(\Omega)$. Thus, $L_s^2(\Omega)$ is itself a Hilbert space.

3.2 The orthonormalization method (ONM)

Let Ω be the N-connected domain introduced in Section 3.1, and assume that the origin 0 lies in the interior of the curve Γ_N, ie. $0 \in \text{Int}\,\Gamma_N$. Also, let f denote a conformal map of Ω onto a slit circular ring $S_{1,N}$, of the form (3.1)-(3.2), so that the outer boundary component Γ_1 and the component Γ_N (which encloses the origin) correspond respectively to the outer and inner circles bounding the ring R. Finally, assume that the other boundary components Γ_j , $j = 2, \ldots , N-1$, correspond respectively to the arcs of the circles $|w| = r_j$, $j = 2, \ldots , N-1$.

The requirements that $\Gamma_1 \to \{ w : |w| < r_1 \}$ and $\Gamma_N \to \{ w : |w| < r_N \}$ determine f uniquely apart from a multiplicative constant. Here we normalize the mapping by imposing a condition of the form

$$f(\zeta) = \zeta, \quad \zeta \in \Gamma_1, \tag{3.8}$$

so that the value of the outer radius r_1 of the ring R is fixed, ie.

$$r_1 = |\zeta| . \tag{3.9}$$

The values r_j , $j = 2, \ldots , N$ are also defined uniquely, but these are not known a priori. That is, the radii r_j , $j = 2, \ldots N$, are unknowns of the conformal mapping problem. These numbers or, more precisely, the ratios

$$M_j := r_1 / r_j , \quad j = 2, \ldots , N , \tag{3.10}$$

are important domain functionals known as the " *radial moduli* " of the domain Ω. In the case $N = 2$, ie. the case where Ω is a doubly-connected domain

$$\Omega = (\text{Int}\,\Gamma_1) \cap (\text{Ext}\,\Gamma_2) \tag{3.11}$$

and the associated canonical domain $S_{1,2}$ reduces to a circular ring R of the form (3.1), the single modulus $M_2 := r_1 / r_2$ identifies completely the conformal equivalence class of the domain. (That is, two doubly-connected domains can be mapped onto each other if and only if they have the same radial modulus.) Similarly, the $N-1$ radial moduli (3.10) together with $2N-5$ " *angular moduli* ", ie. the $2N-5$ angles that determine the lengths and positions of the arcs (3.2), identify completely the conformal equivalence class of a domain of connectivity $N > 2$.

The ONM is an orthonormalization method for approximating the conformal map $f : \Omega \to S_{1,N}$ and the associated radial moduli (3.10). The method is based on the theory contained in in [2, Ch.VII], [3, Ch.V, §5] and [10, Ch.VIII]. It was first studied by Papamichael and Kokkinos [13], in connection with the mapping of doubly-connected domains, and was recently extended to the general N-connected case, with $N \geq 2$, by Kokkinos et al [6].

Let

$$A(z) := \log f(z) - \log z \tag{3.12}$$

so that

$$f(z) = z \exp\{ A(z) \}, \tag{3.13}$$

and observe that the function A is analytic and single-valued in Ω. Also, let

$$H(z) := A'(z) = f'(z) / f(z) - 1/z , \tag{3.14}$$

and recall the definition (3.7) of the space $L_s^2(\Omega)$. Then, the ONM is based essentially on the following two results, concerning the determination of the inner products of the function H and the evaluation of the moduli (3.10) :

— *Inner products of H* · For each function $\eta \in L_s^2(\Omega)$ which is also continuous on $\partial\Omega$ apart, possibly, from a finite number of branch point singularities of the form

$$(z - z_0)^{\alpha}, \quad z_0 \in \partial\Omega, \quad -\tfrac{1}{2} < \alpha < 0, \tag{3.15}$$

$$(\eta, H) = i \int_{\partial\Omega} \eta(z) \log|z| \, dz \tag{3.16}$$

— *Radial moduli* : Let α_j, $j = 2, \ldots, N$ be $N-1$ points lying respectively in the interiors of the boundary curves Γ_j, $j = 2, \ldots N$. Then

$$2\pi \log M_j = \frac{1}{2i} \int_{\partial\Omega} \frac{1}{(z - \alpha_j)} \{ \overline{A(z)} + 2 \log|z| \} \, dz , \quad j = 2, \ldots, N \tag{3.17}$$

In particular, for $j = N$ we can take $\alpha_N = 0$ and can then replace (3.17) by

$$2\pi \log M_N = \frac{1}{i} \int_{\partial\Omega} \frac{1}{z} \log|z| \, dz - \| H \|^2 . \tag{3.18}$$

The first result (3.16) follows by applying Green's formula (1.6) to the inner product $(\eta, H) = (\eta, A')$ and observing that

$$\overline{A(z)} = -A(z) + 2\log|f(z)| - 2\log|z| , \tag{3.19}$$

that η and A are analytic in Ω and that η has a single-valued integral (cf. [3, pp.249-250]). The second result (3.17) follows in a similar manner by applying (1.6) to the inner products

$$\left(\frac{1}{(z-\alpha_j)} , H \right) = \left(\frac{1}{(z-\alpha_j)} , A' \right) . \tag{3.20}$$

Finally, (3.18) is obtained by integration by parts after first applying (3.16) to $\| H \| = (H , H)$ (cf. [3, pp. 250-251]).

Let $\{ \eta_j \}_1^\infty$ be a complete set for the space $L_s^2(\Omega)$ and assume that the functions η_j satisfy the additional continuity requirement needed for the results (1.6) and (3.16). (As will be seen later, it is always possible to construct such basis sets.) Then, by analogy with the BKM procedure of Section 2.1, the results (3.16)-(3.18) suggest the following orthonormalization process for computing approximations to the mapping function f and to the moduli M_j , $j = 2, \dots N$ (cf. [6, p.343]) :

(i) Orthonormalize the set $\{ \eta_j \}_1^n$ by means of the Gram-Schmidt process to obtain the orthonormal set $\{ \eta_j^* \}_1^n$.

(ii) Approximate the function H by the Fourier sum

$$H_n(z) = \sum_{j=1}^{n} (H , \eta_j^*) \eta_j^*(z) , \tag{3.21}$$

where the Fourier coefficients (H , η_j^*) are known by means of (3.16). (That is, H_n is the least squares approximation to H in span$(\eta_1, \dots , \eta_n)$.)

(iii) Approximate the function f by

$$f_n(z) = z \exp\{ A_n(z) \} \quad \text{where} \quad A_n(z) = \int_{t=\zeta}^{z} H_n(t)\, dt ; \tag{3.22}$$

see (3.12) and (3.14) and recall the normalizing condition (3.9).

(iv) Approximate the moduli M_j , $j = 2, \dots , N$, respectively by

$$M_j^{(n)} = \exp\left[\left(\frac{1}{2i} \int_{\partial\Omega} \frac{1}{(z-\alpha_j)} (\overline{A_n(z)} + 2\log|z|)\, dz \right) / 2\pi \right] , \; j = 2, \dots , N ; \tag{3.23}$$

see (3.17). Alternatively, M_N can be approximated by

$$M_N^{(n)} = \exp\left[\left(\frac{1}{i} \int_{\partial\Omega} \frac{1}{z} \log|z|\, dz - \| H_n \|^2 \right) / 2\pi \right] ; \tag{3.24}$$

see (3.18).

It follows from the above that the ONM may regarded as the extension of the BKM to the mapping of multiply-connected domains, where the function H takes the place of the kernel function $K(\cdot , 0)$ and (3.16) plays the role of the reproducing property (2.2). Thus, as is indicated below, the computational details of the two methods are very similar.

3.2.1 Choice of the basis set

As before, let $\alpha_j \in \operatorname{Int}\Gamma_j$, $j = 2, \ldots, N$ and, since $0 \in \operatorname{Int}\Gamma_N$, take $\alpha_N = 0$. Then, it is well-known that the functions

$$z^m, \quad m = 0, 1, \ldots, \tag{3.25a}$$

and

$$(z-\alpha_j)^{-m}, \quad j = 2, \ldots, N, \quad m = 2, 3, \ldots, \tag{3.25b}$$

form a complete set for the space $L_s^2(\Omega)$ (cf. [Ga1, p.244]). However, all the remarks of Section 2.2, concerning the use of the BKM monomial set (2.6), also apply to (3.25). In particular, the use of (3.25) is not recommended when the domain under consideration involves " *singular corners* ", ie. corners where the function H has branch point singularities. For the reasons explained in Section 2.2, in such cases the basis set must be constructed by introducing appropriate " *singular functions* " into (3.25). The functions needed for this purpose can again be deduced (via (3.14)) from the results of Lehman [8]. Here, however, their form depends on whether the corner point z_c lies on the outer component Γ_1 of $\partial\Omega$ or on one of the other components Γ_j, $j = 2, \ldots, N$. Thus, if $z_c \in \partial\Omega$ is a singular corner of interior angle $\omega\pi$, $0 < \omega < 2$ (by interior we mean interior to the domain Ω), then the corresponding singular functions are of the form (2.10) when $z_c \in \Gamma_1$ and, typically, of the form

$$\eta(z) = \frac{d}{dz}\{ [1/(z-\alpha_j) - 1/(z_c-\alpha_j)]^r \}, \quad r = k + l/\omega \text{ or } r = l/\omega, \tag{3.26}$$

when $z_c \in \Gamma_j$, $j = 2, \ldots, N$ (cf. [13, §4], [17, §4.1], [6, §3.1]). (Observe that the form of (3.26) ensures that all the singular functions used have single-valued indefinite integrals in Ω.)

Apart from corner singularities, the function H may also have singularities off $\partial\Omega$ in $\mathbb{C}\setminus\bar{\Omega}$, but the question regarding the location and nature of such singularities has not yet been fully answered. This problem has been studied by Papamichael and Warby [14], but only in connection with the doubly-connected case $N = 2$ (see also [17, §5.2]). Briefly, the results of [14] show that if Ω is a 2-connected domain of the form (3.11), then : (a) The function H may have singularities at the so-called " *common symmetric points* " with respect to the two boundary curves Γ_1 and Γ_2. (b) If $z_1 \in \operatorname{Ext}\Gamma_1$ and $z_2 \in \operatorname{Int}\Gamma_2$ is a pair of such common symmetric points then, for the purposes of the ONM, the corresponding singular behaviour of H may be reflected approximately by introducing into the set (3.25) the two rational functions

$$1/(z-z_1) \quad \text{and} \quad 1/(z-z_2) - 1/z \tag{3.27}$$

Unfortunately, we do not know of any similar results for the N-connected case with $N > 2$.

3.2.2 Computation of inner products and of radial moduli

Since all the basis functions (3.25), (2.10), (3.26) and (3.27) used in the ONM satisfy the boundary continuity requirement needed for the results (1.6) and (3.16), the inner products (η_k, η_l) involved in the orthonormalization step of the method can be computed by Gauss-Legendre quadrature in exactly the same way as in the BKM (cf. Section 2.3). Similarly, the inner products (H, η_j^*) in the approximation (3.21) can be computed by applying the same Gaussian rule to

$$(\eta_j^*, H) = i \int_{\partial\Omega} \eta_j^*(z) \log|z| \, dz \,. \tag{3.28}$$

Finally, the same rule can be used for computing the contour integrals involved in the approximations (3.17), (3.18). (As in the case of the BKM, if the basis set $\{\eta_j\}$ involves singular functions of the form (2.10) or (3.26), then the corresponding integrand singularities must be treated by choosing suitable parametrizations for the boundary curves Γ_j (cf. [13, §5]).)

3.2.3 Stability - Convergence - Error estimates

The BKM stability results of [15] generalize easily to the ONM in the doubly-connected case $N = 2$. Thus, in the case $N = 2$, we have a good understanding of the stability properties of the method in the sense that we again have a geometrical characterization of the level of instability and computable instability indicators (cf. [15, pp. 647-648, pp.666-667]).

The situation regarding the convergence of the ONM is much less satisfactory than that of the BKM. As in the case of the BKM, we again have that

$$\lim_{n\to\infty} \| H_n - H \| = 0 \tag{3.29}$$

and hence that $f_n(z) \to f(z)$ uniformly on each compact subset of Ω. In addition, for the doubly-connected case (3.11) we have a result due to Gaier [3] which establishes the uniform convergence in $\bar{\Omega}$ of the ONM approximations corresponding to the use of the set (3.25), ie. of the set

$$z^m \,,\ j = 0, 1, \dots , \qquad z^{-m} \,,\ j = 2, 3, \dots \,. \tag{3.30}$$

This result is proved in [3, p.250] under the assumption that the two components Γ_1 and Γ_2 of $\partial\Omega$ are both analytic Jordan curves, and also in [15, p.654] under the somewhat less restrictive assumption that the mapping function f is regular on $\partial\Omega$. (We note that the proof given in [3] is not for the ONM but for an equivalent variational method.) Unfortunately, however, even for the doubly-connected case, we do not know of any theoretical results for piecewise analytic boundaries, or for ONM approximations corresponding to the augmented basis sets used in [13] and [6].

From the more practical point of view, we again have a reliable method for estimating the maximum error E_n in $|f_n(z)|$ (cf. [13, p.691], [6, p.349]). This is done by : (a) Recalling that $r_1 = |\zeta|$ and computing the approximations

$$r_j^{(n)} = |\zeta| / M_j^{(n)} \,, \quad j = 2, \dots , N \,, \tag{3.31a}$$

to the radii r_j, $j = 2, \dots , N-1$, of the arcs (3.2) and to the inner radius r_N of the ring R. (b)

Computing the values

$$\varepsilon_1^{(n)} := \max_{z \in \sigma_1} | r_1 - | f_n(z) | | \tag{3.31b}$$

and

$$\varepsilon_j^{(n)} := \max_{z \in \sigma_j} | r_j^{(n)} - | f_n(z) | | , \quad j = 2, \dots , N , \tag{3.31c}$$

where σ_j , $j = 1, \dots , N$ are N sets of " *test points* " on the curves Γ_j , $j = 1, \dots , N$ respectively.

(c) Taking

$$E_n := \max_{1 \le j \le N} \varepsilon_j^{(n)} \tag{3.32}$$

We expect (3.32) to be a reliable estimate, because in general the approximations $M_j^{(n)}$, $j = 2, \dots , N$, are more accurate than $| f_n(z) |$, $z \in \partial\Omega$. (This can be explained by observing that the computation of the $M_j^{(n)}$ involves an averaging process; see (3.17)-(3.18).)

3.2.4 Optimum number of basis functions

The procedure for determining the optimum number n_{opt} of basis functions is essentially the same as that described, for the BKM, in Section 2.5 (cf. [13, p.691], [6, p.350]).

3.2.5 Numerical examples

The results of several numerical experiments illustrating the application of the ONM to the mapping of doubly-connected domains can be found in references [13], [14] and [15]. In addition, [6] contains six numerical examples illustrating the application of the method to the mapping of N-connected domains with $N > 2$. Here, we illustrate the high accuracy that the ONM is capable of achieving, by merely presenting the results of one example taken from [13, pp.692-694].

Let Ω denote the square frame

$$\Omega := \{ (x , y) : | x | < 1 , | y | < 1 \} \cap \{ (x , y) : | x | > 0.5 , | y | > 0.5 \} \tag{3.33}$$

Then, the function H has a serious branch point singularity at each of the four re-entrant corners of the inner square ($\omega = 3/2$ at each of these corners).

The ONM with an augmented basis, including four singular functions of the form (3.26) for dealing with the branch point singularities of H, gives the following results :

$$n_{opt} = 24 , \quad E_{24} = 5.0 \cdot 10^{-8} , \quad M_2^{(24)} = 1.847\,709\,011\,22 .$$

The exact value of M_2 (which in this case is known in terms of elliptic functions) is :

$$M_2 = 1.847\,709\,011\,24 .$$

3.3 Other standard conformal maps

In this section we consider the problem of extending the application of the ONM to the conformal maps that correspond respectively to the other four canonical domains $S^{(d)}, S^{(c)}, S^{(r)}$ and $S^{(p)}$ listed in Section 3.1. It turns out that the modifications needed for this purpose are straightforward and can, again, be deduced easily from the theory contained in [2, Ch.VI], [3, Ch.V, §5.1] and [10, Ch.VII, §5]. In fact, the details of the numerical method remain essentially unchanged, and the only major modifications concern the functions A, H and the formulae for computing the inner products of H and the moduli of Ω.

3.3.1 The slit disc conformal map

Let Ω be the N-connected domain of Section 3.1, but assume now that the origin 0 lies in Ω, ie. assume that $0 \in \Omega$ rather than $0 \in \mathrm{Int}\,\Gamma_N$. Also, let f denote a conformal map of Ω onto a slit disc $S^{(d)}$, so that the boundary components Γ_1 and Γ_j, $j = 2, \dots, N$ correspond respectively to the outer circle $|w| = r_1$ and to the arcs of the circles $|w| = r_j$, $j = 2, \dots, N$. Such a conformal map can be made unique by requiring that

$$f(0) = 0 \quad \text{and} \quad f'(0) > 0, \tag{3.34}$$

and imposing a condition of the form

$$f(\zeta) = \zeta, \quad \zeta \in \Gamma_1. \tag{3.35}$$

As in the case of the map $\Omega \to S_{1N}$, the condition (3.35) determines the value of the outer radius r_1, but the other radii r_j, $j = 2, \dots, N$ are unknowns of the mapping problem $f : \Omega \to S^{(d)}$.

It is easy to see that the ONM can also be applied directly to the above conformal mapping problem. That is, the functions A, H and the inner products of H are again given by (3.12), (3.14) and (3.16), and the radial moduli $M_j := r_j / r_1$, $j = 2, \dots, N$, can again be computed by means of (3.17). A numerical example illustrating such an application of the ONM can be found in [6, pp.357-358].

3.3.2 The circular slit conformal map

Let $0 \in \Omega$, let u be some other fixed point in Ω and let f denote a conformal map of Ω onto a circular slit domain $S^{(c)}$, so that

$$f(0) = 0, \quad \text{and} \quad \lim_{z \to u} = \infty. \tag{3.36}$$

Such a conformal map can be made unique (apart from a multiplicative constant) by requiring that the Laurent series expansion of f near $z = u$ is of the form

$$f(z) = \frac{1}{z - u} + c_0 + c_1(z - u) + \dots. \tag{3.37}$$

We assume that $f : \Omega \to S^{(c)}$ takes the boundary components Γ_j, $j = 1, \ldots N$, respectively onto the arcs of the circles $|w| = r_j$, $j = 1, \ldots, N$, and impose a normalizing condition of the form (3.35) so that the radius r_1 is fixed, ie. $r_1 = |\zeta|$.

For the application of the ONM, we let

$$A(z) := \log f(z) - \log z + \log(z - u), \tag{3.38}$$

which is analytic and single-valued in Ω, and set

$$H(z) := A'(z) = f'(z)/f(z) - 1/z + 1/(z-u). \tag{3.39}$$

Then, corresponding to (3.16) and (3.17), we now have the following two results for computing respectively the inner products of H and the radial moduli $M_j := r_1/r_j$, $j = 2, \ldots, N$.

— *Inner products of H* : For each function $\eta \in L_z^2(\Omega)$ which is also continuous on $\partial\Omega$ apart, possibly, from a finite number of branch point singularities of the form (3.15)

$$(\eta, H) = i \int_{\partial\Omega} \eta(z) \{ \log|z| - \log|z-u| \} dz. \tag{3.40}$$

— *Radial moduli* : Let α_j, $j = 2, \ldots, N$, be $N-1$ points lying respectively in the interiors of the curves Γ_j, $j = 2, \ldots, N$. Then

$$2\pi \log M_j = \frac{1}{2i} \int_{\partial\Omega} \frac{1}{z - \alpha_j} \{ \overline{A(z)} + 2(\log|z| - \log|z-u|) \} dz, \quad j = 2, \ldots, N \tag{3.41}$$

The above results are derived (like (3.16) and (3.17)) by applying Green's formula (1.6) to the inner products $(\eta, H) = (\eta, A')$ and (3.20) and observing that in this case

$$\overline{A(z)} = -A(z) + 2\log|f(z)| - 2\log|z| + 2\log|z-u|. \tag{3.42}$$

3.3.3 The radial slit conformal map

Let f denote a conformal map of Ω onto a radial slit domain $S^{(r)}$, so that

$$f(0) = 0 \quad \text{and} \quad \lim_{z \to u} f(z) = \infty, \tag{3.43}$$

where, as in § 3.3.2, $0 \in \Omega$ and u is some other fixed point in Ω. This conformal map can also be made unique (apart from a multiplicative constant) by requiring, as in § 3.3.2, that f has a Laurent expansion of the form (3.37) near $z = u$.

We assume that $f : \Omega \to S^{(r)}$ takes the boundary components Γ_j, $j = 1, \ldots, N$, respectively onto the segments of the rays $\arg w = \theta_j$, $1, \ldots, N$, and impose again a condition of the form (3.35) so that the angle θ_1 is fixed, ie. $\theta_1 = \arg\zeta$.

For the ONM, we again define the functions A and H by means of (3.38) and (3.39), and again seek to derive formulae for the inner products of H and the angular moduli θ_j, $j = 2, \ldots, N$, by applying Green's formula (1.6) to the inner products (η, H) and (3.20). Now, however, we do this by expressing $\overline{A(z)}$ in the form

$$\overline{A(z)} = A(z) - 2i\,\{\,\arg f(z) - \arg z + \arg(z-u)\,\}\,, \tag{3.44}$$

rather than (3.42). This gives the following :

— *Inner products of H* · For each function $\eta \in L_s^2(\Omega)$ which is also continuous on $\partial\Omega$ apart, possibly, from a finite number of branch point singularities of the form (3.15)

$$(\eta, H) = \int_{\partial\Omega} \eta(z)\,\{\,\arg z - \arg(z-u)\,\}\,dz \tag{3.45}$$

— *Angular moduli* : Let α_j, $j = 2, \ldots, N$, be $N-1$ points lying respectively in the interiors of the curves Γ_j, $j = 2, \ldots, N$. Then

$$4\pi(\theta_1 - \theta_j) = \int_{\partial\Omega} \frac{1}{z-\alpha_j}\,\{\,\overline{A(z)} - 2i\,[\,\arg z - \arg(z-u)\,]\,\}\,dz\,,\quad j = 2, \ldots, N \tag{3.46}$$

It is of interest to note that the inner products (H, η) can also be computed by means of

$$(H, \eta) = -\pi\,\{\,\overline{\mu(0)} - \overline{\mu(u)}\,\} - \frac{1}{2i}\int_{\partial\Omega}\{\,\frac{1}{z} - \frac{1}{z-u}\,\}\,\overline{\mu(z)}\,dz\,, \tag{3.47}$$

where $\mu' = \eta$. The above result, which is given in [10, p.377, Ex.5], can be deduced easily from (3.45) by partial integration. In fact similar alternative formulae that express (H, η) in terms of μ where, $\mu' = \eta$, can also be derived by partial integration from (3.16) (for both the cases of the slit circular ring $S^{(a)}$ and the slit disc $S^{(d)}$) and from (3.40) for the circular slit domain $S^{(c)}$ (cf. [10, p.377, Exs 5,7,8]).

3.3.4 *The parallel slit conformal map*

Let f denote a conformal map of Ω onto a parallel slit domain $S^{(p)}$, so that

$$\lim_{z\to u} = \infty\,, \tag{3.48}$$

where u is some fixed point in Ω. Such a conformal map can be made unique by requiring that the parallel slits form a given angle θ with the positive real axis, and that f has a Laurent series expansion of the form

$$f(z) = \frac{1}{z-u} + c_1(z-u) + \ldots\,, \tag{3.49}$$

near $z = u$.

For the ONM we now let

$$A(z) := f(z) - \frac{1}{z-u}\,, \tag{3.50}$$

which (since the only singularity of f is a simple pole of residue 1 at $z = u$) is analytic and single-valued in Ω. That is, in this case, we take

$$H(z) := A'(z) = f'(z) + \frac{1}{(z-u)^2} . \tag{3.51}$$

Then, we have the following result for computing the inner products of H.

— *Inner products of* H : For each function $\eta \in L_s^2(\Omega)$ which is also continuous on $\partial\Omega$ apart, possibly, from a finite number of branch point singularities of the form (3.15)

$$(H, \eta) = \pi e^{i\theta}\overline{\eta(u)} + \frac{1}{2i}\int_{\partial\Omega} \frac{\overline{\eta(u)}}{z-u}\, d\bar{z} . \tag{3.52}$$

This formula can be derived, as shown in [3, pp. 242-243] and [10, p.374], by applying Green's formula (1.6) to the inner product (H, η) and making use of (3.51).

References

[1] V.V. Andrievskii, *Uniform convergence of Bieberbach polynomials in domains with piecewise-quasiconformal boundary*, Ukrainian Math. J. **35** (1983) 233 - 236 .

[2] S. Bergman, *The Kernel Function and Conformal Mapping*, Math. Surveys, No. **5** (Amer. Math. Soc., Providence, R.I. 1970) .

[3] D. Gaier, *Konstruktive Methoden der konformen Abbildung* (Springer, Berlin 1964) .

[4] D. Gaier, *Lectures on Complex Approximation* (Birkhäuser , Boston 1987) .

[5] D. Gaier, *On the convergence of the Bieberbach polynomials in regions with corners*, Constr. Approx. **4** (1988) 289 - 305 .

[6] C.A. Kokkinos, N. Papamichael and A.B. Sideridis, *An orthonormalization method for the approximate conformal mapping od multiply-connected domains*, IMA J. Numer. Anal. **9** (1990) 343 - 359 .

[7] I.V. Kulikov, W_2^1 , L_∞ *- convergence of Bieberbach polynomials in a Lipschitz domain*, Russian Math. Surveys **36** (1981) 161 - 162 .

[8] R.S. Lehman, *Development of the mapping function at an analytic corner*, Pacific J. Math. **7** (1957) 1437 - 1449 .

[9] D. Levin, N. Papamichael and A. Sideridis, *The Bergman kernel method for the numerical conformal mapping of simply-connected domains*, J. Inst. Maths Applics **22** (1978) 171 - 187

[10] Z. Nehari, *Conformal Mapping* (McGraw-Hill, New York 1952) .

[11] N. Papamichael and C.A. Kokkinos, *Two numerical methods for the conformal mapping of simply-connected domains*, Comput. Meths Appl. Mech. Engrg **28** (1981) 285 - 307 .

[12] N. Papamichael and C.A. Kokkinos, *Numerical conformal mapping of exterior domains*, Comput. Meths Appl. Mech. Engrg **31** (1982) 189 -203 .

[13] N. Papamichael and C.A. Kokkinos, *The use of singular functions for the approximate conformal mapping of doubly-connected domains*, SIAM J. Sci. Stat. Comput. **5** (1984) 684-700 .

[14] N. Papamichael and M.K. Warby, *Pole-type singularities and the numerical conformal mapping of doubly-connected domains*, J. Comput. Appl. Math. **10** (1984) 93 - 106 .

[15] N. Papamichael and M.K. Warby, *Stability and convergence properties of Bergman kernel methods in numerical conformal mapping*, Numer. Math. **48** (1986) 639 - 669 .

[16] N. Papamichael, M.K. Warby and D.M. Hough, *The determination of the poles of the mapping function and their use in numerical conformal mapping*, J. Comput. Appl. Math. **9** (1983) 155 - 166 .

[17] N. Papamichael, M.K. Warby and D.M. Hough, *The treatment of corner and pole-type singularities in numerical conformal mapping techniques*, J. Comput. Appl. Math. **14** (1986) 163-191.

[18] G. Sansone and J. Gerresten, *Lectures on the Theory of Functions of a Complex Variable*, Vol. II (Wolters-Noordhoff, Groningen, 1969) .

[19] I.B. Simonenko, *On the convergence of Bieberbach polynomials in the case of a Lipschitz domain*, Math. USSR-Izv. **13** (1979) 166 - 174 .

[20] P.K. Suetin, *Polynomials orthogonal over a region and Bieberbach polynomials*, Proc. Steklov Inst. Math. **100** (1971). (Amer. Math. Soc. , Providence, RI, 1974) .

[21] M.K. Warby, *BKMPACK User's Guide*, Technical Report, Dept of Maths and Stats, Brunel University, 1990 .

ON THE COMPUTATION OF ZEROS AND TURNING POINTS OF BESSEL FUNCTIONS

R. Piessens
Department of Computer Science
Catholic University of Leuven
Celestijnenlaan 200 A
B-3001 HEVERLEE (BELGIUM)

Abstract

This paper contains a description of a set of FORTRAN subroutines for the accurate computation of the zeros and turning points of the Bessel functions of the first kind.

1 Introduction

The zeros and turning points of the Bessel functions $J_p(x)$ have important applications in mathematical physics, applied mathematics and numerical analysis. We denote the s th positive zero of $J_p(x)$ $(p > -1)$ by $j_{p,s}$ and the s th positive turning point of J_p $(p > 0)$ by $j'_{p,s}$, i.e. $j'_{p,s}$ is the s th positive zero of $J'_p(x)$. For $p = -1$, $x = 0$ is counted as the first positive zero of $J_{-1}(x)$. For $p = 0$, $x = 0$ is counted as the first positive turning point of $J_0(x)$. In this paper we give a description of a set of FORTRAN subroutines for the accurate computation of $j_{p,s}$ and $j'_{p,s}$, $p > -1$ and $s = 1, 2, 3, \ldots$. The computerprograms are available from the author.

2 Chebyshev series approximations

For $s = 1, 2, 3, 4, 5, 6$ and $p \leq 1$, $j_{p,s}$ and $j'_{p,s}$ are approximated by the truncated Chebyshev series :

$$j_{p,1} \cong \sqrt{p+1} \sum_{k=0}^{N_1}{}' a_k^{(1)} T_k(p) \qquad -1 \leq p \leq 1 \tag{1}$$

$$j_{p,s} \cong \sum_{k=0}^{N_s}{}' a_k^{(s)} T_k(p) \quad s = 2,3,4,5,6 \quad -1 \leq p \leq 1 \tag{2}$$

$$j'_{p,s} \cong \sqrt{p}\sum_{k=0}^{M_1}{}' c_k^{(1)} T_k^*(p) \qquad 0 \leq p \leq 1 \tag{3}$$

$$j'_{p,s} \cong \sum_{k=0}^{M_s}{}' c_k^{(s)} T_k^*(p) \quad s = 2,3,4,5,6 \quad 0 \leq p \leq 1 \tag{4}$$

The notation $\sum'$ in (1), (2), (3) and (4) indicates a summation with first term halved. In (3) and (4) T_k^*, $k = 0,1,2,\ldots$, are the shifted Chebyshev polynomials. The factors $\sqrt{p+1}$ in (1) and $\sqrt{p}$ in (3) are introduced in order to take into account the asymptotic behaviour of $j_{p,1}$ for $p \to -1$ and of $j'_{p,1}$ for $p \to 0$ (see [5] and [7]).

The Chebyshev coefficients $a_k^{(s)}$ and $b_k^{(s)}$ are calcutated using an algorithm of Gentleman [1]. They are tabulated in [8].

For $s = 1,2,\ldots 6$ and $p > 1$, $j_{p,s}$ and $j'_{p,s}$ are approximated by

$$j'_{p,s} \cong p + A_s p^{1/3} + p^{-1/3} \sum_{k=0}^{L_s}{}' b_k^{(s)} T_k^*(p^{-2/3}) \tag{5}$$

and

$$j'_{p,s} \cong p + B_s p^{1/3} + p^{-1/3} \sum_{k=0}^{K_s}{}' d_k^{(s)} T_k^*(p^{-2/3}) \tag{6}$$

The form of these approximations is suggested by the asymptotic expansions [5]

$$j_{p,s} \sim p + A_s p^{1/3} + 0(p^{-1/3}) \quad p \to \infty \tag{7}$$

and

$$j'_{p,s} \sim p + B_s p^{1/3} + 0(p^{-1/3}) \quad p \to \infty \tag{8}$$

where $A_s = -2^{-1/3} a_s$ and $B_s = -2^{-1/3} a'_s$, with a_s the s th negative zero of the Airy function $Ai(x)$ and a'_s the s th negative zero of the derivative $Ai'(x)$.

The coefficients A_s and B_s, and the Chebyshev coefficients $b_k^{(s)}$ and $d_k^{(s)}$ of the expansions (5) and (6) are presented in [8].

3 Mc Mahon's asymptotic expansion

In the region $\{7 \leq s \leq 30, -1 \leq p \leq \max(7, s/2)\}$ and in the region $\{s > 30, -1 \leq p \leq 15\}$, Mc Mahon's approximation

$$j_{p,s} \cong \beta - \sum_{r=1}^{7} \frac{A_{2r-1}}{(2r-1)! 2^{3r} \beta^{2r-1}} \tag{9}$$

where $\beta = (2p + 4s - 1)\pi/4$, is used. The coefficients A_{2r-1} can be found in [4].

For $\{7 \leq s \leq 20,\ 0 \leq p \leq max(4.5, s/2)\}$ and for $\{s > 20,\ 0 \leq p \leq 10\}$ Mc Mahon's approximation

$$j'_{p,s} \cong \beta' - \sum_{r=1}^{6} \frac{A'_{2r-1}}{(2r-1)!2^{3r}\beta^{2r-1}} \tag{10}$$

where $\beta' = (2p + 4s - 3)\pi/4$, is used. The coefficients A'_{2r-1} are given in [4].

4 Olver's uniform asymptotic expansion [3]

For $\{7 \le s \le 30,\ p > \max(7, s/2)\}$ and for $\{s > 30,\ p > 15\}$ $j_{p,s}$ is approximated by Olver's uniform asymptotic expansion, truncated after three terms,

$$j_{p,s} \cong pz(\varsigma) + \alpha_1 z'(\varsigma)/p + [\alpha_2 z'(\varsigma) + \alpha_1^2 z''(\varsigma)/2]/p^3 \tag{11}$$

where $\varsigma = p^{-2/3}a_s$, with a_s the s th negative zero of the Airy function $Ai(x)$. The function $z(x)$, $x < 0$ is implicitly defined by

$$\frac{2}{3}(-x)^{3/2} = \sqrt{z^2 - 1} - \arccos(1/z) \tag{12}$$

Formulas for the evaluation of α_1 and α_2 are given in [8]. A similar formula for $j'_{p,s}$ is employed in the regions $\{7 \le s \le 20,\ p > \max(4.5, s/2)\}$ and $\{s > 20,\ p > 10\}$.

5 Iterative refinement

If high accuracy is required, the approximation obtained with the formulas given in sections 2, 3 and 4, has to be refined by an iterative method. If X_1 is an approximation of $j_{p,s}$, then

$$X_2 = X_1 - \frac{J_p(X_1)}{J'_p(X_1)} - \frac{J''_p(X_1)}{2J'_p(X_1)}\left(\frac{J_p(X_1)}{J'_p(X_1)}\right)^2 \tag{13}$$

is a more accurate approximation of $j_{p,s}$. Formula (13) is a third order iterative formula of Newton type.

Similarly, if X'_1 is an approximation of $j'_{p,s}$, then

$$X'_2 = X'_1 - \frac{J'_p(X'_1)}{J''_p(X'_1)} - \frac{J'''_p(X'_1)}{2J''_p(X'_1)}\left(\frac{J'_p(X'_1)}{J''_p(X'_1)}\right)^2 \tag{14}$$

is a more accurate approximation of $j'_{p,s}$. The derivates of the Bessel functions in (13) and (14) can be evaluated efficiently using the relations :

$$J'_p(x) = -J_{p+1}(x) + pJ_p(x)/x$$

$$J''_p(x) = -J'_p(x)/x - (1 + p^2/x^2)J_p(x)$$

$$J'''_p(x) = -3J''_p(x)/x - (1 + (1 - p^2)/x^2)J'_p(x) - 2J_p(x)/x$$

6 Description of the subroutines

1. DOUBLE PRECISION FUNCTION ZERJ (P,NUM)

   This subprogram gives an approximation for the NUM-th zero of $J_p(x)$ (NUM ≥ 1, $P \geq -1$), using the formulas of sections 2, 3 and 4. The relative error of the approximation is less than 0.5×10^{-10}.

2. DOUBLE PRECISION FUNCTION ZERJAC (P,NUM)

   This subprogram gives an approximation for the NUM-th zero of $J'_p(x)$ (NUM ≥ 1, $P \geq 0$), using the formulas of sections 2, 3 and 4. The relative error of the approximation is less than 0.2×10^{-9}.

3. DOUBLE PRECISION FUNCTION Z2J (P,NUM)

   This subprogram computes an high-accuracy approximation of the NUM-th zero of $J_p(x)$ (NUM ≥ 1, $P \geq -1$) by applying formula (13) where X_1 is the approximation given by ZERJ. Z2J needs ZERJ and a subprogram BESJP for evaluating Bessel functions (see ref. [6]).

4. DOUBLE PRECISION FUNCTION Z2JAC (P,NUM)

   This subprogram computes en accurate approximation of the NUM-th zero of $J'_p(x)$ (NUM ≥ 1, $P \geq 0$) by applying formula (14), where X'_1 is the approximation given by ZERJAC. Z2JAC needs ZERJAC and a subprobram BESJP for evaluating Bessel functions (see ref. [6]).

7 Numerical results

In the tables 1 and 2, we show for some values of P and NUM the absolute error of the approximation given by ZERJ and ZERJAC.

Table 1: Absolute errors of the output of ZERJ

P	NUM				
	1	5	10	15	20
-1	0	0.2(-12)	0.1(-15)	0.1(-15)	0.1(-15)
0	0.7(-13)	0.2(-12)	0.4(-15)	0.1(-15)	0.1(-15)
1	0.7(-13)	0.1(-12)	0.9(-15)	0.9(-15)	0.1(-15)
5	0.4(-12)	0.7(-12)	0.7(-12)	0.5(-14)	0.1(-15)
10	0.2(-13)	0.2(-12)	0.1(-9)	0.4(-10)	0.2(-11)
15	0.4(-12)	0.2(-13)	0.4(-10)	0.2(-10)	0.7(-11)
20	0.5(-12)	0.8(-12)	0.2(-10)	0.7(-11)	0.4(-11)

Table 2: Absolute errors of the output of ZERJAC

P	NUM				
	1	5	10	15	20
0	0	0.6(-12)	0.2(-14)	0.1(-15)	0.1(-15)
1	0.5(-12)	0.3(-12)	0.4(-15)	0.1(-15)	0.1(-15)
5	0.7(-13)	0.2(-11)	0.2(-9)	0.2(-11)	0.6(-13)
10	0.2(-12)	0.2(-11)	0.2(-9)	0.6(-10)	0.1(-9)
15	0.2(-12)	0.8(-12)	0.8(-10)	0.3(-10)	0.1(-10)
20	0.3(-13)	0.2(-11)	0.4(-10)	0.1(-10)	0.6(-11)

8 Conclusion

Using Chebyshev series approximations, Mc Mahon's asymptotic expansion and Olver's uniform asymptotic expansion, subroutine ZERJ evaluates $j_{p,s}$ with a relative error less than 0.5×10^{-10} (for $p \geq -1$ and $s = 1, 2, \ldots$) and subroutine ZERJAC evaluates $j'_{p,s}$ with a relative error les than 0.2×10^{-9} (for $p \geq 0$ and $s = 1, 2, \ldots$). If higher accuracy is required, then iterative refinement has to be applied which is implemented in the subroutines Z2J and Z2JAC.

References

[1]. W. M. Gentleman, Algorithm 424 - Clenshaw - Curtis quadrature, Comm. ACM. *15* (1972), 353.

[2]. F. W. J. Olver, A further method for the evaluation of zeros of Bessel functions and some new asymptotic expansions for zeros of functions of large order, Proc. Cambridge - Phil. Soc. *47* (1951) 699.

[3]. F. W. J. Olver, The asymptotic expansion of Bessel functions of large order, Phil. Trans. Royal Society. *A 247* (1954), pp. 328-368.

[4]. F. W. J. Olver, Zeros and associated values, in "Royal Society Mathematical Tables, Bessel functions", Part. III, vol. 7, (Cambridge University Press, London, 1960).

[5]. R. Piessens, A series expansion for the first positive zero of the Bessel functions, Math. Comp. *42* (1984), 195.

[6]. R. Piessens, The computation of Bessel functions on a small computer, Comp. Math with Appls, 10 (1984) 161.

[7]. R. Piessens and S. Ahmed, Approximation for the turning points of Bessel functions, J. Comp. Phys. *64* (1986), 253-257.

[8]. R. Piessens and G. Engelen, The Computation of zeros and turning points of the Bessel functions of the first kind. Report TW72 (Catholic University of Leuven, 1985).

ASYMPTOTIC ERROR EXPANSIONS FOR LINEAR MULTISTEP METHODS FOR A CLASS OF DELAY INTEGRO-DIFFERENTIAL EQUATIONS

Christopher T.H. Baker[1] & Neville J. Ford[2]

[1] The Victoria University of Manchester

[2]. Chester College, Chester

Abstract We describe a class of step-by-step methods for the numerical solution of delay-integro-differential equations of the form

$$y'(t) = G(t, y(t), \int_{t-\tau}^{t} K(t,s,y(s))ds) \quad (t \geq 0)$$

where τ is fixed, subject to an appropriate initial condition. We present a theory giving the order of convergence of such methods and an asymptotic expansion for the error.

1. Introduction

We are concerned with certain numerical methods for the approximate solution of equations of the form

$$(1.1) \qquad y'(t) = G(t, y(t), \int_{t-\tau}^{t} K(t,s,y(s))ds) \quad (t \geq 0),$$

subject to the prescribed initial condition

$$(1.2) \qquad y(t) = \psi(t) \quad (-\tau \leq t < 0).$$

We shall consider (1.1) for $t \in [0,T]$, $T < \infty$. Our aim is to provide a self-contained convergence theory for a class of numerical methods for this problem.

The published literature contains results on the existence and uniqueness of solutions of various Volterra functional equations, in particular those of the form

$$(1.3) \qquad u'(t) = H(t, u(t), \int_{0}^{t} L(t,s, u(s))ds), \quad t \geq 0,$$

with prescribed initial value

$$(1.4) \qquad u(0) = u_0.$$

Various numerical schemes have been proposed for equations (1.3), (1.4); see for example [4,5,6,8,9,10]. By defining $L(t,s,u)$ 0 if $s<t-\tau$ and by considering $0 \leq t < \tau$ and $\tau \leq t < \infty$ separately when defining H in (1.3), it is possible to place (1.1), (1.2) into the context of equations of the form (1.3) subject to (1.4). However, for numerical work, the direct treatment of (1.1), (1.2) presents greater attractions.

We propose to assume that G is continuous on D $\mathbb{R}_+ \times \mathbb{R} \times \mathbb{R}$ and that

$$(1.5) \qquad \begin{array}{ll} \text{(H1)} & |G(t,y_1,z) \quad G(t,y_2,z)| \leq \Lambda_1 \, |y_1 - y_2| \\ \text{(H2)} & |G(t,y,z_1) \quad G(t,y,z_2)| \leq \Lambda_2 \, |z_1 - z_2| \end{array}$$

uniformly in D, and we assume that K is continuous on the domain

$$t \geq 0,\ s \in [t-\tau, t],\ |y| < \infty,$$

whereon we ask that

(1.6) (H3) $|K(t,s,y_1) - K(t,s,y_2)| \leq \Lambda_3 |y_1 - y_2|$.

Then the theory for (1.3) and (1.4) can be employed to establish (for suitable prescribed ψ) the existence of a unique continuously differentiable solution $y(t)$, $t \in (0,\infty)$ to (1.1), (1.2).

For motivation, recall that linear multistep methods [3,7] for approximating u defined by (1.3), (1.4) require a choice of a consistent zero-stable k-step linear multistep formula (LMF) defined by its characteristic polynomials [ibid]:

(1.7) $\hat{\rho}(\mu) = a_0\mu^k + a_1\mu^{k-1} + \cdots + a_k,\ \hat{\sigma}(\mu) = b_0\mu^k + b_1\mu^{k-1} + \cdots + b_k,$

and a family of quadrature rules

(1.8) $\int_0^{nh} \varphi(s)ds \sim h \sum_{j=0}^{n} w_{nj}\ \varphi(jh), \qquad n \in \mathbb{Z}_+.$

The methods with fixed step $h > 0$ are defined by appropriate starting values and the formulae

(1.9) $\sum_{r=0}^{k} a_r\ \tilde{u}_{n-r} = h \sum_{r=0}^{k} b_r\ H(t_{n-r}, \tilde{u}_{n-r}, \tilde{z}_{n-r}) \qquad (n \geq k)$

where

(1.10) $\tilde{z}_n = h \sum_{j=0}^{n} w_{nj}\ L(t_n, t_j, \tilde{u}_j) \qquad (t_n = nh \in [0,T]).$

Then $\tilde{u}_n \simeq u(t_n)$, and, further, $\tilde{z}_n \sim \int_0^{t_n} L(t_n,s,u(s))ds$.

We propose to modify (1.9), (1.10) to establish a method for the direct numerical treatment of (1.1), (1.2). Given $\tau > 0$, we select $h = \tau/N$, $N \in \mathbb{Z}_+$ and we determine $\{\tilde{y}_n \mid n \in \mathbb{Z}_+; t_n = nh \leq T\}$ such that $\tilde{y}_j = \psi(t_j)$, $j \leq 0$, and

(1.11) $\sum_{r=0}^{k} a_r\ \tilde{y}_{n-r} = h \sum_{r=0}^{k} b_r\ G(t_{n-r}, \tilde{y}_{n-r}, \zeta_{n-r}) \ (n \geq k).$

Here, for a certain choice of parameters $A_0, \cdots, A_N$,

(1.12) $\zeta_n = h \sum_{j=0}^{N} A_\ell\ K(t_n, (n-j)h, \tilde{y}_{n-j}),\ (n \geq 0).$

The parameters a_r, b_r of (1.11) are those of (1.7). The weights $\{A_\ell\}$ in (1.12) are those of a quadrature rule

(1.13) $\int_0^{\tau} \psi(s)ds \sim h \sum_{j=0}^{N} A_j\ \psi((N-j)h) \qquad (\tau = Nh),$

and (1.12) provides an approximation to $\int_{t_n-\tau}^{t_n} K(t_n,s,y(s))ds$.

If $k>1$, (1.11) requires starting values $\{\tilde{y}_\ell \mid \ell = 0,1, \ldots, k-1\}$, in addition to the given values $\tilde{y}_\ell = \psi(t_\ell)$, $-N \leq \ell < 0$.

__Lemma 1.1.__ (*Existence*) With the assumptions stated, (a) there exists a unique continuously differentiable solution $y(t)$ on $\mathbb{R}_+$, and (b) there exists $h_* \in \mathbb{R}_+$ such that for $h \in (0,h_*)$, (1.11) and (1.12) define from the starting values a unique sequence $\{\tilde{y}_\ell \mid \ell \geq k\}$.

If (1.11), (1.12) yield implicit equations for $\{\tilde{y}_\ell \mid \ell \geq k\}$, we here assume that they are solved exactly by a predictor-corrector iteration. (The iteration forms a basis for a proof of part (b) of Lemma 1.1.)

The approximation (1.12) for integration over intervals $[t_n-\tau, t_n]$ of fixed length $\tau = Nh$ may be contrasted with (1.10) for integration over $[0, t_n]$. We shall show that the parameters $\{a_\ell, b_\ell; A_\ell\}$ may be chosen to yield high-order convergence in the presence of sufficient differentiability. The differentiability of y depends, inter alia, upon that of G and of K; it also depends upon ψ. Evidently, y' suffers a discontinuity at $t=0$ if $\psi'(0-) \neq G(0,\psi(0), \int_{-\tau}^{0} K(0,s,\psi(s))ds)$. It may then be that y'' suffers a discontinuity at $t=\tau$, y''' suffers a discontinuity at $t=2\tau$ and $y^{(m)}$ may suffer a discontinuity at $t = (m-1)\tau$. Such behaviour of y can have an adverse effect on the order of convergence of the numerical scheme, as our theory will suggest. In consequence, while starting values are required in any event if $k > 1$, special starting procedures may also be required to provide a high-order approximation for $t \in [m^*\tau, (m^*+1)\tau]$ for various $m^* \in \mathbf{Z}_+$ We limit ourselves to the foregoing caveat and do not dwell here upon overcoming any lack of smoothness.

2. Convergence

Various authors (e.g. [4,5,6,8,9,11]) have considered convergence proofs for linear multistep methods applied to (1.3). We shall not attempt to adapt this analysis to the discussion of (1.1). Instead, we shall establish a convergence theorem specific to the treatment of (1.1), using a short sequence of lemmas of a type familiar in discretization theory for evolutionary problems.

For the LMF, we make the hypothesis that

(H4) $\hat{\rho}, \hat{\sigma}$ in (1.7) define a zero-stable LMF of order at least $p > 0$.

In consequence, for any closed and bounded interval I of $\mathbf{R}$ and any function $\varphi \in C^{(p+1)}(I)$

$$(2.1) \quad \sum_{\ell=0}^{k} a_\ell \, \varphi(t-\ell h) \quad h \sum_{\ell=0}^{k} b_\ell \, \varphi'(t-\ell h) = O(h^{p+1})$$

uniformly for t in the interior of I. As a further consequence, $\hat{\rho}$ in (1.7) is simple von Neumann; that is, the zeros of $\hat{\rho}(\mu)$ have modulus at most unity and those of modulus unity are simple. For the quadrature (1.12) we assume that there exist a finite constant a_q such that

$$(\text{H5}) \quad \left| \int_0^\tau \psi(s)ds \quad h \sum_{j=0}^{N} A_\ell \, \psi((N-j)h) \right| \leq \alpha_q \, \|\psi^{(q)}\|_\infty h^q$$

if $\psi \in C^q[0,\tau]$. For example, if $N \quad MR$ and the rule (1.12) is an M times repeated version of a rule for integration on $[0,Rh]$, then (H.5) is valid if the basic rule has degree of precision $Q \geq q-1$. As an alternative example, Gregory's rule incorporating r end-correction terms to the composite trapezium rule allows us to take q upto $2+r$. A consequence of (H5) is that

$$(2.2) \quad \int_0^\tau \Psi(t,s)ds \quad h \sum_{j=0}^{N} A_j \, \Psi(t,(N-j)h) = O(h^q)$$

uniformly for t in a compact interval $I \subset \mathbf{R}$ when $\{\partial/\partial s\}^q \Psi(t,s)$ is continuous on $I \times [0,\tau]$.

Concerning the smoothness of $y(s)$ and $K(t,s,y(s))$ we make, in addition to (H1)-(H3), the following hypotheses

$$(\mathrm{H6}) \quad \begin{cases} \text{(i) } y \in C^{p+1}[-\tau,T]; \\ \text{(ii) } K(t,s,y(s)) \text{ has continuous derivatives } (\partial/\partial s)^q(\partial/\partial t)^{p+1}K(t,s,y(s)) \\ \text{for } s \in [t-\tau,t] \text{ and } t \in [0,T]. \end{cases}$$

Our first result establishes convergence and order of convergence:

<u>Theorem 2.1</u> *(Order of convergence)* Given (H1), (H2), (H3), (H4), (H5), (H6), suppose also that $\tilde{y}_\ell$ $\psi(t_\ell)$ for $-N \le \ell < 0$ and the additional starting values satisfy

$$(\mathrm{H7}) \qquad \lim_{h\to 0} \sum_{\ell=0}^{k-1} |\tilde{y}_\ell \quad y(t_\ell)| = O(h^r).$$

Then, for any $T < \infty$,

$$(2.3) \qquad \lim_{h\to 0} \sup_{t_n \in [0,T]} |\tilde{y}_n \quad y(t_n)| = O(h^s)$$

where

$$(2.4) \qquad s = \min(r,p,q).$$

To establish Theorem 2.1 we must relate the errors

$$(2.5) \qquad \varepsilon_r = \tilde{y}_r - y(t_r)$$

(which are uniquely defined under the conditions (H1)-(H3) and $h \in (0,h_0)$ assumed for Lemma 1.1) to the local truncation errors associated with the scheme. We have:

<u>Lemma 2.2.</u> *(The local truncation error)* Subject to (H1)-(H6), the local truncation error

$$(2.6) \qquad \begin{aligned} \eta(t) := &\sum_{\ell=0}^{k} a_\ell\, y(t-\ell h) \\ &h\sum_{\ell=0}^{k} b_\ell\, G(t-\ell h,\ y(t-\ell h),\ h\sum_{j=0}^{N} A_j\ K(t-\ell h, t-\ell h-jh, y(t-\ell h-jh))) \end{aligned}$$

is such that $|\eta(t)|$ $O(h^{p+1}) + O(h^{q+1})$ uniformly for t in any open subinterval of $[0,T]$ (as $h \to 0$).

<u>Proof:</u> Replace φ by y in (2.1), replace $\Psi(t,s)$ by $K(t,s,y(s))$ in (2.2) and employ (H2), (H5) and (H6). □

A sequence of results is required to establish a relationship between the values of the local truncation errors

$$(2.7) \qquad \eta_n = \eta(t_n)$$

and the approximation errors ε_n of (2.5). The first result is familiar in the study of Volterra integral equations:

<u>Lemma 2.3.</u> *(A discrete Gronwall inequality)* Suppose

$$|e_n| \le h\,\Omega \sum_{j=0}^{n-1} |e_j| + \Delta_n \qquad (n \ge k)$$

and that $\sum_{\ell=0}^{k-1} |e_\ell| \le e$. If, for $k \le n \le K$, $|\Delta_n| \le \Delta$ then

$$|e_n| \le (h\Omega e + \Delta)\exp\{Kh\Omega\} \quad (n=k,k+1,\cdots,K).$$

<u>Proof:</u> See [1], [3] or [8]. (An intermediate result, with $\{(1+h\Omega)^{n-k}\}$ in place of $\exp\{Kh\Omega\}$, viz.,

$$|e_n| \le (h\Omega e + \Delta)(1+hL)^{n-k} \qquad (n=k,k+1,\cdots,K)$$

can be established by induction, and the lemma then follows.) □

<u>Lemma 2.4</u> *(A consequence of zero-stability)* Suppose that, for $n=0,1,2,\cdots$,

$$(2.8) \qquad \sum_{\ell=0}^{k} a_\ell\, \varepsilon_{n-\ell} \quad h\sum_{\ell=0}^{k} b_\ell\, \xi_{n-\ell} + \nu_n$$

where $\varepsilon_{-\ell} = \xi_{-\ell} = 0$ for $\ell \in \mathbb{Z}_+$. Then, subject to (H4),

$$\varepsilon_n = h \sum_{j=0}^{n} w_{n-j}\, \xi_j + \sum_{j=0}^{n} c_{n-j}\, \nu_j$$

wherein $|c_\ell| \le C < \infty$ and $|w_\ell| \le W < \infty$ uniformly for $0 \le \ell \in \mathbb{Z}$.

Proof: Denote by $T(d_0,\cdots,d_m)$ the lower-triangular Toeplitz matrix (isoclinal matrix):

$$\begin{bmatrix} d_0 & 0 & \cdots & & & 0 \\ d_1 & d_0 & 0 & & & 0 \\ d_2 & d_1 & d_0 & 0 & \cdots & 0 \\ & & & & & \vdots \\ d_{m-1} & & \cdots & & d_0 & 0 \\ d_m & d_{m-1} & \cdots & & d_1 & d_0 \end{bmatrix}$$

As the reader may check, for $m > k$, $[T(a_0,\cdots,a_k,0,\cdots,0)]^{-1} = T(c_0,\cdots,c_m)$ where $\sum_{\ell=0}^{\infty} c_\ell \mu^\ell$ is the formal power series for $\{a_0 + a_1\mu+\cdots+a_k\mu^k\}^{-1} \equiv \{\hat\rho(\mu)\}^{-1}$ wherein $a_0 \neq 0$. Also for $m > k$, the product

$$[T(a_0,\cdots,a_k,0,\cdots,0)]^{-1}.T(b_0,\cdots,b_k,0,\cdots,0) = T(w_0,\cdots,w_m)$$

is isoclinal where

$$\sum_{\ell=0}^{\infty} w_\ell \mu^\ell = \{a_0+a_1\mu+\cdots+a_k\mu^k\}^{-1}\{b_0+b_1\mu+\cdots+b_k\mu^k\}$$

is the formal power series for $\{a_0+a_1\mu+\cdots+a_k\mu^k\}^{-1}\{b_0+b_1\mu+\cdots+b_k\mu^k\}$. The uniform boundedness of $\{|c_\ell|\}$ by $C < \infty$, and hence that of $\{|w_\ell|\}$ by $W < \infty$, follows from the binomial expansion using partial fractions to simplify $\{\hat\rho(\mu)\}^{-1}$ and the property that $\hat\rho(\mu)$ is simple von Neumann, since $\{\hat\rho,\hat\sigma\}$ is zero-stable. (The results above can be found in the literature on numerical methods for ordinary differential equations, but are easily verified directly.)

Express (2.8) for $k \le m$ in matrix-vector form

$$T(a_0,\cdots,a_k,0,\cdots,0).\varepsilon = hT(b_0,\cdots,b_k,0,\cdots,0)\xi + \nu$$

where $\varepsilon = [\varepsilon_0, \varepsilon_1,\cdots, \varepsilon_m]^T$ etc., and the Lemma is immediately established.□

Proof of Theorem 2.1. Suppose h sufficiently small ($h < h_0$ where h_0 is given by Lemma 1.1(b)) so that $\{\tilde y_n \mid n \ge k\}$ is uniquely defined. To employ the foregoing lemmas, we note that, from (2.6) and (2.7),

$$\sum_{\ell=0}^{k} a_\ell\, y(t_{n-\ell}) \tag{2.9}$$
$$-h\sum_{\ell=0}^{k} b_\ell\, G(t_{n-\ell}, y(t_{n-\ell}),\ h\sum_{j=0}^{N} A_j\, K(t_{n-\ell}, t_{n-\ell-j}, y(t_{n-\ell-j}))) = \eta_n \quad (n \ge k)$$

and that

$$\sum_{\ell=0}^{k} a_\ell \tilde y_{n-\ell}$$
$$-h\sum_{\ell=0}^{k} b_\ell G(t_{n-\ell}, \tilde y_{n-\ell},\ h\sum_{j=0}^{N} A_j K(t_{n-\ell}, t_{n-\ell-j}, \tilde y_{n-\ell-j})) = 0 \quad (n \ge k). \tag{2.10}$$

It follows that, with ε_ℓ defined as in (2.5),

$$\sum_{\ell=0}^{k} a_\ell\, \varepsilon_{n-\ell} \quad h\sum_{\ell=0}^{k} b_\ell\, \gamma_{n-\ell} + \eta_n \qquad (n \ge k) \tag{2.11}$$

where

$$\gamma_n = G(t_n, y(t_n),\ h\sum_{j=0}^{N} A_j\ K(t_n, t_{n-j},\ y(t_{n-j})))$$

(2.12)

$$G(t_n, \tilde{y}_n,\ h\sum_{j=0}^{N} A_j\ K(t_n, t_{n-j}, \tilde{y}_{n-j})).$$

We intend to augment (2.11) by a further relation valid for $n = 0,1,\cdots,k-1$ and then to apply Lemma 2.4. We define

(2.13) $\xi_\ell = 0\ (\ell \le k-1),\ \xi_\ell\quad \gamma_\ell\ (\ell \ge k);$

(2.14) $v_\ell\quad \sum_{j=0}^{k} a_j \varepsilon_{\ell-j}\ (\ell = 0,\ 1,\ \cdots,\ k-1),\ v_\ell = \eta_\ell\ (\ell \ge k).$

By virtue of (2.14) and (2.11), we have

(2.15) $\sum_{\ell=0}^{k} a_\ell\ \varepsilon_{n-\ell}\quad h\sum_{\ell=0}^{k} b_\ell\ \xi_{n-\ell} + v_n\qquad (n \ge 0).$

Applying Lemma 2.4 to (2.15), we have

(2.16) $\varepsilon_n\quad h\sum_{j=0}^{n} w_{n-j}\ \xi_j + \upsilon_n$

wherein

(2.17) $|\upsilon_n| = |\sum_{j=0}^{k-1} c_{n-j}\ v_j + \sum_{j=k}^{n} c_{n-j}\ v_j\ |.$

By (H7) and (2.6), since $v_\ell = \sum_{\ell=0}^{k} a_j\varepsilon_{\ell-j}$ for $0 \le j < k$ and $v_j = \eta_j$ for $j \ge k$, we have $\upsilon_n = O(h^r) + nC\{O(h^{p+1}) + O(h^{q+1})\}\quad O(h^r) + nhC\{O(h^p) + O(h^q)\}$ where C is the constant appearing in Lemma 2.4. Thus,

(2.18) $\sup |\upsilon_n| = O(h^s)$ as $h \to 0,\ s\quad \min(r,p,q),$

where the supremum is taken for $0 \le n \le K$ with $Kh = T$.

Returning to (2.12) and (2.13), we observe (using $\Lambda_1, \Lambda_2, \Lambda_3$ from (H1)-(H3)) that

(2.19) $|\xi_n| \le \Lambda_1|\varepsilon_n| + \Lambda_4 h\sum_{\ell=0}^{N} |\varepsilon_{n-\ell}|$

for $n \ge k$ (and hence for $n \ge 0$) where Λ_4 is bounded and depends on the Lipschitz constants $\{\Lambda_i\}$ and the quadrature weights $\{A_\ell\}$:

(2.20) $\Lambda_4\quad \{\Lambda_2\Lambda_3 \max_\ell |A_\ell|\}.$

We deduce from (2.16) and (2.19) the result

(2.21) $|\varepsilon_n| \le hW\sum_{j=0}^{n}\{\Lambda_1|\varepsilon_j| + h\Lambda_4\{\sum_{i=j-N}^{j}|\varepsilon_i|\}\} + \upsilon_n$

for $0 \le n \in \mathbb{Z}$. To employ Lemma 2.3 it is convenient to define

(2.22)

$$e_n = \max\{|\varepsilon_j|\ |\ j = n-N, n-N+1, \cdots, n\}$$

$$\delta_n\quad \max\{|\upsilon_j|\ |\ j = n-N, n-N+1, \cdots, n\}$$

where $\varepsilon_{-j}\quad \upsilon_{-j}\quad 0$ if $j \in \mathbb{Z}_+$. Then (2.21), (2.22) give:

$$|\varepsilon_n| \le hW\sum_{j=0}^{n}\{\Lambda_1 e_j + (N+1)h\ \Lambda_4 e_j\} + \delta_n.$$

We deduce

(2.23) $e_n \le hL\sum_{j=0}^{n} e_j + \delta_n$

with $L\quad \{\Lambda_1 + 2\tau\Lambda_4\}W \ge \{\Lambda_1 + (N+1)h\ \Lambda_4\}W$. Thus, for $hL < 1$,

$$(2.24) \qquad e_n \le h\Omega \sum_{j=0}^{n-1} e_j + \Delta_n$$

with $\Omega = L/(1-hL)$ and $\Delta_n = \delta_n/(1-hL)$. Clearly,

$$(2.25) \qquad e := \sum_{\ell=0}^{k-1} |e_\ell| = O(h^r), \quad \Delta := \sup_{0 \le n \le K} |\Delta_n| = O(h^s),$$

as $h \to 0$ with $Kh = T$.

We now apply Lemma 2.3 to (2.24). From the resulting bound we conclude that $\sup e_n = O(h^s)$ and we deduce, since $|\varepsilon_n| \le e_n$ that $|\varepsilon_n| = O(h^s)$, where $s = \min(r,p,q)$, as required. This establishes our theorem. □

The convergence theory presented here can be modified to deal with the discretization of (1.3) by (1.9), (1.10).

3. Asymptotic expansion of the discretisation error

In the following, we provide an asymptotic expansion for the discretisation error which was demonstrated to be $O(h^s)$ by Theorem 2.1. We are able to prove that, under suitable hypotheses defined below, *there exists a function* $e(t)$, *which is independent of* h, *such that* $\tilde{y}_n = y(t_n) = h^m e(t_n) + o(h^m)$. (We shall require, in the notation above, that $r = p+1$; then $m = \min(p,q)$, given certain conditions.) In order to establish this result, we shall require that the starting errors, quadrature errors and LMF errors take a particular form. Stetter [10, (Theorem 4.4.2, p232)] obtains the result for the ordinary differential equation which is a special case of (1.1) with $K=0$. He remarks that an asymptotic expansion for the error up to the term of order J exists only when the starting values satisfy rigorous constraints. The special case where $J=p$ and the order of the starting errors is $p+1$ gives a satisfactory result. Thus the condition on the starting errors imposes a condition on how many terms in the asymptotic expansion we may establish. For the moment, we shall confine our discussion to deriving the dominant term in the expansion, since this corresponds closely to the special case identified by Stetter.

For the LMF we make the hypothesis (strengthening our earlier assumption (H4)) that

(3.1) (H8) $\hat{\rho},\hat{\sigma}$ in (1.7) define a zero-stable linear multistep method of order at least $p>0$

such that, for any closed bounded interval I of $\mathbb{R}$ and any function $\varphi \in C^{R+1}(I)$,

$$(3.2) \qquad \sum_{\ell=0}^{k} a_\ell \varphi(t-\ell h) = h \sum_{\ell=0}^{k} b_\ell \varphi'(t-\ell h) + \sum_{\ell=p+1}^{R} c_\ell(\varphi,t)h^\ell + O(h^{R+1}).$$

For the present, we shall use this result only where $R = p+1$; the values $c_\ell(\varphi,t)$ depend on derivatives of φ.

For an M-times repeated quadrature rule of order q with $N=MR$, we assume that $\tau=Nh$,

$$(3.3) \ \text{(H9)} \qquad \int_0^\tau \Psi(s)\,ds = h \sum_{j=0}^{N} A_j \Psi((N-j)h)$$

$$+ \sum_{\sigma=q}^{R} d_\sigma h^\sigma \{\Psi^{(\sigma-1)}(\tau) - \Psi^{(\sigma-1)}(0)\} + O(h^{R+1}).$$

where the constants d_σ are determined by the basic rule. As a consequence of (3.3) we find that there exist functions $D_\sigma(K,t)$ ($\sigma = q, q+1, \ldots$) such that

$$\int_{t-\tau}^{t} K(t,s,y(s))ds \quad h\sum_{j=1}^{N} A_j \; K(t,t-jh,y(t-jh))$$
$$+\sum_{\sigma=q}^{R} D_\sigma(K,t)h^\sigma + O(h^{R+1})$$

where y is the true solution. For the present, we shall use this result only with R = q.

Stetter's result [10] prompts us to make the following hypothesis regarding the starting errors: The values of $\psi(t)$ $(-\tau \le t \le 0)$ are prescribed exactly and the starting errors are $O(h^r)$ as $h \to 0$ where $r = \min(p+1,q+1)$. We assume:

(3.4) (H10) $\max_{0 \le \ell \le k-1} |y_\ell \quad \tilde{y}_\ell| \quad O(h^r)$ as $h \to 0$ where $r = \min(p+1,q+1)$.

By the results of § 2 we may deduce that under hypotheses (H8) (H9) (H10) (H1) (H2) (H3) and (H6), the result of Theorem 2.1 holds. (This is true since hypotheses (H8) (H9) (H10) are properly stronger than (H4) (H5) (H7).)

We shall need the notation

(3.5) $G_2(t,y,w) = (\partial/\partial y)G(t,y,w)$, $G_3(t,y,w) = (\partial/\partial w)G(t,y,w)$, $K_3(t,s,y) = (\partial/\partial y)K(t,s,y)$.

The result of interest to us in this section takes the following form:

Theorem 3.1 (Asymptotic expansion for the error) Assume (H1), (H2), (H3), (H6), (H8), (H9), (H10) are satisfied. Suppose that

$$w(t) = \int_{t-\tau}^{t} K(t,s,y(s))\, ds,$$

With the notation of § 2, there exists a function e(t), satisfying an equation of the form

$$e'(t) = G_2(t,y(t),w(t))e(t) + \int_{t-\tau}^{t} K_3(t,s,y(s))e(s)ds \; G_3(t,y(t),w(t)) + \chi(t) \quad (t > 0),$$
$$e(t) = 0 \quad (-\tau \leqslant t \leqslant 0),$$

for some function $\chi(t)$, such that

$$\tilde{y}_n \quad y(t_n) \quad h^m e(t_n) + o(h^m).$$

Proof: We may identify the three steps in the proof of this theorem, which we shall present as a series of three lemmas. Define

$$z_h(t) := y(t) + h^m e(t),$$

assuming e(t) to be a known function which we shall define later, and define

$$\eta(z_h,t_n) := \sum_{\ell=0}^{k} a_\ell \; z(t_n - \ell h)$$
$$- h \sum_{\ell=0}^{k} b_\ell \; G(t_n-\ell h,\; z(t_n-\ell h), \sum_{j=0}^{N} hA_j K\{t_n-\ell h, t_n-\ell h-jh, z(t_n-\ell h-jh)\})$$

First we shall obtain an expression for $\eta(z_h,t_n)$ in terms of the quadrature errors given by (H9), the errors in the linear multistep formula given by (H8) and the function e(t). A carefully selected definition of e(t) then gives a simplified function η which is $o(h^{m+1})$. Finally we are able to show that $|\tilde{y}_n - z(t_n)| = o(h^m)$ and the theorem follows. :

Lemma 3.1 (An expression for $\eta(z_h,t_n)$) The function values $\eta(z_h,t_n)$ are given by

$$\eta(z_h,t_n) \quad \sum_{\ell=0}^{k} a_\ell y(t_n-\ell h) - h\sum_{\ell=0}^{k} b_\ell y'(t_n-\ell h) +$$
$$+ h^m\{\sum_{\ell=0}^{k} a_\ell e(t_n-\ell h) \quad h\sum_{\ell=0}^{k} b_\ell e(t_n-\ell h)G_2(t_n-\ell h, y(t_n-\ell h), w(t_n-\ell h))$$
$$-h\sum_{\ell=0}^{k} \{b_\ell \int_{t_n-\ell h-\tau}^{t_n-\ell h} K_3(t_n-\ell h,s,y(s))e(s)ds \; G_3(t_n-\ell h,\; y(t_n-\ell h), w(t_n-\ell h))\}\}$$
$$-h^q(h\sum_{\ell=0}^{k} \; b_\ell d_q(K,t_n-\ell h)G_3(t_n-\ell h,\; y(t_n-\ell h), w(t_n-\ell h)))\} +$$
$$+ o(h^{m+1}) + o(h^{q+1}).$$

Proof: We continue with the notation $w(t) = \int_{t-\tau}^{t} K(t,s,y(s))ds$,

so that (1.1) gives (for $n > \ell$)

$$y'(t_n-\ell h) = G(t_n-\ell h, y(t_n-\ell h), w(t_n-\ell h)).$$

We set

$$\tilde{w}_h(t) = \sum_{j=0}^{N} hA_j\ K(t,s,z_h(t_j)),$$

and

$$\eta(z_h,t_n) := \sum_{\ell=0}^{k} a_\ell z(t_n-\ell h) - h\sum_{\ell=0}^{k} b_\ell G(t_n-\ell h, z(t_n-\ell h), \tilde{w}_h(t_n-\ell h)).$$

Thus,

$$\eta(z_h,t_n) = \sum_{\ell=0}^{k} a_\ell\ y(t_n \quad \ell h) + h^m \sum_{\ell=0}^{k} e(t_n-\ell h) - h\sum_{\ell=0}^{k} b_\ell G(t_n-\ell h, z_h(t_n-\ell h), \tilde{w}_h(t_n-\ell h))$$

$$- \sum_{\ell=0}^{k} a_\ell\ y(t_n \quad \ell h) + h^m \sum_{\ell=0}^{k} a_\ell\ e(t_n-\ell h)$$

$$-h\sum_{\ell=0}^{k} b_\ell G(t_n-\ell h, y(t_n-\ell h), w(t_n-\ell h))$$

$$-h\sum_{\ell=0}^{k} b_\ell \{G(t_n-\ell h, z_h(t_n-\ell h), \tilde{w}_h(t_n-\ell h)) - G(t_n-\ell h, y(t_n-\ell h), w(t_n-\ell h))\}.$$

Now, to discuss the last term in the preceding equation, we write

$$\tilde{w}_h(t) - w(t) = h\sum_{j=0}^{N} A_j K(t,t_j,y(t_j)) + h\sum_{j=0}^{N} A_j h^m e(t_j) K_3(t,t_j,y(t_j)) + o(h^m) - w(t)$$

$$= \tilde{d}_q(K,t)h^q + h^m\ h\sum_{j=0}^{N} A_j e(t_j) K_3(t,t_j,y(t_j)) + o(h^m),$$

by (H9),

$$= \tilde{d}_q(K,t)h^q + h^m\ \{\int_{t-\tau}^{t} K_3(t,s,y(s))e(s)ds + o(1)\} + o(h^m)\} = O(h^m).$$

It follows that

$$h\sum_{\ell=0}^{k} b_\ell \{G(t_n-\ell h, z_h(t_n-\ell h), \tilde{w}_h(t_n-\ell h)) - G(t_n-\ell h, y(t_n-\ell h), w(t_n-\ell h))\}.$$

$$= h \sum_{\ell=0}^{k} b_\ell\ h^m\ e(t_n-\ell h)\ G_2(t_n-\ell h, y(t_n-\ell h), w(t_n-\ell h))$$

$$+ h\sum_{\ell=0}^{k} b_\ell\ \{(\tilde{w}_h(t_n-\ell h) - w(t_n-\ell h)) G_3(t_n-\ell h,\ y(t_n-\ell h), w(t_n-\ell h)) + o(h^m)\}.$$

Thus,

$$\eta(z_h,t_n) = \sum_{\ell=0}^{k} a_\ell y(t_n-\ell h) - h\sum_{\ell=0}^{k} b_\ell y'(t_n-\ell h) + h^m\sum_{\ell=0}^{k} a_\ell e(t_n-\ell h)$$

$$- h^{m+1} \sum_{\ell=0}^{k} b_\ell\ e(t_n-\ell h)\ G_2(t_n-\ell h, y(t_n-\ell h), w(t_n-\ell h))$$

$$- h\sum_{\ell=0}^{k} b_\ell\ \{(\tilde{w}_h(t_n-\ell h) - w(t_n-\ell h)) G_3(t_n-\ell h,\ y(t_n-\ell h), w(t_n-\ell h)) + o(h^m)\}$$

$$= \sum_{\ell=0}^{k} a_\ell y(t_n-\ell h) - h\sum_{\ell=0}^{k} b_\ell y'(t_n-\ell h)$$

$$+ h^m\Big\{\sum_{\ell=0}^{k} a_\ell e(t_n-\ell h) - h\sum_{\ell=0}^{k} b_\ell\ e(t_n-\ell h) G_2(t_n-\ell h, y(t_n-\ell h), w(t_n-\ell h))$$

$$-h\sum_{\ell=0}^{k} b_\ell \int_{t_n-\ell-\tau}^{t_n-\ell} K_3(t_n-\ell h, s, y(s)) e(s)ds\ G_3(t_n-\ell h, y(t_n-\ell h), w(t_n-\ell h))\Big\} \tag{3.6}$$

$$- h\sum_{\ell=0}^{k} h^q d_q(K,t_n-\ell h) G_3(t_n-\ell h,\ y(t_n-\ell h), w(t_n-\ell h))$$

$$+ o(h^{m+1}) + o(h^{q+1}).$$

Lemma 3.2 (An equation for $e(t)$.) Let $m = \min(p,q)$ and let the function $e(t)$ be chosen to satisfy $e(t) = 0$, $-\tau \leq t \leq 0$, and, for $t > 0$,

$$e'(t) = G_2(t,y(t),w(t))e(t) + \{\int_{t-\tau}^{t} K_3(t,s,y(s))e(s)ds\}\{G_3(t,y(t),w(t))\} + \chi(t)$$

where $\chi(t) = c_{p+1}(y,t) \ /\ \sum_{\ell=0}^{k} b_\ell$, if $p < q$,

$\chi(t) = D_q(K,t)G_3(t,y(t),w(t))$, if $q < p$,

$\chi(t) \quad c_{p+1}(y',t) \ /\ \sum_{\ell=0}^{k} b_\ell + D_q(K,t)G_3(t,y(t),w(t))$, if $p \quad q$.

the function w being defined in Theorem 3.1. Suppose, in addition to the preceding hypotheses, that $e(t) \in C^{p+1}[0,T]$. Then, as $h \to 0$,

$$\sup |\eta(z_h, t_n)| = o(h^{m+1})$$

where the supremum is for $t_n \in [0,T]$ and $z_h(t) = y(t) + h^m e(t)$.

Proof Lemma 3.2 follows by applying (H8) to the result of Lemma 3.1. □

Note The function $\chi(t)$ in Lemma 3.2 is dependent on y: $\chi(t) \equiv \chi(t, y(\cdot))$.

Lemma 3.3 **(The asymptotic form of the error)** Assume (H10) and that m   min(p,q). Then $r > m$, and with $e(t)$ defined as in Lemma 3.2, $|\tilde{y}_n \quad z(t_n)| = o(h^m)$ (equivalently,

$$\tilde{y}_n - y(t_n) \quad h^m e(t) + o(h^m) \text{)}$$

provided $e(t) \in C^{p+1}[0,T]$.

Proof By (H10), the starting errors are of sufficiently high order that $r > m$. The remainder of the proof is broadly similar to the proof of Theorem 2.1:

First we may observe that, under the above conditions, $\eta(z_h, t_n) \quad o(h^{m+1})$. Suppose h sufficiently small ($h < h_0$ where h_0 is given by Lemma 1.1(b))) so that $\{\tilde{y}_n \mid n \geq k\}$ is uniquely defined. To employ the foregoing lemmas, we note that,

$$\sum_{\ell=0}^{k} a_\ell z(t_{n-\ell}) \tag{3.7}$$
$$-h\sum_{\ell=0}^{k} b_\ell G(t_{n-\ell}, z(t_{n-\ell}), h\sum_{j=0}^{N} A_j K(t_{n-\ell}, t_{n-\ell-j}, z(t_{n-\ell-j})))$$
$$= \eta(z_h, t_n) \quad (n \geq k)$$

and that

$$\sum_{\ell=0}^{k} a_\ell \tilde{y}_{n-\ell} \tag{3.8}$$
$$-h\sum_{\ell=0}^{k} b_\ell G(t_{n-\ell}, \tilde{y}_{n-\ell}, h\sum_{j=0}^{N} A_j K(t_{n-\ell}, t_{n-\ell-j}, \tilde{y}_{n-\ell-j})) \quad 0 \quad (n \geq k).$$

It follows that, with $\bar{\varepsilon}_\ell$ defined by

$$\bar{\varepsilon}_\ell := z(t_\ell) \quad \tilde{y}_\ell,$$

$$\sum_{\ell=0}^{k} a_\ell \bar{\varepsilon}_{n-\ell} = h\sum_{\ell=0}^{k} b_\ell \bar{\gamma}_{n-\ell} + \eta(z_h, t_n) \quad (n \geq k) \tag{3.9}$$

where

$$\bar{\gamma}_n \quad G(t_n, z(t_n), h\sum_{j=0}^{N} A_j K(t_n, t_{n-j}, z(t_{n-j}))) \tag{3.10}$$
$$G(t_n, \tilde{y}_n, h\sum_{j=0}^{N} A_j K(t_n, t_{n-j}, \tilde{y}_{n-j})).$$

We intend to augment (3.9) by a further relation valid for n   0,1,⋯,k-1 and then to apply Lemma 2.4. We define

(3.11) $\quad \bar{\xi}_\ell \quad 0 \ (\ell \leq k-1), \ \bar{\xi}_\ell \quad \bar{\gamma}_\ell \ (\ell \geq k);$

(3.12) $\quad \bar{v}_\ell \quad \sum_{j=0}^{k} a_j \bar{\varepsilon}_{\ell-j} \ (\ell = 0, 1, \cdots, k-1), \ \bar{v}_\ell = \eta(z_h, t_\ell) \ \ell \geq k.$

We note that

$$\bar{v}_\ell \quad \sum_{j=0}^{k} a_j\{y(t_{\ell-j}) + h^m e(t_{\ell-j}) \quad \bar{y}_{\ell-j}\} \quad (\ell \quad 0, 1, \cdots, k-1)$$

and hence

$$\bar{v}_\ell \quad \sum_{j=0}^{k} a_j\{h^m e(t_{\ell-j}) \quad \varepsilon_{\ell-j}\} = O(h^{m+1}) \quad (\ell \quad 0, 1, \cdots, k-1)$$

since $e(0) = 0$ and $e(rh) = rhe'(\theta_r rh)$ for some $\theta_r \in (0,1)$.

By virtue of (3.11) and (3.12), we have

$$(3.13) \quad \sum_{\ell=0}^{k} a_\ell \, \bar{\varepsilon}_{n-\ell} = h\sum_{\ell=0}^{k} b_\ell \, \xi_{n-\ell} + \bar{v}_n \qquad (n \geq 0).$$

Applying Lemma 2.4 to (3.13), we have

$$(3.14) \qquad \bar{\varepsilon}_n = h\sum_{j=0}^{n} w_{n-j} \, \xi_j + \bar{\upsilon}_n$$

wherein

$$(3.15) \qquad |\bar{\upsilon}_n| = \Big| \sum_{j=0}^{k-1} c_{n-j} \, \bar{v}_j + \sum_{j=k}^{n} c_{n-j} \, \bar{v}_j \Big| .$$

By (H10) and Lemma 3.2, since $\bar{v}_\ell = \sum_{j=0}^{k} a_j \bar{\varepsilon}_{\ell-j}$ for $0 \leq j < k$ and $\bar{v}_j = \eta(z_h, t_j)$ for $j \geq k$, we have $\bar{\upsilon}_n = O(h^r) + nC\{O(h^{m+1})\} = O(h^r) + nhC\{O(h^m)\}$ where C is the constant appearing in Lemma 2.4. Thus,

$$(3.16) \qquad \sup |\bar{\upsilon}_n| \quad o(h^m) \text{ as } h \to 0$$

where the supremum is taken for n with $t_n \in [0,T]$.

Returning to (3.10) and (3.11), we observe (using $\Lambda_1, \Lambda_2, \Lambda_3$ from (H1)-(H3)) that

$$(3.17) \qquad |\xi_n| \leq \Lambda_1|\bar{\varepsilon}_n| + \Lambda_4 h\sum_{\ell=0}^{N} |\bar{\varepsilon}_{n-\ell}|$$

for $n \geq k$ (and hence for $n \geq 0$) where Λ_4 is bounded and depends on the Lipschitz constants $\{\Lambda_i\}$ and the quadrature weights $\{A_\ell\}$:

$$(3.18) \qquad \Lambda_4 \quad \{\Lambda_2\Lambda_3 \max_\ell |A_\ell|\}$$

as before. We deduce from (3.14) and (3.17) the result

$$(3.19) \qquad |\bar{\varepsilon}_n| \leq hW \sum_{j=0}^{n}\Big\{\Lambda_1|\bar{\varepsilon}_j| + h\Lambda_4 \Big\{\sum_{\ell=j-N}^{j} |\bar{\varepsilon}_\ell|\Big\}\Big\} + \bar{\upsilon}_n$$

for $0 \leq n \in \mathbf{Z}$. To employ Lemma 2.3 it is convenient to define

$$(3.20) \qquad \begin{aligned} \bar{e}_n &= \max\{|\bar{\varepsilon}_\ell| \mid \ell = n-N, n-N+1, \cdots, n\} \\ \bar{\delta}_n &= \max\{|\bar{\upsilon}_\ell| \mid \ell = n-N, n-N+1, \cdots, n\} \end{aligned}$$

where $\bar{\varepsilon}_{-\ell} = \bar{\upsilon}_{-\ell} = 0$ if $\ell \in \mathbf{Z}_+$. Then (3.19), (3.20) give:

$$|\bar{\varepsilon}_n| \leq hW \sum_{j=0}^{n}\Big\{\Lambda_1\bar{e}_j + (N+1)h \, \Lambda_4\bar{e}_j\Big\} + \bar{\delta}_n .$$

We deduce

$$(3.21) \qquad \bar{e}_n \leq hL \sum_{j=0}^{n} \bar{e}_j + \bar{\delta}_n$$

with $L = \{\Lambda_1 + 2\tau\Lambda_4\}W \geq \{\Lambda_1 + (N+1)h \, \Lambda_4\}W$. Thus, for $hL < 1$,

$$(3.22) \qquad \bar{e}_n \leq h\Omega \sum_{j=0}^{n-1} \bar{e}_j + \bar{\Delta}_n$$

with $\Omega = L/(1-hL)$ and $\bar{\Delta}_n = \bar{\delta}_n/(1-hL)$. Clearly,

$$(3.23)\qquad \bar{e} := \sum_{\ell=0}^{k-1} |\bar{e}_\ell| = O(h^r), \quad \bar{\Delta} := \sup_{0 \leq n \leq K} |\bar{\Delta}_n| = O(h^m),$$

as $h \to 0$ with $Kh = T$.

We now apply Lemma 2.3 to (3.22). From the resulting bound we conclude that $\sup \bar{e}_n = o(h^m)$ and we deduce, since $|\bar{\varepsilon}_n| \leq \bar{e}_n$ that $|\bar{\varepsilon}_n| \quad o(h^m)$, as required. This establishes Lemma 3.3. □

Remarks (1) It may happen that $D_{q+r}(K,t)$ vanishes identically for $r = 0, 1, 2, \ldots, Q \leq R$. In this case we should replace $D_q(K,t)$ by $D_Q(K,t)$ and substitute Q for q in §3, in particular in Lemma 3.2.

(2) We may consider the question of whether the above analysis may be extended to give an expression for the next term in the asymptotic expansion. In the proof of Lemma 3.3 above, we used the fact that the starting errors were such that

$$h^m e(t_\ell) \quad \varepsilon_\ell = O(h^{m+1}) \quad (\ell = 0, 1, \cdots, k-1).$$

In order to be able to obtain successive terms in the series expansion

$$\varepsilon_n = \tilde{y}_n \quad y(t_n) \quad h^m e_0(t_n) + h^{m+1} e_1(t_n) + \ldots + h^{m+R} e_R(t_n) + o(h^{m+R})$$

uniformly for $t_n \in [0,T]$, we should need the starting errors themselves to satisfy this relation to appropriate order accuracy:

$$h^m e_0(t_\ell) + h^{m+1} e_1(t_\ell) + \ldots + h^{m+R} e_R(t_\ell) - \varepsilon_\ell = O(h^{m+R+1}) \quad (\ell \quad 0, 1, \cdots, k-1)$$

This restriction which the starting errors impose on the number of terms which can be obtained corresponds to the restriction identified by Stetter in the case of ordinary differential equations which we referred to above. It does not apply, of course, in the case that no starting values are required ($k = 1$).

4. Illustrative examples.

The convergence theorem of Section 2 gives the order of convergence in terms of the integers p, q, r. If the latter are varied, it is possible to demonstrate the realism of Theorem 2.1 by simple numerical experiments. Further, the asymptotic error expansion which we discussed in section 3, Theorem 3.1, suggests that we may employ the deferred approach to the limit to improve on the approximations, and this may also be demonstrated numerically. Finally, graphical plots may be presented of the values $(\tilde{y}_n \quad y(t_n))/h^m$ for varying values of h. This yields successive approximations to the function e(t).

In the present work, we shall present one such example. More details and further examples appear in [2].

Consider the equation:

$$(4.1)\quad y'(t) = \cos(t) \quad \cos(t-1) \quad \int_{t-1}^{t} (1-(t-s))y(s)\, ds,$$

with the initial conditions

$$(4.2)\qquad y(t) = \cos(t), \quad t \leq 0.$$

The equation (4.1) subject to (4.2) has smooth solution $y(t) \quad \cos(t)$, $t \geq 0$.

Suppose we use the trapezium rule in (1.11) as the linear multistep method, i.e. we set

$$(4.3)\qquad \tilde{y}_{n+1} \quad \tilde{y}_n = \frac{h}{2}\left\{G(t_{n+1},\, \tilde{y}_{n+1},\, \tilde{\zeta}_{n+1}) + G(t_n,\, \tilde{y}_n,\, \tilde{\zeta}_n)\right\} \qquad (h = 1/N),$$

and we approximate the integral in (1.13) using the composite explicit Euler rule, i.e. we use the rule

$$\int_0^1 \psi(s)ds \approx h \sum_{\ell=0}^{N-1} \psi((N-\ell)h).$$

to replace $\int_{t-1}^{t} y(s)\, ds$ by $h\sum_0^{N-1}\tilde{y}_\ell$. Table 1 gives the absolute values of the errors and relative errors

$$|\varepsilon_n^{\#}| \qquad |\{\tilde{y}_n \quad y(nh)\}/\tilde{y}_n|$$

for $nh = 5$ and $nh = 10$ and varying h. Theorem 2.1 predicts convergence of O(h), which is borne out by the values in Table 1.

h	n	absolute error	relative error	extrapolated error	relative extrapolated error
(a) $t = nh = 5$					
0.05	100	$0.48_{10}-2$	$0.17_{10}-1$	$0.11_{10}-3$	$0.39_{10}-3$
0.025	200	$0.24_{10}-2$	$0.85_{10}-2$	$0.28_{10}-4$	$0.98_{10}-4$
0.0125	400	$0.12_{10}-2$	$0.42_{10}-2$	$0.69_{10}-5$	$0.24_{10}-4$
0.00625	800	$0.62_{10}-3$	$0.22_{10}-2$		
(b) $t = nh = 10$					
0.05	200	$0.91_{10}-2$	$0.11_{10}-1$	$0.15_{10}-3$	$0.18_{10}-3$
0.025	400	$0.46_{10}-2$	$0.55_{10}-2$	$0.37_{10}-4$	$0.44_{10}-4$
0.0125	800	$0.23_{10}-2$	$0.27_{10}-2$	$0.90_{10}-5$	$0.11_{10}-4$
0.00625	1600	$0.12_{10}-2$	$0.14_{10}-2$		

Table 1 Errors and extrapolated errors for the low order formula (4.3) nh=5, 10.

Figure 1 shows the plots of $(\tilde{y}_n \quad y(t_n))/h^m$ for varying h, which indicate convergence to the function e(t). In the figure, the bold line is caused by superimposition of multiple lines for successive small h values.

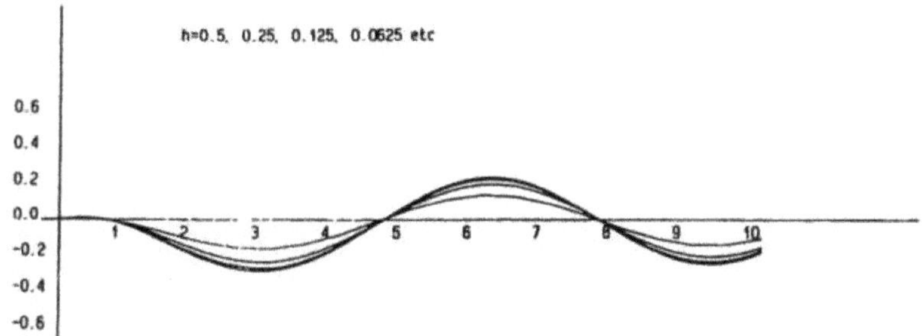

Plots of the approximations to the function e(t) for varying h, using the Trapezium rule as LMF, and explicit Euler for the integral

Figure 1

5. Further work.

As is well known in the case of initial value problems in ordinary differential equations, stability restrictions impose limitations on the choice of stepsize h for the correct simulation of qualitative properties. A similar observation applies to the problems considered here, for which the analysis covers asymptotic features as $h \to 0$. Stability restrictions on the choice of h will be discussed in a forthcoming report.

References

[1] Baker, C.T.H. (1978) ***The numerical treatment of integral equations.*** Clarendon Press, Oxford.

[2] Baker, C.T.H. and N.J.Ford (1989) Asymptotic error expansions for linear multistep methods for a class of delay integro-differential equations. Numer. Anal. Tech. Report 174, Mathematics Dept., The University of Manchester.

Brunner, H. and P.J. van der Houwen (1986) ***The numerical solution of Volterra equations.*** North-Holland, Amsterdam.

[3] Feldstein, A. and J.R. Sopka. (1974) Numerical methods for nonlinear Volterra integro-differential equations. **SIAM J. Numer. Anal.** 11 pp. 826-846.

[4] Jackiewicz, Z. and M. Kwapisz. (1983) Convergence of multistep methods for Volterra integro-differential equations. **Annales Polonici Mathematici** 43 pp.121-139.

[5] Jackiewicz, Z. (1979) Convergence of multistep methods for Volterra functional differential equations. **Numer. Math.** 37 pp.307-322.

[6] Lambert, J.D. (1973) ***Computational methods in ordinary differential equations.*** Wiley, London

[7] Linz, P. (1985) ***Analytical and numerical methods for Volterra functional equations.*** SIAM Publications, Philadelphia.

[8] Linz, P. (1969) Linear multistep methods for Volterra integro-differential equations. **Journal Assoc. Comput. Mach.** 16 pp. 295-301.

[9] Stetter, H. (1973) ***Analysis of discretisation methods for ordinary differential equations.*** Springer-Verlag, New York-Heidelberg-Berlin.

[10] Wolkenfelt, P.H.M. (1981) ***The numerical analysis of reducible quadrature methods for Volterra integral and integro-differential equations.*** Math. Centrum, Amsterdam.

APPROXIMATE SOLUTION OF O.D.E.S IN BANACH SPACE - RATIONAL APPROXIMATION OF SEMIGROUPS

VIDAR THOMÉE

ABSTRACT. We survey some stability and convergence results concerning approximation of a semigroup e^{tA} in a Banach space by operators of the form $r(kA)^n, nk = t$, where r is an A-acceptable rational function. In particular, we present a recent result concerning stability in the case e^{tA} is analytic and $|r(\infty)| = 1$.

0. Introduction.

Let A be a densely defined, closed linear operator in a Banach space X, and assume that A generates a bounded strongly continuous semigroup $E(t) = e^{At}$ with $\|E(t)\| \le C_0$ for $t > 0$ Then, for any $v \in X$, the initial value problem

$$u' = Au, \quad \text{for } t > 0, \quad u(0) = v, \tag{0.1}$$

has the unique (generalized) solution $u(t) = E(t)v$, and

$$\|E(t)v\| \le C_0\|v\|, \quad \text{for } t > 0. \tag{0.2}$$

Let k be a time step and $t_n = nk$. An approximate solution $U^n \approx u(t_n)$ of (0.1) may be defined by

$$U^{n+1} = r(kA)U^n, \quad \text{for } n \ge 0, \qquad U^0 = v, \tag{0.3}$$

where r is a rational function that satisfies certain conditions. For example, $r(z) = 1/(1-z)$ and $r(z) = (1+z/2)/(1-z/2)$ correspond to the backward Euler and Crank-Nicolson methods, respectively. The scheme (0.3) then provides an approximation $r(kA)^n$ of $E(t_n)$ and one may ask the following two questions:

1. Stability: Is $r(kA)^n$ bounded for $n \ge 0$?
2. Convergence: Does $r(kA)^n v$ converge to $E(t)v$ as $k \to 0$, $nk = t$?

In the scalar case where A is defined as multiplication by $z \in C$, we have $E(t) = e^{zt}$, and (0.2) holds if and only if $\mathrm{Re} z \le 0$ (and C_0 may then be chosen as 1, i.e., $E(t)$ is a contraction semigroup).

In this case $r(kA)^n = r(kz)^n$, and for this to be bounded for all $n \ge 0$ it is necessary that r is A-acceptable, i.e.,

$$|r(z)| \le 1, \qquad \text{for } \mathrm{Re} z \le 0.$$

We shall therefore assume throughout that this condition holds. We shall also assume that r is accurate of order p for some $p \geq 1$, i.e.,

$$r(z) = e^z + O(z^{p+1}), \qquad \text{as } z \to 0. \tag{0.4}$$

In Section 1 below we shall give answers to the two questions posed above under these assumptions, following Brenner and Thomée [1]. It turns our that $\|r(kA)^n\| = O(\sqrt{n})$, uniformly in k, and that this result is best possible in the present generality. It is also shown that convergence of order $O(k^p)$ takes place even when $r(kA)^n$ is not bounded, provided $v \in D(A^{p+1})$. The proofs utilize Laplace and Fourier transformation, and, in particular, the so called Carlsson's inequality. Typical examples of the results in this section are provided by approximation methods for well-posed initial-value problems for partial differential equations, i.e., when A in (0.1) denotes a partial differential operator with respect to a space variable x, or a spatially discrete version of such and operator.

In Section 2 we shall make the additional assumption that $E(t)$ is an analytic semigroup. This applies to situations when (0.1) is a parabolic equation, but also other cases are possible, cf., e.g., [5] where a strongly damped wave equation and its spatially discrete analogue are treated within this framework.

We shall then first show that the convergence of order $O(k^p)$ under a somewhat weaker assumption than above, namely that $v \in D(A^p)$. The improvement in this result is related to the smoothing property of the analytic semigroup. Assuming in addition that $|r(\infty)| < 1$ it is also possible to derive a $O(k^p)$ convergence for $t > 0$, without any smoothness assumptions on v, or, more precisely,

$$\|r(kA)^n v - E(t_n)v\| \leq Ck^p t_n^{-p}\|v\|, \quad \text{for } t_n > 0. \tag{0.5}$$

This, in particular, immediately implies the stability of $r(kA)^n$ in the case considered.

The condition $|r(\infty)| < 1$ is not satisfied, e.g., for the Crank-Nicolson method, for which $r(\infty) = -1$. Moreover, by considering the case of an unbounded selfadjoint operator A in a Hilbert space, it is easy to see that the error bound of (0.5) does not hold for the Crank-Nicolson method, as this method does not have the smoothing property required. However, as has recently been proved, stability still holds in the case $|r(\infty)| = 1$, and we present a proof in the special case of the Crank-Nicolson method.

The analysis in this section depend on spectral representations of operators in terms of integrals in the complex plane, see Le Roux [6], Larsson, Thomée, and Wahlbin [5], Fujita and Suzuki [3]. The proof of stability in the case $|r(\infty)| = 1$ was given in Crouzeix, Larsson, Piskarev, and Thomée [2], cf. also Palencia [7].

1. Bounded semigroups.

In this section we shall consider the two questions posed in the introduction for a bounded strongly continuous semigroup $E(t) = e^{tA}$. We begin by presenting the following result.

Theorem 1.1. *Assume that (0.2) holds. Then there is a constant $C = C(r)$ such that if r is A-acceptable, then*

$$\|r(kA)^n\| \leq CC_0\sqrt{n}, \qquad \text{for } n \geq 0,\ k > 0.$$

Sketch of proof. The proof depends on three lemmas. We begin by introducing some notation. Let M be the set of bounded measures on R. This is a Banach space with norm $\|\mu\| = \int_{-\infty}^{\infty} d|\mu|(t)$. Let further $\hat{M}$ denote the Fourier transforms of elements $\mu \in M$,

$$\hat{\mu}(\xi) = \int_{-\infty}^{\infty} e^{it\xi} d\mu(t).$$

This is again a Banach space with norm $m(\hat{\mu}) = \|\mu\|$. Finally, let $\check{M}$ be the Laplace transforms of measures $\mu \in M$, with supp $\mu \subset [0, \infty)$, i.e.,

$$\check{\mu}(z) = \int_{0}^{\infty} e^{zt} d\mu(t).$$

Lemma 1.1. *Assume that $f(z)$ is bounded and analytic for $Re\, z \leq 0$, and set $f_I(\xi) = f(i\xi)$ for $\xi \in R$. Then, if $f_I = \hat{\mu} \in \hat{M}$ we have $f = \check{\mu}$, i.e.,*

$$f(z) = \int_{0}^{\infty} e^{zt} d\mu(t), \qquad \text{for } Re\, z \leq 0.$$

Ex. (i): Let $f(z) = e^{zt}$, $t > 0$. Then $f = \check{\mu}$ with $\mu = \delta_t$.

In order to use this result we need following version of Carlsson's lemma. Here and below $\|\cdot\|_p$ denotes the norm in $L_p = L_p(R)$.

Lemma 1.2. *Assume that $g, g' \in L_2$. Then $\hat{g} \in L_1$ and hence $g \in \hat{M}$. Moreover,*

$$m(g) \leq \sqrt{2}\|g\|_2^{1/2}\|g'\|_2^{1/2}.$$

Ex. (ii): Let $f(z) = r(z)$ be an A–acceptable, rational function. Now, with $r_\infty = r(\infty)$, there are polynomial P and Q, with deg(P) <deg(Q), such that

$$f_I(\xi) = r_\infty + \frac{P(\xi)}{Q(\xi)}.$$

Here, obviously $r_\infty \in \hat{M}$. Further, $P/Q, (P/Q)' \in L_2$ and thus $P/Q \in \hat{M}$. Thus $f \in \check{M}$.

Ex. (iii): Let

$$f(z) = \frac{r(kz)^n - e^{nkz}}{z^{p+1}},$$

where r is an A–acceptable rational function satisfying (0.4). Again $f \in \check{M}$.

Lemma 1.3. *Let A generate a bounded strongly continuous semigroup $E(t)$ on X satisfying (0.2), and let $f = \tilde{\mu} \in \tilde{M}$. Then*

$$f(A) = \int_0^\infty e^{tA} d\mu(t) = \int_0^\infty E(t) d\mu(t).$$

defines a bounded linear operator on X, and

$$\|f(A)\| \leq C_0 \|\mu\| = C_0 m(f_I).$$

Moreover, the mapping $f \to f(A)$ is a homomorphism of $\tilde{M}$ into the algebra of bounded linear operators on X.

We refer to [1] for details concerning Lemmas 1.1 1.3. Assuming them we now have the tools to show Theorem 1.1: It follows from Lemma 1.3 that for $f = r$ as in Example (ii), $r(A)$ as defined in Lemma 1.3 coincides with its definition in terms of resolvants, and further that

$$\|r(kA)^n\| \leq C_0 m(r(ki\xi)^n) = C_0 m(r_I^n).$$

Here the latter step follows by the easily proven fact that $m(f(k\xi)) = m(f(\xi))$. The proof then uses a partition of unity, $1 \equiv \sum_{j=0}^\infty \varphi_j(\xi), \varphi_j \in C^\infty(R)$, *with*

$$\operatorname{supp}(\varphi_j) \subset \begin{cases} \{\xi; 2^{j-1} < |\xi| < 2^{j+1}\}, & j > 0, \\ \{\xi; |\xi| < 2\}, & j = 0. \end{cases}$$

One may now write $r_I^n = \sum_{j=0}^\infty \varphi_j r_I^n$. Here, by Lemma 1.2,

$$m(\varphi_0 r_I^n) \leq \sqrt{2} \|\varphi_0 r_I^n\|_2^{1/2} \|(\varphi_0 r_I^n)'\|_2^{1/2}.$$

Since $|r_I(\xi)| \leq 1, |r_I'(\xi)| \leq C$, we have $\|\varphi_0 r_I^n\|_2 \leq C$ and $\|(\varphi_0 r_I^n)'\|_2 \leq Cn$, and hence $m(\varphi_0 r_I^n) \leq C\sqrt{n}$. Similarly, $m(\varphi_j r_I^n) \leq C\sqrt{n}\, 2^{-j/2}$, for $j > 0$ and hence

$$m(r_I^n) \leq \sum_{j=0}^\infty m(\varphi_j r_I^n) \leq \sum_{j=0}^\infty C\sqrt{n} 2^{-j/2} = C\sqrt{n},$$

which completes the proof.

The $O(\sqrt{n})$ stability estimate of Theorem 1.1 is best possible, as can be seen by the following example: Let $X = L_1(R)$, $A = -\partial/\partial x$, i.e. $(E(t)v)(x) = v(x-t)$. Here $E(t)$ is a contraction semigroup, i.e., satisfies (0.2) with $C_0 = 1$. If $f = \tilde{\mu} \in \tilde{M}$, then, by Lemma 1.3,

$$(f(A)v)(x) = \int_0^\infty (E(t)v)(x) d\mu(t) = \int_0^\infty v(x-t) d\mu(t) = (v * \mu)(x),$$

and thus $\mathcal{F}(f(A)v) = \hat{\mu}\hat{v} = f(i\xi)\hat{v} = f_I\hat{v}$. We now consider the Crank-Nicolson method,

$$r(z) = \frac{1+\frac{1}{2}z}{1-\frac{1}{2}z}, \tag{1.1}$$

and we want to show

$$\|r(A)^n\| = \|\mathcal{F}^{-1}(r_I^n\hat{v})\|_{L_1\to L_1} \geq c\sqrt{n}, \quad \text{with } c>0,$$

In this case $|r_I(\xi)| \equiv 1$ and we may write $r_I(\xi) = e^{i\psi(\xi)}$, where $\psi(\xi) = \xi - \frac{1}{2}\xi^3 + \cdots$ is realvalued. Let v be such that $\mathcal{F}v = \hat{v} \in C_0^\infty(R)$ and $\psi''(\xi) \neq 0$ in $\operatorname{supp}(\hat{v})$. By Parseval's relation we have

$$\|\hat{v}\|_2^2 = \|r_I^n\hat{v}\|_2^2 = 2\pi\|\mathcal{F}(r_I^n\eta)\|_2^2 \leq 2\pi\|\mathcal{F}(r_I^n\hat{v})\|_1\|\mathcal{F}(r_I^n\hat{v})\|_\infty.$$

Here

$$\|\mathcal{F}(r_I^n\hat{v})\|_1 = 2\pi\|\mathcal{F}^{-1}(r_I^n\hat{v})\|_1 = 2\pi\|r(A)^nv\|_1 \leq 2\pi\|r(A)^n\|\|v\|_1.$$

Further, as will be shown below,

$$\|\mathcal{F}(r_I^n\hat{v})\|_\infty \leq \frac{C}{\sqrt{n}}. \tag{1.2}$$

Assuming this for a moment, we have

$$\|\hat{v}\|_2^2 \leq \frac{C}{\sqrt{n}}\|r(A)^n\|\|v\|_1$$

and hence

$$\|r(A)^n\| \geq c\sqrt{n}, \quad \text{with } c>0,$$

which is the result stated.

Recalling that $r_I(\xi) = e^{i\psi(\xi)}$ with $\psi''(\xi) \neq 0$ in $\operatorname{supp}\hat{v}$, (1.2) follows from the following lemma.

Lemma 1.4. *Let $\eta \in C_0^1$ and let $\psi \in C^2$ be realvalued, with $|\psi''| \geq \delta > 0$ in supp(η). Then*

$$\|\mathcal{F}(e^{i\psi}\eta)\|_\infty \leq \frac{8}{\sqrt{\delta}}\|\eta'\|_1.$$

Proof. We have, with x_0 arbitrary,

$$\mathcal{F}(e^{i\psi}\eta) = \int_{-\infty}^{\infty} e^{i\psi(x)+i\xi x}\eta(x)dx = -\int_{-\infty}^{\infty}\eta'(x)\int_{x_0}^{x} e^{i(\psi(y)+\xi y)}dydx.$$

By van der Corput's lemma (cf. [1]) we find

$$\left|\int_{x_0}^{x} e^{i(\psi(y)+\xi y)}dy\right| \leq \frac{8}{\sqrt{\delta}},$$

which completes the proof, and thus the proof that Theorem 1.1 is sharp.

For special choices of $r(z)$ better results are possible. We have the following:

Theorem 1.2. *Assume that the following conditions are satisfied for* r*:*

(a) $$|r_I(\xi)| < 1, \quad \text{for } 0 \neq \xi \in R, \text{ and } |r(\infty)| < 1.$$

There exist positive integers p, q *with* q *even,* $q \geq p+1$ *such that*

(b) $$r_I(\xi) = e^{i\xi + \psi(\xi)}, \quad \text{with } \psi(\xi) = O(|\xi|^{p+1}), \quad \xi \to 0.$$

and $\gamma > 0$ *such that*

(c) $$Re\,\psi(\xi) \leq -\gamma\xi^q, \quad \text{for } |\xi| \leq 1.$$

Then there is a constant $C = C(r)$ *such that*

$$\|r(kA)^n\| \leq Cn^{\frac{1}{2}(1-\frac{p+1}{q})}.$$

This result is also best possible. Note that the accuracy condition (0.4) implies condition (b). But then $Re\,\psi(\xi) = \gamma_0\xi^q(1+o(1))$, as $\xi \to 0$, for some q with $q \geq p+1$. Now it follows from (a) that $\gamma_0 < 0$ and q even, and hence that (c) holds. Thus, if (0.4) holds together with (a), there are p and q for which the conclusion of theorem is valid.

We give some examples:

Ex. For the Padé approximants of e^z we have

$$r(z) = r_{jl}(z) = \frac{P_l(z)}{Q_j(z)}, \quad \text{where } \deg(P_l) = l, \ \deg(Q_j) = j,$$

and (0.4) is satisfied with $p = j + l$. It is known (see Hairer and Wanner [4]) that $r_{jl}(z)$ is A–acceptable if and only if $0 \leq j - l \leq 2$ and that (c) then holds with $q = 2j$. Theorem 1.2 now shows that

$$\|r_{j,j-1}(kA)^n\| \leq C \quad \text{and} \quad \|r_{j,j-2}(kA)^n\| \leq Cn^{1/(4j)}.$$

Ex. The restricted Padé approximants

$$R_j(z) = (-1)^j(1-\gamma z)^{-j}\sum_{m=0}^{j} L_j^{(j-m)}(\gamma^{-1})(\gamma z)^m,$$

where L_j denotes the Laguerre polynomials of degree j. For suitable γ, $r = R_j$ satisfies (0.4) with $p = j+1$. It is known (see [4]) that R_j is A–acceptable if and only if $j = 1,2,3,5$, that $|r(\infty)| < 1$ for $j = 2,3,5$, and that (c) holds with $q = 4$ for $j = 2$ and with $q = 6$ for $j = 3$. It follows, in particular, from Theorem 1.2 that

$$\|R_2(kA)^n\| \leq C \quad \text{and} \quad \|R_3(kA)^n\| \leq Cn^{1/12}.$$

We remark that it is essential in the in above example showing the sharpness of Theorem 1.1 that X is not a Hilbert space. In fact, if A generates a contraction semigroup $E(t)$ on a Hilbert space X, and r is A-acceptable, then $\|r(kA)^n\| \le 1$. To see this, we note that in this case $\|u(t)\| = \|E(t)v\|$ is nonincreasing and hence

$$\frac{d}{dt}\|u\|^2 = (Au, u) + (u, Au) = 2Re\,(Au, u) \le 0, \quad \text{for } t > 0,$$

so that $Re\,A \le 0$. We recall the following definition by von Neumann: A closed set $Z \subset C$ is called a spectral set for the linear operator A in the Hilbert space X if for any rational function $R(z)$ with $|R(z)| \le 1$ in Z one has $\|R(A)\| \le 1$. (We note that, in general, the spectrum $sp(A)$ of A is a subset of Z, and that $sp(A)$ is a spectral set for A if A is a normal operator.) Von Neumann showed (cf. [9]) that the set $Z = \{z; Re\,z \le 0\}$ is a spectral set for A if and only if $Re\,A \le 0$. Thus, in particular, if r is A-acceptable, we have $|r(kz)^n| \le 1$ for $Re\,z \le 0$, and thus $\|r(kA)^n\| \le 1$.

We close this section with the following convergence result:

Theorem 1.3. *There is a constant $C = D(r)$ such that if r is A-acceptable and accurate of order p, then*

$$\|r(kA)^n v - E(t_n)v\| \le C C_0 t_n k^p \|A^{p+1}v\|, \qquad \text{for } n \ge 0,\ k > 0,\ v \in D(A^{p+1}).$$

In particular, optimal order convergence occurs for v sufficiently smooth, even in ustable cases. The proof uses Ex.(iii) above, see [1] for details.

2. Analytic semigroups.

In this section we make the additional assumption that the bounded semigroup $E(t)$ generated by A is analytic, and assume (without loss of generality, cf. [2]) that $0 \in \rho(A)$, where $\rho(A)$ denotes the resolvent set of A. This is equivalent (cf. [8, Theorem 2.5.2]) to saying that there is an angle $\alpha \in (0, \pi/2)$ such that

$$\rho(A) \supset \overline{\Sigma}_\alpha, \qquad \text{where } \Sigma_\alpha = \{z = \rho e^{i\varphi} : \rho > 0,\ |\varphi| < \pi - \alpha\}, \tag{2.1}$$

and a constant $M \ge 1$ such that, with $R(z; A) = (zI - A)^{-1}$ denoting the resolvent of A

$$\|R(z; A)\| \le \frac{M}{|z|}, \qquad \text{for } z \in \Sigma_\alpha. \tag{2.2}$$

In this case, it will be sufficient in the analysis of the rational approximations $r(kA)^n$ of $E(t_n)$ to assume that r is $A(\theta)$-acceptable for some angle $\theta \in (\alpha, \pi/2]$, i.e., that

$$|r(z)| \le 1, \quad \text{for } z \in \mathbf{C} \setminus \Sigma_\theta.$$

We first show the following convergence result in which the regularity assumption on the initial data has been weakened as compared to Theorem 1.3.

Theorem 2.1. *There is a constant $C = C(r, \alpha)$ such that if A satisfies (2.1) and (2.2), and if r is $A(\theta)$-acceptable with $\theta \in (\alpha, \pi/2]$ and accurate of order p, then*

$$\|r(kA)^n v - E(t_n)v\| \le CMk^p\|A^p v\|, \qquad \text{for } t_n \ge 0,\ k > 0,\ v \in D(A^p).$$

Proof. With $F_n(z) = r(z)^n - e^{nz}$ we want to show that, in operator norm,

$$\|F_n(kA)A^{-p}\| \le CMk^p.$$

To demonstrate this we shall use the Dunford-Taylor spectral representation of operators, and denote $\Gamma = \Gamma_+ \cup \Gamma_-$, $\Gamma_\pm = \{\rho e^{\pm i\alpha'}, 0 \le \rho < \infty\}, \alpha' = \pi - \alpha$. Since for $z \in \Gamma$

$$F_n(z)z^{-p} = \begin{cases} O(|z|^{-p}) & \text{as } z \to \infty, \\ O(|z|) & \text{as } z \to 0, \end{cases}$$

we may write

$$F_n(kA)A^{-p} = \frac{1}{2\pi i}\int_\Gamma F_n(kz)z^{-p}R(z;A)dz,$$

Using (2.2) it follows that

$$\begin{aligned} \|F_n(kA)A^{-p}\| &\le CM\int_0^\infty (|F_n(k\rho e^{i\alpha'})| + |F_n(k\rho e^{-i\alpha'})|)\frac{d\rho}{\rho^{p+1}} \\ &= CMk^p\int_0^\infty (|F_n(\rho e^{i\alpha'})| + |F_n(\rho e^{-i\alpha'})|)\frac{d\rho}{\rho^{p+1}}. \end{aligned}$$

Since $r(z)^n$ and e^{tz} are bounded on Γ we find

$$\int_1^\infty |F_n(\rho e^{\pm i\alpha'})|\frac{d\rho}{\rho^{p+1}} \le C\int_1^\infty \frac{d\rho}{\rho^{p+1}} \le C.$$

One shows easily

$$|F_n(z)| \le n|z|^{p+1}e^{-cn|z|}, \quad \text{for } |z| \le 1, z \in \Gamma, \quad \text{with } c > 0. \tag{2.3}$$

Hence

$$\int_0^1 |F_n(\rho e^{i\pm\alpha'})|\frac{d\rho}{\rho^{p+1}} \le Cn\int_0^1 e^{-cn\rho}d\rho \le C.$$

Together these estimates complete the proof.

We now turn to an estimate for initial data which do not satisfy any regularity assumption in addition to $v \in X$. To accomplish this we need to make the additional assumption $|r(\infty)| < 1$, which will secure a certain smoothing property for the discrete solution operator $r(kA)^n$.

Theorem 2.2. *There is a constant $C = C(r,\alpha)$ such that if A satisfies (2.1) and (2.2), and if r is $A(\theta)$-acceptable with $\theta \in (\alpha, \pi/2]$, accurate of order p, and satisfies $|r(\infty)| < 1$, then*

$$\|r(kA)^n v - E(t_n)v\| \le CMk^p t_n^{-p}\|v\|, \qquad \text{for } t_n > 0,\ k > 0,\ v \in X.$$

Proof. This time we need to show

$$\|F_n(kA)\| \le CM\frac{k^p}{t_n^p} = \frac{CM}{n^p}.$$

In this presentation we assume for simplicity that $r(\infty) = 0$, and refer to our references for the general case. Now $F_n(z) = O(|z|^{-1})$ as $z \to \infty$ on Γ and thus

$$F_n(kA) = \frac{1}{2\pi i}\int_\Gamma F_n(kz)R(z;A)dz,$$

so that

$$\|F_n(kA)\| \le CM\int_0^\infty (|F_n(\rho e^{i\alpha'})| + |F_n(\rho e^{-i\alpha'})|)\frac{d\rho}{\rho}.$$

Since $r(\infty) = 0$ we have

$$|r(z)| \le e^{-c}, \quad \text{and} \quad |r(z)| \le \frac{C}{|z|}, \qquad \text{for } z \in \Gamma,\ |z| > 1,$$

so that

$$|F_n(z)| \le |e^{nz}| + |r(z)^n| \le e^{-cn|z|} + C\frac{e^{-cn}}{|z|} \le C\frac{e^{-cn}}{|z|},$$

and hence, for any $p > 0$,

$$\int_1^\infty |F_n(\rho e^{\pm i\alpha'})|\frac{d\rho}{\rho} \le Ce^{-cn}\int_1^\infty \frac{d\rho}{\rho^2} \le \frac{C}{n^p}.$$

This time we obtain from (2.3)

$$\int_0^1 |F_n(\rho e^{\pm i\alpha'})|\frac{d\rho}{\rho} \le Cn\int_0^1 \rho^p\, e^{-cn\rho}d\rho \le \frac{C}{n^p}.$$

The proof is now complete in the case considered.

We now turn to the stability result for the general case $|r(\infty)| \le 1$:

Theorem 2.3. *There is a constant $C = C(r,\alpha)$ such that if A satisfies (2.1) and (2.2), and if r is $A(\theta)$-acceptable with $\theta \in (\alpha, \pi/2]$, then*

$$\|r(kA)^n\| \le CM, \qquad \text{for } n \ge 0,\ k > 0.$$

We shall only outline the proof in the special case of the Crank-Nicolson method (1.1), and note that then $r(\infty) = -1$. Setting $w = 1/z$ we have

$$r(z) = -\frac{1+2w}{1-2w} = -r(4w) = -e^{4w} + O(w^3), \quad \text{as } w \to 0.$$

The idea is now to modify the function $F_n(z)$ used in the proofs of Theorems 2.1 and 2.2 by subtracting a function which matches the behavior of $r(kz)^n$ for large z and such that the corresponding operator is bounded. We thus introduce $\tilde{F}_n(z) = r(z)^n - e^{nz} - (-1)^n e^{4n/z}$. We then have for $z \in \Gamma$

$$\tilde{F}_n(z) \leq \begin{cases} C|z| & \text{for } |z| \leq 1, \\ C/|z|, & \text{for } |z| \geq 1, \end{cases}$$

and we may therefore define

$$\tilde{F}_n(kA) = \frac{1}{2\pi i}\int_\Gamma \tilde{F}_n(kz)R(z;A)\,dz. \tag{2.4}$$

To show that this operator is bounded we now note that, for $|z| \leq 1, z \in \Gamma$, one has

$$|r(z)^n - e^{nz}| \leq |r(z) - e^z| \sum_{j=0}^{n-1} |r(z)|^{n-j-1}||e^{jz}| \leq C|z|^2 n e^{-cn|z|} \leq C|z|,$$

and

$$|e^{4n/z}| \leq e^{-cn/|z|} \leq e^{-c/|z|} \leq C|z|,$$

and, for $|z| \geq 1, z \in \Gamma$,

$$\begin{aligned} |r(z)^n - (-1)^n e^{4n/z}| &\leq |r(4w) + e^{4w}| \sum_{j=0}^{n-1} |r(4w)|^{n-j-1}|e^{4jw}| \\ &\leq C|w|^2 n e^{-cn|w|} \leq C|w| = C/|z|, \end{aligned}$$

and

$$|e^{nz}| \leq e^{-cn|z|} \leq e^{-c|z|} \leq C/|z|. \tag{2.5}$$

It follows that, uniformly in n and k,

$$\|\tilde{F}_n(kA)\| \leq \frac{M}{2\pi}\int_0^\infty \sum_{\pm} |\tilde{F}_n(\rho e^{\pm i\alpha'})| \frac{d\rho}{\rho} \leq CM\Big\{\int_0^1 d\rho + \int_1^\infty \frac{d\rho}{\rho^2}\Big\} \leq CM.$$

To complete the proof it now suffices to show that the operator $\tilde{F}_n(kA)$ equals $r(kA)^n - e^{nkA} - (-1)^n e^{4n(kA)^{-1}}$, and that $e^{tA^{-1}}$ is uniformly bounded for $t > 0$. For this purpose we rewrite (2.4) as

$$\begin{aligned} \tilde{F}_n(kA) = &\frac{1}{2\pi i}\int_\Gamma (r(kz)^n - (-1)^n)R(z;A)\,dz - \frac{1}{2\pi i}\int_\Gamma e^{nkz}R(z;A)\,dz \\ &- (-1)^n \frac{1}{2\pi i}\int_\Gamma (e^{4n/z} - 1)R(z;A)\,dz. \end{aligned}$$

Here we see that the first term on the right equals $r(kA)^n - (-1)^n I$, since the rational function $r(kz)^n - (-1)^n$ vanishes at $z = \infty$, and the second term is $E(t_n)$. The proof is therefore completed by the following lemma:

Lemma 2.1. *There is a constant $C = C(r, \alpha)$ such that*

$$\|e^{tA^{-1}}\| \le CM, \qquad \text{for } t \ge 0. \tag{2.6}$$

Moreover, with Γ as above, we have

$$\frac{1}{2\pi i}\int_{\Gamma}(e^{tz^{-1}} - 1)R(z; A)\,dz = e^{tA^{-1}} - I. \tag{2.7}$$

Proof. It suffices to consider $t > 0$. We then have

$$e^{tA^{-1}} = \frac{1}{2\pi i}\int_{\tilde{\Gamma}} e^{tw}R(w; A^{-1})\,dw, \tag{2.8}$$

for a suitable contour $\tilde{\Gamma}$ enclosing the spectrum of the bounded operator A^{-1} We may choose $\tilde{\Gamma} = \Gamma_{\epsilon,R} = \{w = -Re^{i\varphi} : |\varphi| \le \alpha\} \cup \{w = -\rho e^{\pm i\alpha} : \epsilon \le \rho \le R\} \cup \{w = \epsilon e^{i\varphi} : |\varphi| \le \pi - \alpha\}$, with counter-clockwise orientation, and with R sufficiently large and arbitrary $\epsilon > 0$. The contribution to (2.8) from the arc with radius R is 0 in the limit as $R \to \infty$. The formula (2.8) therefore holds with $\tilde{\Gamma} = \Gamma_\epsilon = \{w = -\rho e^{\pm i\alpha} : \rho \ge \epsilon\} \cup \{w = \epsilon e^{i\varphi} : |\varphi| \le \pi - \alpha\}$.

Using the identity $R(w; A^{-1}) = -w^{-1}AR(w^{-1}; A)$, we now have

$$e^{tA^{-1}} = -\frac{1}{2\pi i}\int_{\Gamma_\epsilon} e^{tw}AR(w^{-1}; A)\,\frac{dw}{w},$$

and, setting $y = tw$,

$$e^{tA^{-1}} = -\frac{1}{2\pi i}\int_{\Gamma_{t\epsilon}} e^{y}AR(ty^{-1}; A)\,\frac{dy}{y}. \tag{2.9}$$

Since $AR(z; A) = -I + zR(z; A)$, we have $\|AR(ty^{-1}; A)\| \le 1 + M \le 2M$ by (2.2). Hence (2.6) follows if we take $\epsilon = t^{-1}$, so that $\Gamma_{t\epsilon}$ is independent of t.

Using $AR(z; A) = -I + zR(z; A)$ again, we rewrite (2.9) as

$$e^{tA^{-1}} = \frac{1}{2\pi i}\int_{\Gamma_\epsilon} e^{tw}\frac{dw}{w}I - \frac{1}{2\pi i}\int_{\Gamma_\epsilon} e^{tw}R(w^{-1}; A)\,\frac{dw}{w^2}.$$

Noting that the first integral equals 1, and setting $z = w^{-1}$ and $\Gamma'_\epsilon = \{z = -\rho e^{\pm i\alpha} : 0 \le \rho \le \epsilon^{-1}\} \cup \{z = \epsilon^{-1}e^{i\varphi} : |\varphi| \le \pi - \alpha\}$ with clockwise orientation in the second integral, we obtain, since the integral of $R(z; A)$ over a closed contour vanishes, the representation

$$e^{tA^{-1}} = I + \frac{1}{2\pi i}\int_{\Gamma'_\epsilon}(e^{t/z} - 1)R(z; A)\,dz.$$

Here the integral over the arc of radius ϵ^{-1} contributes 0 to the last term in the limit as $\epsilon \to 0$. This proves (2.7).

REFERENCES

1. P. Brenner and V. Thomée, *On rational approximation of semigroups*, SIAM J. Numer. Anal. **16** (1979), 683–694.
2. M. Crouzeix, S. Larsson, S. Piskarev, and V. Thomée, *The stability of rational approximations of analytic semigroups*, Preprint, Department of Mathematics, Chalmers University of Technology, Göteborg **1991-28**.
3. H. Fujita and T. Suzuki, *Evolution problems*, Handbook of Numerical Analysis, Vol. II. Finite Element Methods (Part 1) (P. G. Ciarlet and J. L. Lions, eds.), North-Holland, 1991, pp. 789–928.
4. E. Hairer and G. Wanner, *Solving Ordinary Differential Equations II, Stiff and Differential-Algebraic Problems*, Springer-Verlag, 1991.
5. S. Larsson, V. Thomée, and L. B. Wahlbin, *Finite element methods for a strongly damped wave equation*, IMA J. Numer. Anal. **11** (1991), 115–142.
6. M.-N. Le Roux, *Semidiscretization in time for parabolic problems*, Math. Comp. **33** (1979), 919–931.
7. C. Palencia, *A stability result for sectorial operators in Banach spaces*, report 1991/4, Departamento de Matemática Aplicada y Computación, Universidad de Valladolid.
8. A. Pazy, *Semigroups of Linear Operators and Applications to Partial Differential Equations*, Springer-Verlag, 1983.
9. F. Riesz and B. Sz.-Nagy, *Leçons d'Analyse Fonctionelle*, deuxième édition, Académie des Sciences de Hongrie, Budapest, 1953.

DEPARTMENT OF MATHEMATICS, CHALMERS UNIVERSITY OF TECHNOLOGY, S-412 96 GÖTEBORG, SWEDEN

E-mail address: thomee@math.chalmers.se

GALERKIN–FINITE ELEMENT METHODS FOR THE NONLINEAR SCHRÖDINGER EQUATION*

Georgios D. Akrivis[1], Vassilios A. Dougalis[2], Ohannes A. Karakashian[3], and William R. McKinney[4]

ABSTRACT

We consider the initial- and boundary-value problem for the Nonlinear (cubic) Schrödinger Equ tion posed with zero Dirichlet boundary conditions on a bounded domain Ω in $\mathbf{R}^d$, $d = 1, 2$, o 3. We discretize the problem in space by a standard Galerkin–finite element method and in the temporal variable by Crank–Nicolson type and also by higher order, algebraically stable, implicit Runge–Kutta procedures. We review the available theoretical results on error estimates for these schemes with particular emphasis on the question of possible temporal order reduction on general domains. Time stepping with the implicit Runge–Kutta schemes requires solving a nonlinear system of equations at each time step. We analyze Newton's method for solving these nonlinear systems and show how modified Newton schemes may be applied to decouple and solve them in efficient ways, especially effective for parallel implementation. Finally, we show how such fully discrete schemes may be applied with adaptive space and time meshes to approximate radial solutions of the nonlinear Schrödinger equation for $d = 2$ and 3, which blow up at the origin in finite time, and present the results of relevant blow-up rate calculations.

1. INTRODUCTION

The Nonlinear Schrödinger Equation with cubic nonlinearity (henceforth referred to as the "NLS equation")

$$u_t = i\Delta u + i|u|^2 u, \quad x \in \mathbf{R}^d, \quad t \geq 0, \tag{1.1a}$$

where u is a complex-valued function of the 'spatial' variable $x \in \mathbf{R}^d$ (we shall consider $d = 1, 2, \text{or } 3$), and of the 'temporal' variable $t \geq 0$, occurs frequently in various areas in Mathematical Physics. For example, the initial-value problem for the NLS, consisting of (1.1a) supplemented by the initial condition

---

* Work supported by a joint travel grant from the General Secretariat of Research and Technology, Greece and the N.S.F., U.S.A..

[1] Mathematics Department, University of Crete, 71409 Heraklion, Greece, and Institute of Applied and Computational Mathematics, FORTH, Greece.

[2] Mathematics Department, National Technical University, Zographou, 15780 Athens, Greece, and Institute of Applied and Computational Mathematics, FORTH, Greece.

[3] Department of Mathematics, University of Tennessee, Knoxville, TN 37996, USA.

[4] Department of Mathematics, North Carolina State University, Raleigh, NC 27695, USA.

$$(1.1b) \qquad u(x,0) = u^0(x), \quad x \in \mathbf{R}^d,$$

where u^0 is a given complex function, arises, for $d = 1$, as an envelope equation in water wave theory, [25], [42]. In two space dimensions it occurs in nonlinear optics, governing, under certain hypotheses, the propagation of electromagnetic beams in media whose index of refraction depends on the amplitude of the field in some simple nonlinear way, [9], [35]. For $d = 3$ it is obtained e.g. as a limiting case of the Zakharov model of Langmuir waves, cf. [44]. For reviews of various physical interpretations and mathematical properties of the equation, cf. e.g. the surveys [31], [28].

It is not hard to see that for $d = 1$ the initial-value problem is globally well-posed for smooth enough data that decays sufficiently fast at infinity. In this case, it is also well-known that it can be solved by the inverse scattering transform, [25]. The case $d = 2$ is the critical exponent case for which global existence occurs under suitable smallness conditions in the initial data, [41]. It is also well-known that, in this case, singular solutions exist that blow up in finite time, [14]; this blow-up in two dimensions is usually referred to as "self-trapping" or "self-focusing", [9], [35]. The three-dimensional case is super-critical; the problem (1.1) is locally well-posed and there exist solutions that blow up ("collapse") in finite time, [44], [13].

We shall be interested in the numerical solution of the NLS equation by fully discrete Galerkin-finite element methods. Specifically, we shall analyze the discretization of the following initial- and boundary-value problem:

$$(1.2) \qquad \begin{cases} u_t = i\Delta u + i|u|^2 u & \text{in} \quad \bar{\Omega} \times [0,T], \\ u = 0 & \text{on} \quad \partial\Omega \times [0,T], \\ u(x,0) = u^0(x) & \text{in} \quad \bar{\Omega}. \end{cases}$$

Here $\Omega \subset \mathbf{R}^d, d = 1, 2,$ or 3, will be a bounded domain with boundary $\partial\Omega$, $0 < T < \infty$ is given, and u^0 is a given complex-valued function defined on $\bar{\Omega}$. For the purposes of proving error estimates, we shall assume that the problem (1.2) possesses a unique classical solution in $C^\mu(\bar{\Omega} \times [0,T])$, where μ is sufficiently large. It is our intention however to use the numerical schemes to investigate the development of blow-up of solutions of (1.2).

Employing standard notation we shall use the symbols $H^s = H^s(\Omega)$, $H_0^1 = H_0^1(\Omega)$ to denote the usual complex (Hilbert) Sobolev spaces. We shall let $(\cdot,\cdot)$ be the inner product on $L^2 = H^0$ defined by $(u,v) = \int_\Omega u(x)\bar{v}(x)dx$ for $u, v \in L^2$ and denote by $\|\cdot\|$ the associated L^2 norm. By $\|\cdot\|_s$ we denote the usual norm on H^s, whereas by $|\cdot|_p$, $1 \le p \le \infty$, $p \ne 2$, that on $L^p = L^p(\Omega)$.

It is quite straightforward to check that the L^2 norm of the solution $u(\cdot,t) = u(t)$ of (1.1) is invariant, i.e. that the equality

$$(1.3) \qquad \|u(t)\| = \|u^0\|, \qquad 0 \le t \le T,$$

holds. A second well-known identity involves the Hamiltonian of the equation:

$$\|\nabla u(t)\|^2 - \frac{1}{2}|u(t)|_4^4 = \|\nabla u^0\|^2 - \frac{1}{2}|u^0|_4^4, \tag{1.4}$$

where we let $(\nabla u, \nabla v) = \sum_{i=1}^{d}(\partial_i u, \partial_i v)$. These invariants will play a significant role in the analysis of numerical schemes that will be presented in the sequel and also in the construction of efficient adaptive Galerkin methods for (1.2). Of course, they also play a major role in the existence–uniqueness theory of solution of (1.2), cf. e.g. [7], [26], [38]. Although most of the literature on the blow-up of solutions of the NLS concerns the pure initial-value problem, there is ample computational and theoretical evidence, cf. e.g. [33], [18], [27], that blow-up also occurs for problems posed on bounded domains, such as the problem (1.2).

A significant number of papers have appeared in the literature in recent years on the numerical approximation of various aspects of the solution of the NLS equation. We refer the reader, for example, to [12], [15], [21], [29], [33], [35], [36], [40], [1], [37] and to the reference lists of these papers.

In this paper we shall first give a brief review of previous theoretical work of the authors on error estimates for Galerkin–finite element methods for (1.2). Specifically, in section 2 we discuss the semidiscrete problem, whereas in sections 3, resp. 4, we review some error estimates on fully discrete schemes of low, resp. high, temporal accuracy. In section 5 we address the issue of efficient implementation of our high order accurate schemes, which are based on temporal discretizations via high order, implicit Runge–Kutta schemes. Finally, in section 6 we present some numerical experiments, performed with adaptive versions of our fully discrete schemes, as a preliminary report of an ongoing study of the blow-up in finite time of radial solutions of the NLS in two and three dimensions by direct methods.

2. GALERKIN SEMIDISCRETIZATION

We shall discretize (1.2) in space by the *standard Galerkin method*. To this effect, for $0 < h < 1$, let S_h be a finite-dimensional subspace of complex-valued continuous functions in H_0^1 (typically a space of piecewise polynomial functions of fixed degree, defined on suitable partitions of $\bar{\Omega}$ and endowed with bases with elements of small support). Assume that there exists an integer $r \geq 2$, a constant c independent of h, and for each $v \in H^r \cap H_0^1$, a $\chi \in S_h$ such that

$$\|v - \chi\| + h\|v - \chi\|_1 \leq ch^r\|v\|_r, \tag{2.1}$$

which is such that, if in addition $v \in C^2(\bar{\Omega})$, then

$$|v - \chi|_\infty \leq Ch^2\|v\|_{C^2(\bar{\Omega})}. \tag{2.2}$$

We shall also assume that the elements φ of S_h satisfy the inverse property

$$|\varphi|_\infty \leq Ch^{-d/2}\|\varphi\|, \tag{2.3}$$

for some constant C independent of φ and h.

The *semidiscrete approximation* of the solution of (1.2) is then defined as the map $u_h : [0,T] \longrightarrow S_h$ satisfying:

$$(2.4) \qquad \begin{cases} (u_{ht}, \varphi) + i(\nabla u_h, \nabla\varphi) = i(|u_h|^2 u_h, \varphi) & \forall \varphi \in S_h, \quad 0 \le t \le T, \\ u_h(0) = u_h^0, \end{cases}$$

where u_h^0 is an approximation of u^0 in S_h, chosen so that

$$(2.5) \qquad \|u^0 - u_h^0\| \le ch^r$$

It is not hard to see, cf. [1, section 2], that u_h exists uniquely on $[0,T]$, that it satisfies the discrete analog $\|u_h(t)\| = \|u_h^0\|$ of (1.3), and, if u is sufficiently smooth on $[0,T]$, the optimal-order L^2 error estimate

$$(2.6) \qquad \max_{0 \le t \le T} \|u(t) - u_h(t)\| \le ch^r$$

3. CRANK-NICOLSON TYPE SCHEMES

In this section we shall review some full discretizations of the system of ordinary differential equations represented by (2.4), which have second-order temporal accuracy. Let $k > 0$ be the (constant) time step and $t^n = nk$, $n = 0, 1, \dots, J$, where $t^J = T$. For example, one may discretize (2.4) by a Crank-Nicolson type method (more precisely by the 'midpoint' scheme) and approximate $u(t^j) = u^j$ by $U^j \in S_h$, $0 \le j \le J$, defined by

$$(3.1) \qquad \begin{cases} (U^{n+1} - U^n, \varphi) + \dfrac{ik}{2}(\nabla(U^{n+1} + U^n), \nabla\varphi) \\ \qquad = \dfrac{ik}{8}(|U^{n+1} + U^n|^2(U^{n+1} + U^n), \varphi) \quad \forall \varphi \in S_h,\ 0 \le n \le J-1, \\ U^0 = u_h^0. \end{cases}$$

This scheme has often been used for numerical computations in the literature, coupled with various iterative techniques for solving at each time step the nonlinear system of equations that it represents, cf. e.g. [30], [36], [1]. It is easily seen that it is L^2*-conservative*, i.e. that its solutions satisfy

$$(3.2) \qquad \|U^n\| = \|U^0\|, \qquad 0 \le n \le J,$$

mimicking (1.3); the scheme however does not conserve the discrete analog of the second invariant (1.4). An application of Brouwer's fixed point theorem yields the *existence* of a solution U^{n+1} of (3.1) given $U^n \in S_h$. The uniqueness question is somewhat more involved. Assume that U^0 is bounded in L^2 uniformly with respect to h, e.g. assume that $U^0 = Pu^0$, where P is the L^2-projection operator onto S_h. Then, making use of the Gagliardo-Nirenberg inequality

$$|v|_4^4 \le C_d \|v\|^{4-d} \|\nabla v\|^d, \qquad \forall v \in H_0^1, \tag{3.3}$$

valid for $d = 1, 2, 3$, it can be seen, cf. [1, section 3], that the solution U^{n+1} of (3.1) is unique if $d = 1$ and k is sufficiently small. For $d = 2$, we recall, [7], that (1.2) has unique solutions in H^2 for all $t \ge 0$, provided $u^0 \in H^2 \cap H_0^1$ is chosen so that $C_2 \|u^0\|^2 < 2$, where C_2 is the constant occurring in (3.3) for $d = 2$, whilst if $C_2 \|u^0\|^2 > 2$, the solution may blow up in finite time. (In [26], it is shown that $C_2 \le \pi^{-1}$.) In [1] it is proved that the solutions of the scheme (3.1) are unique for $d = 2$ if $C_2 \|U^0\|^2 < 0.22\ldots$, a restriction that agrees qualitatively with the bound imposed for global existence of the continuous problem. (For uniqueness in $d = 3$ one should either impose a stringent mesh condition on k and h or prove suitable a-priori estimates on the solution of (3.1) in L^4.)

This state of affairs should be contrasted with that pertaining to the scheme

$$\begin{cases} (U^{n+1} - U^n, \varphi) + \dfrac{ik}{2}(\nabla(U^{n+1} + U^n), \nabla\varphi) \\ \qquad = \dfrac{ik}{4}((|U^{n+1}|^2 + |U^n|^2)(U^{n+1} + U^n), \varphi) \quad \forall \varphi \in S_h,\ 0 \le n \le J-1, \\ U^0 = u_h^0, \end{cases} \tag{3.4}$$

introduced by Delfour, Fortin and Payre [12], following an idea of [32] and widely used in computations, cf. e.g. [12], [29], [33]. As was pointed out in [12], solutions of (3.4) satisfy, in addition to (3.2), the discrete analog of (1.4) obtained by replacing $u(t)$ and u^0 by U^n and U^0, respectively. This latter fact enables one to prove an interesting uniqueness result for (3.4), if $d = 2$; namely that uniqueness is guaranteed if U^0 is bounded in H_0^1 uniformly in h and $C_2 \|U^0\|^2 < 2$, exactly as in the condition for global existence of solutions to the continuous problem. (Uniqueness is again unconditional if $d = 1$ but requires stringent conditions if $d = 3$.)

Both schemes are second-order accurate in the temporal variable. Indeed, if u is *sufficiently smooth*, $k = o(h^{\frac{d}{4}})$ and $U^0 = u_h^0$ is chosen so that (2.5) holds, it is shown in [1] that both have *unique* solutions satisfying

$$\max_{0 \le n \le J} \|u^n - U^n\| \le c(k^2 + h^r). \tag{3.5}$$

(The uniqueness part of this theorem is thus a local property of the discrete schemes guaranteed in a "tube" around smooth solutions of the p.d.e., as opposed to the previously stated uniqueness results which did not presuppose any smoothness on the part of u.)

In [1] we also analyzed the convergence of Newton-type methods for solving, for each n, the nonlinear systems represented by (3.1) and (3.4). We shall discuss such Newton methods in a more general context in section 5 below.

4. HIGHER ORDER TEMPORAL DISCRETIZATIONS

In this section, we shall review some results of [16] on higher-order temporal discretizations of (2.4) by Implicit Runge-Kutta (IRK) methods. It is well-known that approximating smooth solutions of initial- and boundary-value problems for some p.d.e's by high-order Runge-Kutta methods results sometimes in observed temporal rates of convergence lower than the (classical) order of the methods, cf. [10], [11]. This is sometimes referred to as 'order reduction due to stiffness'. In the specific instance of the NLS we shall see that such order reduction does not occur in some special cases of considerable practical interest.

For $q \geq 1$, a q-stage IRK method is defined by a set of constants arranged in tableau form, [8]:

$$\begin{array}{ccc|c} a_{11} & \dots & a_{1q} & \tau_1 \\ \vdots & & \vdots & \vdots \\ a_{q1} & \dots & a_{qq} & \tau_q \\ \hline b_1 & \dots & b_q & \end{array} ,$$

whose meaning is as follows: Given the initial-value problem

$$dy/dt = f(t,y), \quad 0 \leq t \leq T, \quad y(0) = y^0, \tag{4.1}$$

IRK methods generate approximations y^n to $y(t^n), 1 \leq n \leq J$, by the equations

$$\left\{ \begin{aligned} y^{n,j} &= y^n + k\sum_{m=1}^{q} a_{jm} f(t^{n,m}, y^{n,m}), \quad 1 \leq j \leq q, \\ t^{n,j} &= t^n + k\tau_j, \quad 1 \leq j \leq q, \\ y^{n+1} &= y^n + k\sum_{j=1}^{q} b_j f(t^{n,j}, y^{n,j}). \end{aligned} \right. \tag{4.2}$$

For our application, we shall assume that these constants satisfy certain ***stability*** and ***consistency*** conditions. Specifically, we shall require that the methods be ***algebraically stable***, i.e. that $b_i \geq 0$, $1 \leq i \leq q$, and that the $q \times q$ matrix with entries $m_{ij} = a_{ij}b_i + a_{ji}b_j - b_ib_j$ is positive semidefinite. In addition we shall assume that certain order ***simplifying assumptions*** hold. Specifically that there exist integers $\nu, p, \varrho \geq 1$ with $\nu \leq \varrho + p + 1$, $\nu \leq 2p + 2$, such that

$$\sum_{j=1}^{q} b_j \, \tau_j^\ell = \frac{1}{\ell + 1}, \; 0 \leq \ell \leq \nu - 1,$$

$$\sum_{j=1}^{q} a_{ij} \, \tau_j^\ell = \frac{\tau_i^{\ell+1}}{\ell + 1}, \; 1 \leq i \leq q, \quad 0 \leq \ell \leq p - 1,$$

$$\sum_{i=1}^{q} a_{ij}\, \tau_i^{\ell}\, b_i = \frac{b_j}{\ell+1}(1-\tau_j^{\ell+1}), \quad 1 \le j \le q, \quad 0 \le \ell \le \varrho - 1.$$

We shall refer to p and ν as the *stage order* and the *(classical) order* of the IRK scheme, respectively. In addition, we postulate the *positivity* condition that the matrix $A = (a_{ij})$ is invertible and that there exists a positive diagonal matrix D such that the matrix $DA^{-1}D^{-1}$ is positive definite. Examples of IRK schemes that satisfy the above conditions are given in [16, section 2]. They include the class of q-stage Gauss-Legendre collocation-type methods for which $\nu = 2q$, $p = \varrho = q$. The 1-stage Gauss-Legendre method coincides with the midpoint scheme (3.1), while the 2-stage member of the class is given by the tableau

$$\begin{array}{cc|c} \frac{1}{4} & \frac{1}{4} - \frac{1}{2\sqrt{3}} & \frac{1}{2} - \frac{1}{2\sqrt{3}} \\ \frac{1}{4} + \frac{1}{2\sqrt{3}} & \frac{1}{4} & \frac{1}{2} + \frac{1}{2\sqrt{3}} \\ \hline \frac{1}{2} & \frac{1}{2} & \end{array}$$

All Gauss-Legendre methods are *conservative* in the sense that they satisfy $m_{ij} = 0$, where (m_{ij}) is the $q \times q$ matrix that intervenes in the definition of algebraic stability. The fully discrete Galerkin method that corresponds to the time-stepping procedure (4.2) is then the following:

$$(4.3) \qquad \begin{cases} U^0 = u_h^0 \\ \text{For } n = 0, 1, \ldots, J-1: \\ (a)\ U^{n,j} = U^n + k \sum_{m=1}^{q} a_{jm} f_h(U^{n,m}), \qquad 1 \le j \le q, \\ (b)\ U^{n+1} = U^n + k \sum_{j=1}^{q} b_j f_h(U^{n,j}) \end{cases}$$

where we have put $f_h = i(\Delta_h + g_h)$, with the maps $\Delta_h, g_h : S_h \longrightarrow S_h$ defined for $\varphi \in S_h$ by the equations $(\Delta_h \varphi, \chi) = -(\nabla\varphi, \nabla\chi)$, $(g_h(\varphi), \chi) = (|\varphi|^2\varphi, \chi)$, $\forall \chi \in S_h$. In [16, section 3] the existence of solutions of (4.3) is proved to be a simple consequence of the *positivity* property of the IRK methods. Moreover, the solutions $\{U^n\}$ are uniformly bounded in L^2; indeed they satisfy

$$(4.4) \qquad \|U^n\| \le \|U^0\|,$$

which holds as an equality for *conservative* IRK schemes such as the Gauss-Legendre methods.

It is also shown in [16] that if $U^0 = u_h^0$ is chosen so that (2.5) is satisfied, then, if u is sufficiently smooth in $\bar{\Omega} \times [0, T]$, we have the error estimate

$$(4.5) \qquad \max_{0 \le n \le J} \|u^n - U^n\| \le c(k^\sigma + h^r),$$

where c is some constant depending only on u, while $\sigma = \min\{p+3, \nu\}$ if Ω is a general domain with piecewise smooth curved boundary, and $\sigma = \nu$ if Ω is any polyhedral domain (or any finite interval if $d = 1$). (For (4.5) to hold, the weak mesh condition $k = o(h^{\frac{d}{2\sigma}})$ should be imposed.)

From (4.5) we infer that, under these hypotheses, no reduction of the (classical) order ν occurs if Ω is a polyhedral domain or if $p+3 \geq \nu$, as would be the case e.g. with the practically important Gauss–Legendre schemes of up to 3 stages. The heart of the proof involves checking up to which integer s do solutions u of (1.2), which are smooth up to the boundary of $\Omega \times (0,T)$, satisfy the *compatibility conditions* that $\Delta^j u = 0$, $0 \leq j \leq s$, on $\partial\Omega \times [0,T]$. For the particular nonlinearity $|u|^2 u$ and for a general curved domain it is not hard to see that s is in general equal to 2 while such a property holds for all s if the boundary is piecewise linear, or if $d = 1$. The compatibility conditions imply that some special smooth approximations to the values $u(t^{n,j})$ of the solution of (1.2) vanish on $\partial\Omega$ and thus may be approximated to optimal order in space by elements of S_h. For details we refer the reader to section 4 of [16].

5. EFFICIENT SOLUTION OF THE NONLINEAR SYSTEMS

Whenever an IRK method is used to generate approximations to nonlinear evolution equations, there arises the issue of solving, at each time step, the nonlinear system that defines the intermediate stages, e.g. in our case the $q \dim S_h \times q \dim S_h$ nonlinear system (4.3a) that defines the functions $U^{n,j}$, $1 \leq j \leq q$. One realizes the importance of solving this system efficiently by recognizing the fact that the computational work is almost entirely concentrated there.

One may start by applying Newton's method to (4.3a). This involves approximating $U^{n,i}$ for each n and $1 \leq i \leq q$, by a sequence of iterants $U^{n,i}_\ell$, $0 \leq \ell \leq \ell_n$, where $\ell_n \geq 1$, obtained as follows: Let $\bar{p} = \min(p, \sigma - 1)$, and $U^{n,i}_0 \in S_h$ be *starting values* computed by extrapolation from previous time steps by the formulas

$$U^{n,i}_0 = \sum_{j=1}^{\bar{p}} L^{\bar{p},n}_j (t^n + k\tau_i) U^{n-j}, \quad 1 \leq i \leq q, \quad \bar{p} \leq n \leq J-1, \tag{5.1}$$

where $L^{\bar{p},n}_i$ are Lagrange interpolating polynomials of degree $\leq \bar{p}$ satisfying

$$L^{\bar{p},n}_i (t^{n-j}) = \delta_{ij}, \quad 0 \leq i,\, j \leq \bar{p}.$$

Then the $U^{n,j}_\ell$, $1 \leq \ell \leq \ell_n$, are defined recursively as solutions of the $q \dim S_h \times q \dim S_h$ linear systems

$$U^{n,j}_{\ell+1} - k \sum_{m=1}^{q} a_{jm} Df_h(U^{n,m}_\ell)(U^{n,m}_{\ell+1} - U^{n,m}_\ell) = U^n + k \sum_{m=1}^{q} a_{jm} f_h(U^{n,m}_\ell), \tag{5.2}$$

$$1 \leq i \leq q, \ \ \ell = 0, \dots, \ell_n - 1.$$

The approximation U^{n+1} is then defined by (4.3b) wherein the $U^{n,j}$ are replaced by $U^{n,j}_{\ell_n}$. This formula may also be written as

$$U^{n+1} = (1 - b^T A^{-1} e) U^n + b^T A^{-1} [U_{\ell_n}^{n,1}, \ldots, U_{\ell_n}^{n,q}]^T, \tag{5.3}$$

where $e = (1, 1, \ldots, 1)^T \in \mathbf{R}^q$.

In (5.2), for each $V \in S_h$ the map $Df_h(V) : S_h \longrightarrow S_h$ is defined as follows: First take real and imaginary parts in the nonlinear system (4.3a). Then, by applying Newton's method to this real-valued system, it is possible to combine terms and write it in complex form as in (5.2). In variational form (5.2) is explicitly given by

$$\begin{aligned}(U_{\ell+1}^{n,j}, \chi) - ik \sum_{m=1}^{q} a_{jm} [(\Delta_h U_{\ell+1}^{n,m}, \chi) + (2|U_\ell^{n,m}|^2 U_{\ell+1}^{n,m} + (U_\ell^{n,m})^2 \bar{U}_{\ell+1}^{n,m}, \chi)] \\ = (U^n, \chi) - 2ik \sum_{m=1}^{q} a_{jm} (|U_\ell^{n,m}|^2 U_\ell^{n,m}, \chi), \ \forall \chi \in S_h, \ 1 \le j \le q, \ 0 \le \ell \le \ell_n - 1.\end{aligned}$$

In [2] we show that one Newton iteration is sufficient to preserve the stability and order of accuracy of the scheme (4.3). In particular, let U^j, $0 \le j \le \bar{p}$, be functions in S_h satisfying $\|u^j - U^j\| \le c(k^\sigma + h^r)$, $0 \le j \le \bar{p}$, with σ as in (4.5). (One may construct such approximations by applying again Newton's method for the first few steps albeit with more iterations per step to compensate for the lower accuracy of the starting values since fewer previous iterants are available in this case.) Then, if $k = o(h^{\frac{4}{7}})$ and $\ell_n \ge \log_2(\sigma - \bar{p} + 1)$, all n, the U^n, $\bar{p} + 1 \le n \le J$, are defined uniquely as solutions of (5.2)-(5.3) and satisfy

$$\max_n \|u^n - U^n\| \le c(k^\sigma + h^r). \tag{5.4}$$

Solving for $U_{\ell+1}^{n,j}$ in Newton's method (5.2) requires forming and decomposing for each n and ℓ a $q \dim S_h \times q \dim S_h$ matrix. This could be prohibitively costly when $\dim S_h$ is large. We seek therefore *modified* versions of Newton's method that could be solved in efficient ways. In one such modified method, we evaluate $Df_h(\cdot)$ in (5.2) at the known value U^n. The corresponding system is

$$\begin{aligned}U_{\ell+1}^{n,j} - k \sum_{m=1}^{q} a_{jm} Df_h(U^n)(U_{\ell+1}^{n,m} - U_\ell^{n,m}) \\ = U^n + k \sum_{m=1}^{q} a_{jm} f_h(U_\ell^{n,j}), \ \ 1 \le j \le q, \ \ 0 \le \ell \le \ell_n - 1,\end{aligned} \tag{5.5}$$

with $U_0^{n,j}, U^{n+1}$ given again by (5.1) and (5.3), respectively. The Jacobian now is the linear operator $J = I - kADf_h(U^n)$ on $(S_h)^q$, i.e. it is independent of the stage index j. Assume now that A has distinct eigenvalues $\lambda_1, \ldots, \lambda_q$ (such is the case e.g. with the Gauss-Legendre methods, cf. [11]). Then, cf. [17], the similarity transformation $A = S^{-1} \Lambda S$, where $\Lambda = \mathrm{diag}(\lambda_1, \ldots, \lambda_q)$, induces a decomposition on the system $Jz = b$, which decouples into q $\dim S_h \times \dim S_h$ linear systems $(I - k\lambda_j Df_h(U^n))\tilde{z}_j = \tilde{b}_j$, $1 \le j \le q$, that are to be solved instead. These q systems are independent of each other and can be solved simultaneously on a computer with at least q independent processors.

Unlike the exact Newton's method, the modified scheme (5.5) is no longer quadratically convergent. As a consequence, in our setting it requires more iterations ℓ_n at each step in order to achieve an error bound of the form (5.4). Specifically, it is proved in [2] that if for $0 \le j \le \bar{p}$, U^j are $O(k^\sigma + h^r)$ approximations to u^j, $k = O(h)$ and $\ell_n \ge 2(\sigma - \bar{p} + 1)$ iterations are performed at each time step, then U^j, $\bar{p}+1 \le j \le J$, exist uniquely and satisfy (5.4).

One may in fact abandon Newton's method altogether and adopt a much simpler iterative scheme, an "explicit-implicit" method in which $f_h(U)$ is split into its linear, $i\Delta_h U$, and nonlinear, $ig_h(U)$, parts, that are evaluated at the iteration levels $\ell+1$ and ℓ, respectively. The scheme is

$$U_{\ell+1}^{n,j} - ik\sum_{m=1}^{q} a_{jm}\Delta_h U_{\ell+1}^{n,m} = U^n + ik\sum_{m=1}^{q} a_{jm} g_h(U_\ell^{n,m}), \qquad 1 \le j \le q, \quad 0 \le \ell \le \ell_n - 1. \tag{5.6}$$

The corresponding linear operator $I - ikA\Delta_h$ on $(S_h)^q$ is now independent of n. Again using the decomposition $A = S^{-1}\Lambda S$, it may naturally be split into q independent systems with operators $I - ik\lambda_j\Delta_h$ that may be then solved simultaneously on a multiprocessor computer. Some results of numerical experiments performed with these parallelizable schemes on machines with distributed and shared memory architecture, are reported in [3]. For example, on the shared memory, 8-processor CRAY Y-MP/832 the speedups achieved (i.e. the ratios of the execution times of the serial vs. the parallel code) for a 1-dimensional test problem on $\Omega = (0,1)$ with $h = 0.01$, using the q-stage Gauss–Legendre schemes coupled with cubic ($r = 4$) splines in space, were equal to 1.81, 2.69, and 3.50 for $q = 2, 3$, and 4, respectively. For quintic splines ($r = 6$) the parallel schemes were even more efficient, achieving speedups of 1.92, 2.81 and 3.65, respectively.

In [2] it is proved that if the U^j are close to u^j in L^2 to $O(k^\sigma + h^r)$ for $0 \le j \le \bar{p}$, then, provided $k = o(h^{\frac{d}{4}})$ and $\ell_n \ge 2(\sigma - \bar{p} + 1)$ for $\bar{p}+1 \le n \le J-1$, the U^j, $\bar{p}+1 \le j \le J$, are uniquely defined by (5.6) and satisfy the error estimate (5.4) of the (exact) fully discrete scheme (4.3).

6. COMPUTATIONS OF BLOW-UP FOR RADIAL SOLUTIONS

In this section we shall use one of the numerical schemes analyzed in previous sections to investigate numerically some aspects of the development of point blow-up singularities of the Cauchy problem for the NLS in $d = 2$ or 3 dimensions. We make the hypothesis that the initial condition and the solution of (1.1) are radially symmetric ('isotropic case'), i.e. that $u = u(r,t)$, where $r = |x| = (x_1^2 + \cdots + x_d^2)^{\frac{1}{2}}$. In this case (1.1) becomes

$$\begin{cases} u_t = i(u_{rr} + \dfrac{d-1}{r}u_r) + i|u|^2u, & r > 0, \ t > 0, \\ u(r,0) = u^0(r), \quad r > 0, & \end{cases} \tag{6.1}$$

with invariants

$$\int_0^\infty |u(r,t)|^2 r^{d-1} dr = \text{const.} \quad \text{for } t \ge 0, \tag{6.2}$$

$$\int_0^\infty (|u_r(r,t)|^2 - \frac{1}{2}|u(r,t)|^4) r^{d-1} dr = \text{const.} \quad \text{for } t \geq 0. \tag{6.3}$$

As already stated in the Introduction, for $d = 2$ and 3 solutions of the Cauchy problem for the NLS may blow up in L^∞ as t tends to some finite value t^*. In recent years there has appeared a considerable amount of work in the direction of a detailed description of the characteristics of this blow-up. We refer the reader e.g. to a series of publications, [33], [24], [21], [22], [19], [23], [20] by a group including Papanicolaou, C. Sulem, P.L. Sulem, McLaughlin and their coworkers. We also refer to the bibliographies of these works for other recent literature on the subject, including references to work of the Russian School of Zakharov and his coworkers.

In [24] the characteristics of the blow-up singularity are studied (in the case $d = 3$ and in the presence of radial symmetry) by a numerical technique which employs "dynamic rescaling", a time-dependent change of scales of the solution and the radial and temporal variables. The scaling factors are chosen so that suitable functionals of the transformed solution and its derivatives are preserved. It turns out that the transformed dependent variable satisfies an equation with global smooth solutions. This equation is then solved numerically and the details of the blow-up singularity are characterized by the long-time asymptotics of the scale factors. In this manner, it was concluded in [24] that singular radial solutions in $d = 3$ dimensions blow up with a peak amplitude that grows like $(t^* - t)^{-\frac{1}{2}}$ as $t \uparrow t^*$ The computations also provide additional information by which further conjectures on the details of the self-similar structure and the phase of the solution as it blows up are plausibly advanced. (Some of these features of the blow-up had been predicted earlier by Zakharov, [44].)

For $d = 2$, still for radially symmetric solutions, earlier conclusions in the literature on the blow-up rate of the amplitude, based on numerical computations and some asymptotic reasoning, varied somewhat. This is not surprising as $d = 2$ is the critical dimension case for the cubic nonlinearity and the blow-up slows considerably down, making, for instance, the numerical integration of the equation harder. For example, a $(t^* - t)^{-\frac{2}{3}}$ law was put forth for the blow-up of the amplitude in [45], [33] and elsewhere, while [39] and [43] favored $\left[\ln(\frac{1}{t^*-t})/(t^* - t)\right]^{\frac{1}{2}}$ With numerical ("dynamical rescaling") evidence and asymptotic techniques, Papanicolaou et al. suggested, cf. e.g. [21], the form $F(t^* - t)/(t^* - t)^{\frac{1}{2}}$, where, as $s \to 0, F(s)$ tends to infinity more slowly than $(\ln \frac{1}{s})^\gamma$ for any $\gamma > 0$. Finally, in [19] and [23] it is concluded, by asymptotic analysis, that the rate is $\left[\ln\ln(\frac{1}{t^*-t})/(t^* - t)\right]^{\frac{1}{2}}$ In [20], dynamic rescaling was extended to the general (i.e. not radially symmetric) Schrödinger equation for $d = 2$ and $d = 3$, and it was demonstrated numerically that initial data with a single peak evolve rapidly into locally radially symmetric, self-similar singularities that blow up, presumably with the last-mentioned rate.

In this section we construct and implement fully discrete, adaptive in space and time Galerkin methods for the purpose of approximating the blow-up singularities of the NLS equation *directly*, i.e. without using dynamic rescaling or any other a-priori knowledge of the asymptotics of the blow-up. As a first, preliminary step we consider here blow-up rate calculations in the radially symmetric case in $d = 2$ and 3 dimensions with initial data of Gaussian type that decays very fast as r increases. It is our intention to extend these adaptive techniques to more general initial data in the future, and, eventually, to the full nonisotropic case in two space dimensions.

In constructing these adaptive schemes we were guided by the experience recently gained (cf. e.g. [4], [5], [6]), from applying similar methods for approximating the development of instabilities (leading to point blow-up) of solutions of the generalized Korteweg-de Vries (GKdV) equation. In particular, the spatial mesh- and temporal step-cutting strategies are motivated from analogous techniques that have been successfully applied to the GKdV.

In (6.1) we shall take as initial value (following [24]) the Gaussian $u^0(r) = 6\sqrt{2}e^{-r^2}$ It is known then, cf. [44], [33], [24] that the solution blows up at $r = 0$ in finite time. We shall approximate (6.1) by a problem posed on a finite spatial interval $[0, R]$, where R will be chosen so that $u(R,t)$ is negligibly small for $t \geq 0$, enabling us to impose the Dirichlet boundary condition $u = 0$ at R. Specifically, after rescaling in r, we approximate the solution of (6.1) by the solution of the problem

$$\begin{cases} u_t = i\varepsilon(u_{rr} + \frac{d-1}{r}u_r) + i|u|^2u, & 0 < r \leq 1, \quad t \geq 0, \\ u(r,0) = u^0(r) = 6\sqrt{2}e^{-25r^2}, & 0 \leq r \leq 1, \\ u(1,t) = 0, \quad t \geq 0, \end{cases} \tag{6.4}$$

where $\varepsilon = 1/25$. The solution of (6.4) should also satisfy $u_r(0,t) = 0$ but this natural boundary condition is not imposed and disappears from the variational formulation.

As our base time-stepping scheme we shall use the simplest member of Gauss-Legendre family, namely the 1-stage method (3.1). S_h will simply consist of those continuous complex-valued functions on [0,1], which vanish at $r = 1$ and are piecewise linear polynomials on each subinterval of an arbitrary partition $0 = r_0 < r_1 < \cdots < r_Q = 1$ of [0,1]. The fully discrete base scheme amounts then to determining $U^n \in S_h, n = 0, 1, 2, \ldots$, satisfying

$$\begin{aligned} \int_0^1 (U^{n+1} - U^n)\bar{\chi}r^{d-1}dr + \frac{ik}{2}\int_0^1 (U^{n+1} + U^n)_r\bar{\chi}_r r^{d-1}dr \\ = \frac{ik}{8}\int_0^1 |U^{n+1} + U^n|^2(U^{n+1} + U^n)\bar{\chi}r^{d-1}dr, \quad \forall\chi \in S_h, \end{aligned} \tag{6.5}$$

where U^0 is taken to be the L^2-projection of u^0 on S_h. The resulting nonlinear system for U^{n+1} is solved by Newton's method using one iteration per step, cf. [1].

Anticipating that the solution blows up at $r = 0$ we implemented (6.5) with variable time step and space mesh that can change with n. As the blow-up time is approached and the solution grows in amplitude near the origin, the adaptive mechanism refines the spatial mesh drastically near $r = 0$ and cuts the time step according to two refinement criteria that appeared to be successful for the problem at hand and will be described below. If the time step reduction criterion is met, then the time step is halved. If the spatial mesh has to be refined, then the number of points r_i is increased in the vicinity of the origin as follows:

Suppose that the spatial grid has already been refined $NSPLIT$ times. Then, [0,1] is partitioned in $NSPLIT + 2$ adjacent successive regions (intervals) $I_0, I_1, \ldots, I_{NSPLIT+1}$, such that the left-hand boundary of I_0 is 0, the right-hand boundary of $I_{NSPLIT+1}$ is 1, and, on each I_j the mesh size is uniform. Specifically, if N and $M < 2N$ are given integers and $h = 1/N$ is the initial spatial meshlength, then:

I_0 consists of M subintervals of constant meshlength $h/2^{NSPLIT+1}$

I_1 consists of $\frac{M}{2}$ subintervals of constant meshlength $h/2^{NSPLIT}$

I_2 consists of $\frac{M}{2}$ subintervals of constant meshlength $h/2^{NSPLIT-1}$

$\vdots$

I_{NSPLIT} consists of $\frac{M}{2}$ subintervals of constant meshlength $h/2$.

$I_{NSPLIT+1}$ consists of $N - \frac{M}{2}$ subintervals of constant meshlength h.

Thus, each time the grid is refined, I_0 is cut in half into two new intervals I_0 and I_1 and all other regions are redefined so that I_1 becomes I_2, I_2 becomes I_3, etc.. U^n is imbedded in the new mesh by linear interpolation.

The empirical criterion for this spatial grid refinement to take place (i.e. for increasing $NSPLIT$ by 1) is that at a certain time step the grid is refined if on the interval I_0 with the finest mesh

$$\frac{|u|_{L^\infty(I_0)} h_{\min}^{\frac{1}{2}}}{\|u\|_{L^2(I_0)}} > TOL_h, \tag{6.6}$$

where $h_{min} = h/2^{NSPLIT+1}$ Thus, motivated by a local $L^\infty - L^2$ "inverse" property on I_0, we reduce h_{min} to allow $|u|_{L^\infty(I_0)}$ to grow. TOL_h is an empirically determined tolerance that was usually taken equal to 0.12 for $d = 2$ and equal to 0.14 for $d = 3$.

The time step reduction is motivated by a need to control, from time step to time step, the change in a weighted version of the second invariant of the equation. Specifically, the time step is halved whenever

$$\frac{|I_2(U^n) - I_2(U^{n+1})|}{1 + \int_0^1 |u_r|^2 r^{d-1} dr} > TOL_k/2, \tag{6.7}$$

where $I_2(U) = \int_0^1 (\epsilon |U_r|^2 - \frac{1}{2}|U|^4) r^{d-1} dr$, is the second invariant associated with the p.d.e. in (6.4). For $d = 2$ the value of TOL_k used was 3×10^{-8}, while for $d = 3$ using $TOL_k = 4 \times 10^{-8}$ resulted in reducing the time step in a satisfactory manner.

With this mesh refinement mechanism in place, we first approximated the solution of (6.4) for $d = 3$ using an initial spatial meshlength $h = 1/1200$ (i.e. taking $N = 1200$), and an initial time step equal to $k = 10^{-4}$ (The constant multiplying the exponential in $u^0(r)$ was actually taken for

$d = 3$ to be equal to 8.485281374, an approximation to $6\sqrt{2}$). The parameter M in the space mesh refinement procedure was taken to be equal to 200. The evolution of the magnitude of the solution is shown in Figures 1a, 1b, 1c. (Note that the radial axis, labelled x in the figure, and the $|u|$ axis, labelled u, are being rescaled in the successive plots of Figures 1b and 1c and that the self-similar nature of the blow-up profile is evident.) By the final ("blow-up") time $t^* = 0.03429981$ the amplitude at zero had risen to approximately 3.3×10^7, the code had refined $NSPLIT = 20$ times the spatial mesh according to the scheme previously described, and the final time step was about $.23 \times 10^{-16}$. The amplitude magnification achieved is thus seen to be considerably larger than the ones previously reported in the literature and which, to our knowledge, are in the range of $O(10^3)$ to $O(10^4)$, cf. [24], [37], and gives us confidence that the blow-up rates, which will be presently tabulated, are accurate enough.

In Table 1 we show some observed rates of blow-up for this problem, i.e. approximations generated by the output of the code to the numbers ϱ such that

$$M(t) \cong (t^* - t)^{-\varrho}, \quad t \uparrow t^*, \tag{6.8}$$

where $M(t)$ are the values at time t of several *norms* of the solution u of (6.4) for $d = 3$ and its radial derivative u_r. We tabulate the blow-up rates at several times t_i, that get very close to t^* as i increases, for the L^4 and L^∞ norms of u and the L^2 and L^∞ norms of the derivative (the latter are shown under the headings L^2_D and L^∞_D in the Table.) The t_i's are the times (their values, being extremely close to that of t^*, are not actually shown) at which the ith spatial mesh refinements occurs. It should be noted that the computation of the rates is quite delicate and should be done with care to avoid roundoff error due to cancellation; cf. e.g. [6] for details.

We conclude from Table 1 that the computed rates are quite robust. The amplitude (L^∞) blow-up rate is equal to $\frac{1}{2}$ to three digits, thus confirming estimates obtained in previously cited references. Table 1 shows that the results of the computations suggest to a high degree of accuracy, that $|u(t)|_4$ blows up at a rate of $\frac{1}{8}$, while $\|u_r\|$ and $|u_r|_\infty$ blow up at rates of $\frac{1}{4}$ and 1, respectively.

i	L^4	L^∞	L^2_D	L^∞_D
7	.1190	.5013	.2415	1.0024
8	.1217	.5001	.2452	1.0006
9	.1232	.5006	.2473	1.0005
10	.1240	.5003	.2485	1.0014
11	.1245	.4997	.2492	0.9999
12	.1247	.5000	.2495	1.0000
13	.1249	.5004	.2497	0.9999
14	.1247	.4998	.2497	0.9993

Table 1

Blow-up rates of various norms. Gaussian initial profile, d=3.

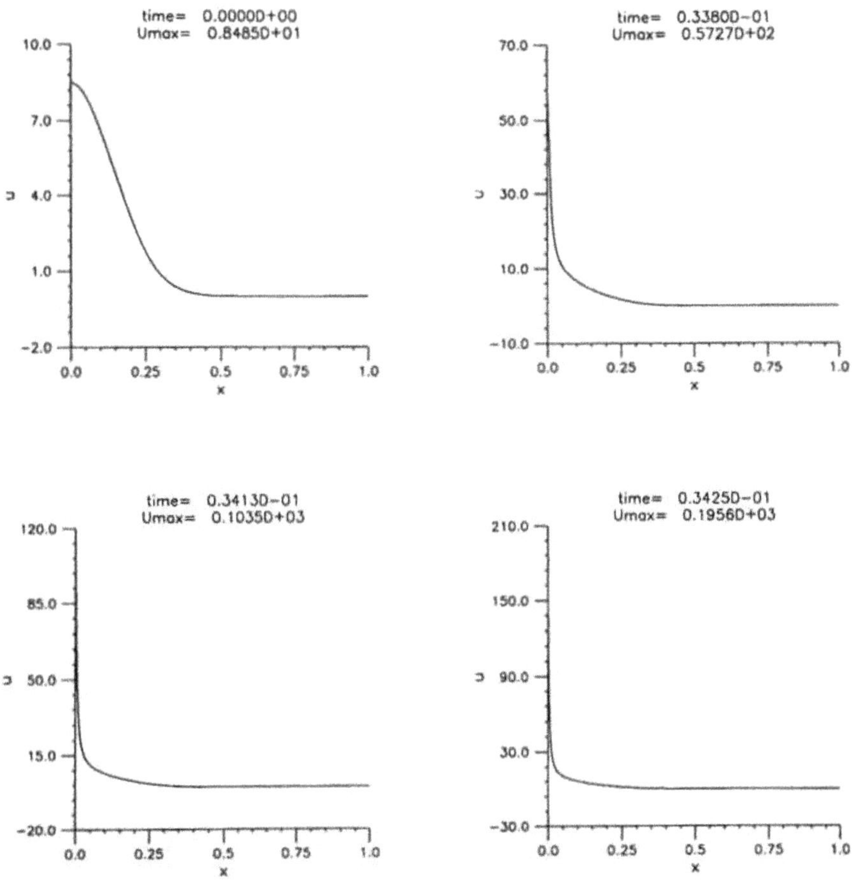

Figure 1a

Temporal evolution of Gaussian initial profile, d=3

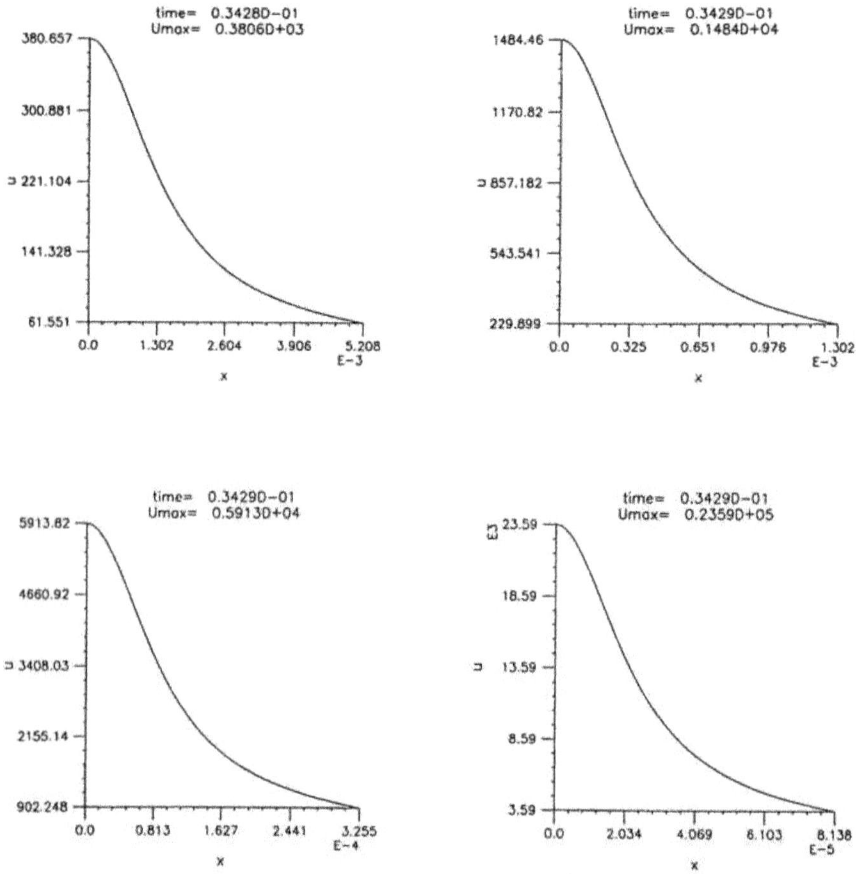

Figure 1b

Continuation of Fig. 1a

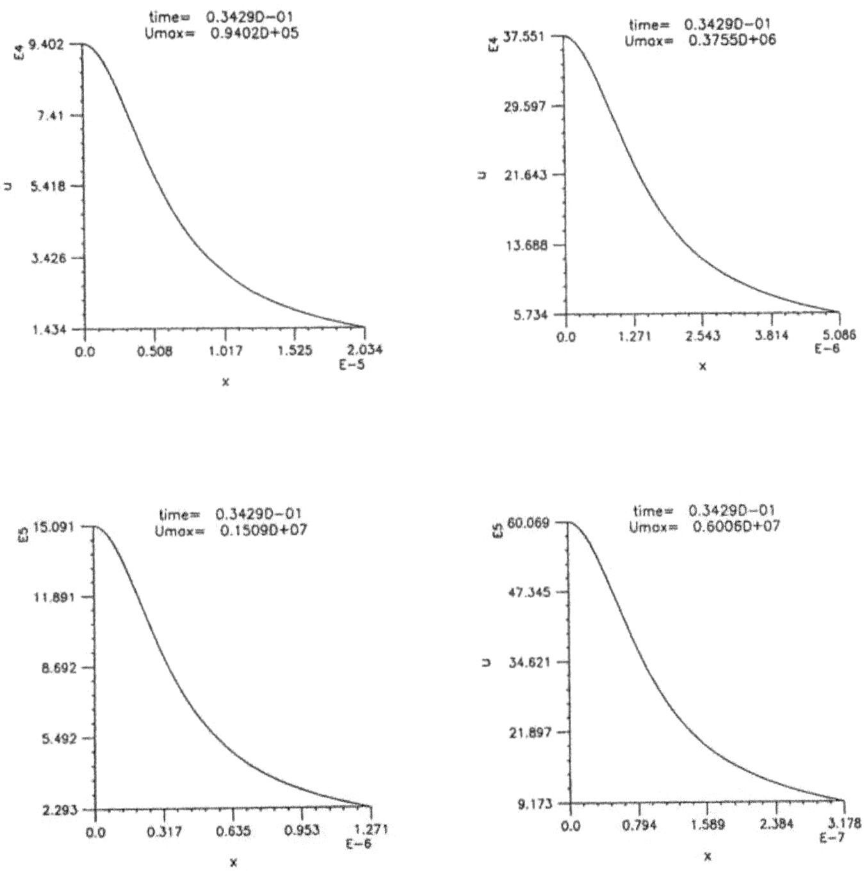

Figure 1c

Continuation of Fig. 1b

We proceed now to the harder case $d = 2$. We solved again the problem (6.4) with the adaptive code previously described. Initially we took $N = 1600$, i.e. $h = 1/1600$, and $k = 0.8 \times 10^{-4}$, and computed with $M = 200$ points in the region of finest mesh. Our code was able to perform 34 refinements in space, at which point it had achieved an amplitude at the origin equal to $.26 \times 10^{12}$, while the time step had been reduced to $.11 \times 10^{-23}$ The final ("blow-up") time reached was $t^* = 0.040289799896$.

The high value of the amplitude reached at the end of this computation tempted us to compute and tabulate blow-up rates of norms including log and log log factors. As an example, in Table 2 we show three columns of rates of blow-up of the L^∞ norm of u at the times t_i of the ith spatial refinement, defined as follows: In the first column ϱ is computed from the output of the code assuming that $M(t) = \|u(t)\|_\infty = |u(0,t)|$ is of the form

$$M_1(t) \cong 1/(t^* - t)^\varrho, \quad \text{as } t \uparrow t^*. \tag{6.9}$$

In the second column, ϱ is computed assuming that

$$M_2(t) \cong \left[\ln(\frac{1}{t^* - t})/(t^* - t)\right]^\varrho, \quad \text{as } t \uparrow t^*, \tag{6.10}$$

while in the third, ϱ corresponds to the assumption that

$$M_3(t) \cong \left[\ln\ln(\frac{1}{t^* - t})/(t^* - t)\right]^\varrho, \quad \text{as } t \uparrow t^* \tag{6.11}$$

i	$M_1(t)$	$M_2(t)$	$M_3(t)$
15	.5076	.4880	.5013
16	.5071	.4886	.5013
17	.5065	.4889	.5011
18	.5060	.4892	.5009
19	.5055	.4895	.5007
20	.5051	.4898	.5006
21	.5047	.4900	.5005
22	.5045	.4903	.5004
23	.5042	.4906	.5003
24	.5039	.4908	.5003
25	.5037	.4910	.5002
26	.5035	.4912	.5001
27	.5032	.4914	.5000

Table 2

Computed rates of blow-up of $|u(0,t)|$ *according to* (6.9)–(6.11).

Gaussian initial profile, d=2.

In Table 2 we tabulated the rates up to the time of the 27th refinement. After that, the difference between the t_i and t^* is so small that the numerical calculation of the rates degenerates. The results suggest of course that ϱ is about $\frac{1}{2}$ under all three assumptions. However, the rates corresponding to M_3, i.e. to the log log modification factor, seem to be more robust and stable for all values of i, whereas in the first two columns the values seem to be still changing in a perceptible, almost uniform rate, certainly faster than the values of the last column. This leads us to conclude tentatively that, on the basis of this experiment, it seems that the hypothesis $M_3(t)$ is most compatible with the data, confirming the results of [19].

Using again a log log factor to modify $(t^* - t)^{-1}$ we can also obtain blow-up rates for other norms of the solution and its radial derivative. In Table 3 we show the computed rates at t_i for the L^3 and L^4 norms of u and the L^2 and L^∞ norms of u_r; the columns of the latter are labelled as L^2_D and L^∞_D as before.

i	L^3	L^4	L^2_D	L^∞_D
15	.1667	.2504	.5009	1.0029
16	.1668	.2505	.5009	1.0030
17	.1667	.2504	.5007	1.0024
18	.1667	.2503	.5006	1.0017
19	.1667	.2502	.5005	1.0013
20	.1667	.2502	.5004	1.0016
21	.1667	.2501	.5003	1.0008
22	.1666	.2501	.5002	1.0014
23	.1666	.2501	.5001	1.0005
24	.1666	.2500	.5001	1.0008
25	.1666	.2500	.5000	1.0005
26	.1666	.2500	.5000	1.0003
27	.1666	.2500	.4999	0.9996

Table 3.

Blow-up rates (modified by log log *factors) of various norms.*

Gaussian initial profile, d=2.

On the basis of these calculations it seems sensible to conclude that the blow-up rates in two dimensions (modified by log log factors) are as follows: The rate of the L^3 norm of u is equal to $\frac{1}{6}$, while that of the L^4 norm of u equals $\frac{1}{4}$. The L^2 norm of u_r blows up at a rate of $\frac{1}{2}$, while the L^∞ norm of u_r seems to be blowing up at a rate equal to 1.

In conclusion, it seems that these blow-up computations are quite encouraging and suggest that the direct numerical integration of the NLS with adaptive Galerkin methods, such as the one proposed herein, should be seriously pursued and extended to cover other cases of interest such as more general initial conditions and nonisotopic cases. Such extensions are currently the object of study by the authors.

REFERENCES

1. G.D. Akrivis, V.A. Dougalis and O.A. Karakashian, "*On fully discrete Galerkin methods of second-order temporal accuracy for the nonlinear Schrödinger equation*", *Numer. Math.* **59** (1991), pp. 31–53.
2. G. Akrivis, V.A. Dougalis and O. Karakashian, "*On the efficient implementation of implicit Runge–Kutta methods*", submitted.
3. G.D. Akrivis, V.A. Dougalis and O. Karakashian, "*Numerical methods for the nonlinear Schrödinger equation*", to appear in the *Proceedings of the 7th IMACS International Conference on Computer methods for P.D.E.'s*, June 1992.
4. J.L. Bona, V.A. Dougalis, O.A. Karakashian and W.R. McKinney, "*Fully-discrete methods with grid refinement for the generalized Korteweg–de Vries equation*", in *Viscous profiles and numerical methods for shock waves*, M. Shearer, ed., pp. 1–11, SIAM, Philadelphia 1991.
5. J.L. Bona, V.A. Dougalis, O.A. Karakashian and W.R. McKinney, "*Computations of blow-up and decay for periodic solutions of the generalized Korteweg–de Vries–Burgers equation*", *Appl. Numer. Math.* **10** (1992), pp. 335–355.
6. J.L. Bona, V.A. Dougalis, O.A. Karakashian and W.R. McKinney, "*Conservative, high-order numerical schemes for the generalized Korteweg–de Vries equation*", submitted.
7. H. Brezis and T. Gallouet, "*Nonlinear Schrödinger evolution equations*", *Nonlinear Anal.* **4** (1980), pp. 677–681.
8. J.C. Butcher, *The numerical analysis of ordinary differential equations. Runge–Kutta and general linear methods*, J. Wiley, Chichester 1987.
9. R.Y. Chiao, E. Garmire and C. Townes, "*Self-trapping of optical beams*", *Phys. Rev. Lett.* **73** (1964), pp. 479–482.
10. M. Crouzeix, "*Sur l' approximation des équations différentielles opérationelles linéaires par des méthodes de Runge–Kutta*", Thèse, Université Paris VI, 1975.
11. K. Dekker and J.G. Verwer, *Stability of Runge–Kutta methods for stiff nonlinear differential equations*, North-Holland, Amsterdam 1984.
12. M. Delfour, M. Fortin and G. Payre, "*Finite-difference solutions of a non-linear Schrödinger equation*", *J. Comp. Phys.* **44** (1981), pp. 277–288.
13. J. Ginibre and G. Velo, "*On a class of nonlinear Schrödinger equations I: the Cauchy problem, general case*", *J. Funct. Anal.* **32** (1979), pp. 1–32.
14. R.T. Glassey, "*On the blowing up of the solutions in the Cauchy problem for nonlinear Schrödinger equations*", *J. Math. Phys.* **18** (1977), pp. 1974–1977.
15. D.F. Griffiths, A.R. Mitchell and J.Ll. Morris, "*A numerical study of the nonlinear Schrödinger equation*", *Comput. Methods Appl. Mech. Eng.* **45** (1984), pp. 177–215.
16. O. Karakashian, G.D. Akrivis and V.A. Dougalis, "*On optimal-order error estimates for the Nonlinear Schrödinger Equation*", to appear in *SIAM J. Numer. Anal.*.

17. O. Karakashian and W. Rust, *"On the parallel implementation of implicit Runge–Kutta methods"*, *SIAM J. Sci. Stat. Comput.* **9** (1988), pp. 1085–1090.

18. O. Kavian, *"A remark on the blowing-up of solutions to the Cauchy problem for nonlinear Schrödinger equations"*, *Trans. A.M.S.* **299** (1987), pp. 193–203.

19. M.J. Landman, G.C. Papanicolaou, C. Sulem and P.L. Sulem, *"Rate of blowup for solutions of the nonlinear Schrödinger equation at critical dimensions"*, *Phys. Rev. A* **38** (1988), pp. 3837–3843.

20. M.J. Landman, G.C. Papanicolaou, C. Sulem, P.L. Sulem and X.P. Wang, *"Stability of isotropic singularities for the nonlinear Schrödinger equation"*, *Physica D* **47** (1991), pp. 393–415.

21. B. Le Mesurier, G. Papanicolaou, C. Sulem and P.-L. Sulem, *"The focusing singularity of the nonlinear Schrödinger equation"*, in *Directions in Partial Differential Equations*, M.G. Crandall, P.H. Rabinowitz and R.E. Turner, eds., pp. 159–201, Academic Press, N.Y. 1987.

22. B.J. Le Mesurier, G. Papanicolaou, C. Sulem and P.L. Sulem, *"Focusing and multi-focusing solutions of the nonlinear Schrödinger equation"*, *Physica D* **31** (1988), pp. 78–102.

23. B.J. Le Mesurier, G. Papanicolaou, C. Sulem and P.L. Sulem, *"Local structure of the self-focusing singularity of the nonlinear Schrödinger equation"*, *Physica D* **32** (1988), pp. 210–226.

24. D.W. McLaughlin, G.C. Papanicolaou, C. Sulem, P.L. Sulem, *"Focusing singularity of the cubic Schrödinger equation"*, *Phys. Rev. A* **34** (1986), pp. 1200–1210.

25. A.C. Newell, *Solitons in Mathematics and Mathematical Physics, CBMS Applied Mathematics Series* **48**, SIAM, Philadelphia, 1988.

26. T. Ogawa, *"A proof of Trudinger's inequality and its application to nonlinear Schrö- dinger equations"*, *Nonlinear Anal.* **14** (1990), pp. 765–769.

27. T. Ogawa and Y. Tsutsumi, *"Blow-up of solutions for the Nonlinear Schrödinger Equation with quartic potential and periodic boundary condition"*, in *Functional Analytic Methods for Partial Differential Equations*, A. Fujita, T. Ikebe and S.T. Kuroda, eds., pp. 236–251, Lecture Notes in Math. **v. 1450**, Springer Verlag 1990.

28. J. Rasmussen and K. Rypdal *"Blow-up in Nonlinear Schrödinger Equations"*, *Parts I and II, Physica Scripta* **33** (1986), pp. 481–504.

29. J.M. Sanz-Serna, *"Methods for the numerical solution of the nonlinear Schrödinger equation"*, *Math. Comp.* **43** (1984), pp. 21–27.

30. J.M. Sanz-Serna and J.G. Verwer, *"Conservative and nonconservative schemes for the solution of the nonlinear Schrödinger equation"*, *IMA J. Numer. Anal.* **6** (1986), pp. 25–42.

31. W.A. Strauss *"The Nonlinear Schrödinger Equation"*, in *Contemporary developments in continuum mechanics and p.d.e.'s*, ed. by G.M. de la Penha and L.A.J. Medeiros, pp. 452–465, North Holland, 1978.

32. W.A. Strauss and L. Vazquez, "*Numerical solution of a nonlinear Klein–Gordon equation*", *J. Comp. Phys.* **28** (1978), pp. 271–278.

33. P.-L. Sulem, C. Sulem and A. Patera "*Numerical simulation of singular solutions to the two-dimensional cubic Schrödinger equation* ", *Comm. Pure Appl. Math.* **37** (1984), pp. 755–778.

34. T.R. Taha and M.J. Ablowitz, "*Analytical and numerical aspects of certain nonlinear evolution equations II. Numerical, nonlinear Schrödinger equation*", *J. Comp. Phys.* **55** (1984), pp. 203–230.

35. V.I. Talanov, "*Self-focusing of wave beams in nonlinear media*" *JETP Lett.* **2** (1965), pp. 138–141.

36. Y. Tourigny and J.Ll. Morris, "*An investigation into the effect of product approximation in the numerical solution of the cubic nonlinear Schrödinger equation*", *J. Comp. Phys.* **76** (1988), pp. 103–130.

37. Y. Tourigny and J.M. Sanz-Serna, "*The numerical study of blow-up with application to a nonlinear Schrödinger equation*", *Oxford Univ. Computing Laboratory Report* 90/2, July 1990.

38. M.V. Vladimirov, "*On the solvability of mixed problem for a nonlinear equation of Schrödinger type*", *Soviet Math. Dokl.* **29** (1984), pp. 281–284.

39. S.N. Vlasov, L.V. Piskunova and V.I. Talanov, "*Structure of the field near a singularity arising from self-focusing in a cubically nonlinear medium*", *Soviet Physics JETP* **48** (1978), pp. 808–812.

40. J.A.C. Weideman and B.M. Herbst, "*Split-step methods for the solution of the nonlinear Schrödinger equation*", *SIAM J. Numer. Anal.* **23** (1986), pp. 485–507.

41. M.I. Weinstein "*Nonlinear Schrödinger equation and sharp interpolation estimates*", *Communs. in Math. Phys.* **87** (1983), pp. 567–576.

42. G.B. Whitham, *Linear and Nonlinear Waves*, J. Wiley, New York 1974.

43. D. Wood, "*The self-focusing singularity in the nonlinear Schrödinger equation*", *Studies in Appl. Math.* **71** (1984), pp. 103–115.

44. V.E. Zakharov, "*Collapse of Langmuir waves*" *JETP* **35** (1972), pp. 908–922.

45. V.E. Zakharov and V.S. Synakh, "*The nature of the self-focusing singularity*", *Soviet Physics JETP* **41** (1976), pp. 465–468.

EFFICIENT RUNGE–KUTTA METHODS FOR HAMILTONIAN EQUATIONS

A. Iserles
DAMTP
University of Cambridge
Cambridge, England

Abstract. Considerable effort has been expanded in the last few years in developing numerical methods for Hamiltonian ordinary differential equations, focusing on algorithms that maintain symplectic structure. Specific Runge–Kutta processes emerged in this context as possibly the most promising general-purpose methods. Unfortunately, such methods are computationally expensive to implement. In the present paper we explore ways and means of designing Runge–Kutta processes that are considerably cheaper computationally, while retaining symplecticity. This is done by marrying the technique of perturbed collocation, due to Nørsett and Wanner [13], with the symplecticity test of Sanz-Serna [15].

1. Symplectic methods

This paper is concerned with the numerical solution of Hamiltonian equations

$$\frac{\mathrm{d}\mathbf{p}}{\mathrm{d}t} = -\frac{\partial H(\mathbf{p},\mathbf{q})}{\partial \mathbf{q}}, \qquad \frac{\mathrm{d}\mathbf{q}}{\mathrm{d}t} = \frac{\partial H(\mathbf{p},\mathbf{q})}{\partial \mathbf{p}}, \tag{1.1}$$

accompanied by initial values $\mathbf{p}(0) = \mathbf{p}_0$, $\mathbf{q}(0) = \mathbf{q}_0$. Here both $\mathbf{p}$ and $\mathbf{q}$ are d-dimensional vectors, whereas H is a C^2 function. Equations of this form describe the evolution of many dynamical systems, of great importance in classical and quantum mechanics ([1], [5], [12]).

It is a major feature of Hamiltonian systems, known as the Liouville theorem, that the flow is symplectic, i.e. that the differential form $\mathrm{d}\mathbf{p} \wedge \mathrm{d}\mathbf{q}$ is invariant under the flow. An application of the Kolmogorov–Arnold–Moser theorem demonstrates that symplecticity implies stability of elliptic equilibria of (1.1) in the underlying phase plane. The price of symplecticity (and of the preservation of other Poincarè invariants) is increasing complexity in the phae-plane geometry of the flow. For example, provided that $d \geq 2$, the flow may be chaotic.

Major features of (1.1) – not just chaotic behaviour but also limit cycles and homoclinic orbits, – follow from its symplectic structure and it comes as no surprise that the maintenance of symplecticity is crucial to reliable numerical solution. The aforementioned features are very delicate and will be the first casualities of an incautious numerical procedure. The concepts so central to classical numerical analysis of ordinary differential equations (convergence, zero-stability, A-stability, algebraic stability *etc.*) are of little use in the present framework, since they either concern behaviour on compact time-intervals – and we are interested in the asymptotics – or performance for dissipative systems – and (1.1) is conservative, not dissipative.

Symplectic integrators for specific Hamiltonian systems were first presented by Ruth [13]. They were further developed by Feng Kang [4], who presented an approach, based on generating functions, that yields symplectic integrators for a wide variety of systems (1.1). However, the crucial breakthrough was the discovery in 1988, simultaneously by Lasagni [11], Sanz-Serna [15] and Suris [18], that Runge–Kutta methods are symplectic, subject to a simple algebraic test. Specifically, given the initial-value problem

$$\mathbf{y}' = \mathbf{f}(t, \mathbf{y}),$$

a general s-stage Runge–Kutta method is

$$\mathbf{y}_{n+1} = \mathbf{y}_n + h \sum_{\ell=1}^{s} b_\ell \mathbf{k}_\ell,$$

where

$$\mathbf{k}_\ell = \mathbf{f}(t_n + c_\ell h, \mathbf{y}_n + h \sum_{j=1}^{s} a_{\ell,j} \mathbf{k}_j), \quad \ell = 1, \ldots, s.$$

It is convenient to denote this in the Butcher *tableau* [6]

$$\begin{array}{c|cccc} c_1 & a_{1,1} & a_{1,2} & \cdots & a_{1,s} \\ c_2 & a_{2,1} & a_{2,2} & \cdots & a_{2,s} \\ \vdots & \vdots & \vdots & & \vdots \\ c_s & a_{s,1} & a_{s,2} & \cdots & a_{s,s} \\ \hline & b_1 & b_2 & \cdots & b_s \end{array} . \tag{1.2}$$

Letting $A = (a_{\ell,j})_{\ell,j=1}^{s}$, $\mathbf{b} = (b_\ell)_{\ell=1}^{s}$ and $B = \mathrm{diag}(b_1, b_2, \ldots, b_s)$, the condition equivalent to symplecticity is

$$M := BA + A^T B - \mathbf{b}\mathbf{b}^T = O. \tag{1.3}$$

We note in passing that the familiar condition for algebraic stability of Runge–Kutta schemes is that M is non-negative definite and $\mathbf{b} \geq \mathbf{0}$ [10]. This is of coincidental significance only, since algebraic stability is all about dissipativity, whereas central feature of Hamiltonian systems is that they conserve, rather than dissipate, energy. It is an important point to grasp, since numerical analysts are more familiar with equations displaying trajectories that locally bunch together and tend to stable hyperbolic equilibria.

For any $s \geq 1$ there exists a unique Runge–Kutta method of the maximal order $2s$, the Gauss–Legendre scheme. It is not difficult to demonstrate that it obeys (1.3). Maximal order in tandem with symplecticity - on the face of it, an unbeatable combination! Unfortunately, further considerations place obstacles in the implementation of Gauss–Legendre schemes. It is important to discuss these obstacles in some detail, since this procedure hints at their possible remedies.

Firstly, we note that considerable precision is required in order to produce correct Hamiltonian dynamics. Recall that, for $d \geq 2$, the exact solution of (1.1) may display chaos. The numerical solution will typically produce chaotic behaviour as well, for all $d \geq 1$, not only modelling generic behaviour of the Hamiltonian system but also *as a numerical artifact*. The distinction between the 'numerical' and the 'exact' chaos is crucial in deriving the dynamics of (1.1) and it calls for numerical methods of high order.

Secondly, the dimension d is likely to be substantial for practical problems. In Hamiltonian dynamics each particle 'contributes' three units to d (hence six variables: three position and three momentum coordinates). If (1.1) originates in semi-discretized Hamiltonian partial differential equations (e.g. the sine-Gordon equation), d is likely to be very large indeed.

Thirdly, numerical symplecticity requires an exact (practically, up to machine accuracy) solution of underlying nonlinear algebraic equations by a Newton–Raphson-type method. Given that all s Runge–Kutta stages are coupled together, we obtain a nonlinear system of dimension sd. It is crucial to realise that the overwhelming portion of computational effort in any implementation of (1.3) is expanded on this task.

All the stages of a Gauss–Legendre scheme are coupled together, and it follows from the aforementioned three observations that its implementation (for 'good' values of s) is likely to be very expensive indeed. Moreover, it is quite clear that the computational bottleneck can be resolved by choosing, instead, a Runge–Kutta method with 'weaker' coupling among the stages.

It is well known to practitioners of Runge–Kutta methods that substantial simplification of nonlinear algebraic equations occurs when the spectrum of A is real, since then the stages can be uncoupled by linear transformations [3]. Unlike Gauss–Legendre (where, in fact, $2[s/2]$ eigenvalues of A are in $\mathcal{C} \setminus \mathcal{R}$), methods with $\sigma(A) \subset \mathcal{R}$ can be implemented with reasonable expansion of computational effort. This more than offsets the underlying reduction in order. The present paper is an initial foray into the design and analysis of symplectic Runge–Kutta methods with real spectra.

The linear stability function (i.e. the function $R(z)$ such that the numerical solution of $y' = \lambda y$, $y(0) = 1$, is $y_n = (R(h\lambda))^n$) of (1.3) is

$$R(z) = 1 + z\mathbf{b}^T(I - zA)^{-1}\mathbf{1},$$

where $\mathbf{1}^T = [1, \ldots, 1] \in \mathcal{R}^s$. Thus, it is a rational function with the denominator $\det(I - zA)$. Clearly, symplecticity implies that $|R(it)| \equiv 1$ for all $t \in \mathcal{R}$, otherwise it fails already for the linear Hamiltonian equation with $d = 1$, $H(p,q) = \frac{1}{2}(p^2 + q^2)$. Therefore, subject to symplecticity, we have

$$R(z) = \frac{\det(I + zA)}{\det(I - zA)}. \tag{1.4}$$

Section 2 is devoted to the investigation of functions (1.4) with real zeros and poles. By using the theory of order stars, we prove that for every odd $s \geq 1$ the largest possible number p such that $R(z) - \exp z = \mathcal{O}(z^{p+1})$ is $s + 1$. This is important, since the order of the Runge–Kutta method is at most p. Moreover, the optimal order can be obtained only if enough of the poles of R are distinct.

In Section 3 we exploit the theory of perturbed collocation of Nørsett and Wanner [13] to analyze in detail the cases $s = 2$ and $s = 3$, obtaining *inter alia* methods of order $s + 1$ with real spectra. We also lie the groundwork for the investigation of more general s-stage symplectic Runge–Kutta methods of order $s + 1$.

2. Rational approximants to $\exp z$

Let $R(z) = Q(-z)/Q(z)$, $Q(0) = 1$, be an s/s real rational function that approximates the exponential to order p, i.e.

$$R(z) = e^z + cz^{p+1} + \mathcal{O}(z^{p+1}), \quad c \neq 0, \quad z \to 0.$$

It can be easily observed that p must be even, since

$$Q(-z)e^{\frac{1}{2}z} - Q(z)e^{-\frac{1}{2}z}$$

is an even function. Moreover, note that $|R(it)| \equiv 1$ for all real t. We further assume that Q has only real zeros.

Following [7] and [9] (and adopting the notation in [9]), we let

$$\rho(z) := e^{-z}R(z), \quad z \in \mathcal{C},$$

and define the *order star* of $\{R, \exp z\}$ as the factorization $\{\mathcal{A}_-, \mathcal{A}_0, \mathcal{A}_+\}$ of C, where

$$\begin{aligned} \mathcal{A}_- &:= \{z \in C : |\rho(z)| < 1\}, \\ \mathcal{A}_0 &:= \{z \in C : |\rho(z)| = 1\}, \\ \mathcal{A}_+ &:= \{z \in C : |\rho(z)| > 1\}. \end{aligned}$$

The connected components of $\mathcal{A}_-$ and $\mathcal{A}_+$ are named $\mathcal{A}_-$-regions and $\mathcal{A}_+$-regions respectively. Note that the order star is symmetric with respect to the real axis, since the coefficients of Q are real. Moreover, it is anti-symmetric with respect to the imaginary axis:

$$z \in \mathcal{A}_\pm \quad \Leftrightarrow \quad -z \in \mathcal{A}_\mp. \tag{2.1}$$

The following propositions of the theory of order stars are pertinent to our argument. Their proofs are provided in [7] and [9], except that in the proof of Proposition 3 we trivially exploit the fact that $|R(it)| \equiv 1$.

Proposition 1 The origin lies on $\mathcal{A}_0$ and it is approached by $p+1$ sectors of $\mathcal{A}_-$ and $p+1$ sectors of $\mathcal{A}_+$, each with an asymptotic angle of $\pi/(p+1)$. □

Proposition 2 The number of zeros inside a bounded $\mathcal{A}_-$-region is identical to the number of interpolation points (that is, solutions of the equation $R(z) = \exp z$) along its directed boundary. Likewise, the number of poles inside an $\mathcal{A}_+$-region coincides with the number of interpolation points along its boundary. □

Proposition 3 There exist precisely one unbounded $\mathcal{A}_-$-region and one unbounded $\mathcal{A}_+$-region, $\mathcal{A}_-^\infty$ and $\mathcal{A}_+^\infty$, say. Moreover, for $|z| \gg 1$, $\mathrm{Re}\, z < 0$ implies that $z \in \mathcal{A}_+^\infty$, whereas $\mathrm{Re}\, z > 0$ means that $z \in \mathcal{A}_-^\infty$. □

Figure 1 displays the order star for

$$R(z) = \frac{1 + \frac{1}{2}z - \frac{239}{344}z^2 - \frac{803}{2064}z^3}{1 - \frac{1}{2} - \frac{239}{344}z^2 + \frac{803}{2064}z^3},$$

with $p = 4$ and poles at $1.5039524\ldots$, $1.4558135\ldots$ and $-1.1739627\ldots$ – all real, as required. This function will play a rôle in the next section. Examination of the figure (where zeros and poles are denoted by '□' and '◇' respectively) illuminates the statements of the aforementioned propositions.

It is known that the maximal order of any s/s real approximant with real poles is $s+1$ [7]. In the present case we have imposed a further requirement,

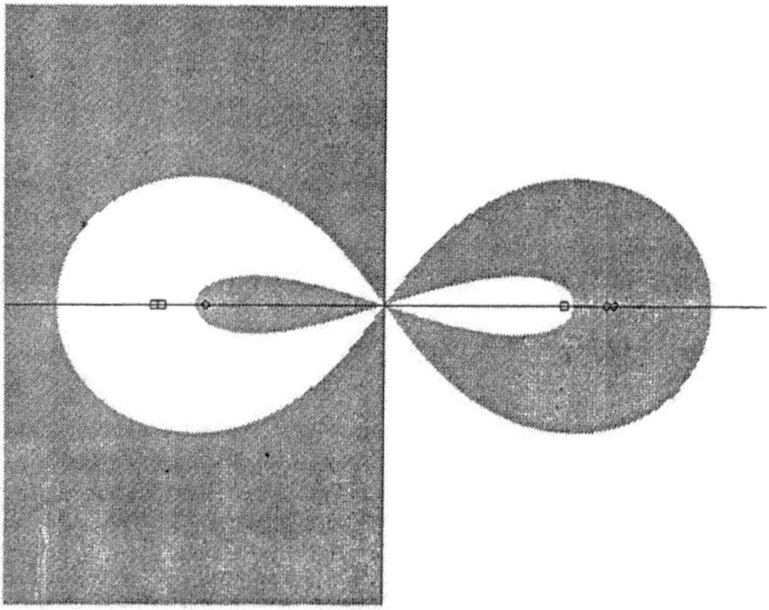

Figure 1 An order star.

namely that $R(z)$ is of the form $Q(-z)/Q(z)$. Consequently, we have a preliminary upper bound

$$p \le 2\left[\frac{s+1}{2}\right].$$

Herewith we restrict our attention to odd s and examine the conditions on Q that are necessary for order $s+1$. This procedure does not result in a loss of generality, since an even s follows as a degenerate case of $s-1$.

Because $i\mathcal{R} \subset \mathcal{A}_0$, the number of sectors of $\mathcal{A}_+$ approaching the origin from the left and from the right is well defined. We denote them by μ_- and μ_+ respectively. Note that, by virtue of Proposition 1,

$$\mu_- + \mu_+ = s+2, \quad |\mu_+ - \mu_-| \le 1. \tag{2.2}$$

s being odd, it follows at once from (2.2) and Proposition 2 that

$$\mu_- = \left[\frac{s+1}{2}\right] + 1, \quad \mu_+ = \left[\frac{s+1}{2}\right]. \tag{2.3}$$

Because of Propositions 2 and 3, at most two sectors of $\mathcal{A}_+$ to the left, and none to the right, might belong to $\mathcal{A}_+^\infty$: Clearly, there are no points of

$\mathcal{A}_+^\infty$ to the right of $i\mathcal{R}$. Suppose that more than two sectors of $\mathcal{A}_+$ to the left belong to $\mathcal{A}_+^\infty$. Then, by symmetry with respect to $\mathcal{R}$, they must enclose two bounded $\mathcal{A}_-$-regions that adjoin the origin and do not intersect with $\mathcal{R}$. It follows from Proposition 2 that these regions must contain zeros of R. This contradicts the fact that all the poles (and hence the zeros) are real.

It follows that s sectors of $\mathcal{A}_+$ at the origin belong to bounded $\mathcal{A}_+$-regions. Suppose that ω such sectors belong to a single $\mathcal{A}_+$-region $\mathcal{U}$, say. Then, by Proposition 2, there must be at least ω poles in $\mathcal{U}$. Moreover, $\mathcal{U}$ must envelop $\omega - 1$ sectors of $\mathcal{A}_-$ that lie in bounded $\mathcal{A}_-$-regions. Using again Proposition 2, it follows that at least $\omega - 1$ zeros are enclosed by $\mathcal{U}$. By symmetry of the order star, at least $\omega - 2$ zeros lie away from the real axis. However, this is forbidden, hence necessarily $\omega \leq 2$. Exploiting again the symmetry of the order star, it follows that *either* $\omega = 1$ and $\mathcal{U}$ adjoins the origin via the real axis *or* $\omega = 2$ and the two sectors adjoin the origin from within $\operatorname{Im} z > 0$ and $\operatorname{Im} z < 0$ (cf. Figure 1). Since $\mu_- - \mu_+ = s + 2$, an odd number, clearly there is exactly one $\mathcal{A}_+$-region with $\omega = 1$ and $[s/2]$ $\mathcal{A}_+$-regions with $\omega = 2$. In particular, it is true that there must be at least $[(s+1)/2]$ distinct poles.

Theorem 4 Suppose that the rational function $R(z) = Q(-z)/Q(z)$, where $\deg Q = s$, approximates the exponential of order $s + 1$ and Q has only real zeros. Then s is odd, exactly $[s/2]$ of the zeros reside in the left half-plane and $[(s+1)/2]$ of them are in the right half-plane. Moreover, there are at least $[(s+1)/2]$ distinct zeros, of which at least $[(s+1)/4]$ lie to the left and $[(s+3)/4]$ to the right of $i\mathcal{R}$.

Proof. The first assertion follows at once from (2.3) and Proposition 2: at most two sectors of $\mathcal{A}_+$ to the left belong to $\mathcal{A}_+^\infty$, hence we need at least $\mu_- - 2 = [s/2]$ poles there to 'support' the sectors from bounded $\mathcal{A}_+$-regions. Likewise, we require at least $\mu_+ = [(s+1)/2]$ poles to the right. Since exactly s poles are available, it follows that all the 'at least' and 'at most' statements are fulfilled as equalities. In particular, there are precisely $[s/2]$ poles to the left and $[(s+1)/2]$ poles to the right of the imaginary axis.

We have already proved that there are at least $[(s+1)/2]$ distinct poles. If $s \bmod 4 = 1$ then the unique $\mathcal{A}_+$-region with $\omega = 1$ lies to the right, whereas in the remaining case, $s \bmod 4 = 3$, it is to the left of $i\mathcal{R}$. The final assertion of the theorem follows by straightforward counting. □

An immediate consequence of Theorem 4 is that, unless $s = 1$, R cannot be A-stable, since it possesses poles in the left half-plane. This is not detrimental at the least to the solution of Hamiltonian equations! Moreover, suppose that a symplectic Runge–Kutta method admits R as its linear stability function. Then it follows that some of the weights b_ℓ must be negative.

For suppose that $b_\ell \geq 0$, $\ell = 1, \ldots, s$. This, in tandem with $M = O$, implies algebraic stability [10]. Algebraic stability, in turn, implies A-stability and a contradiction ensues.

More important implication of the theorem has to do with the number of distinct zeros. The most popular Runge–Kutta methods with real spectra are the *singly-diagonally implicit* (SDIRK) schemes, originally due to Crouzeix and Nørsett, and the *singly-diagonally implicit* (SIRK) methods of Burrage. In both cases the spectrum consists of a single eigenvalue of multiplicity s (for SDIRK, in addition, A is lower-triangular). This is advantageous in specific circumstances. In particular, if Gaussian elimination is used in the nonlinear algebraic solver then only a single LU factorization is necessary *per* time-step. However, this becomes far less significant for very large dimensions d, since then the method of choice in solving linear systems is iterative and no special advantage accrues from the SDIRK or SIRK structure, as opposed to general methods with real spectra.

3. Perturbed collocation and symplectic Runge–Kutta

Order conditions for symplectic Runge–Kutta methods were derived by Sanz-Serna and Abia [17], implementing Butcher's approach that identifies elementary differentials with rooted trees. In the present paper we adopt a different approach, that of *perturbed collocation*, as introduced in the pioneering work of Nørsett and Wanner [13]. First, however, we derive some critical information by investigating order conditions due to Butcher [2]. Given a method (1.2), we say that it is $B(m)$ if

$$\mathbf{b}^T \mathbf{c}_{k-1} = \frac{1}{k}, \quad k = 1, \ldots, m$$

and $D(n)$ if

$$A\mathbf{c}_{k-1} = \frac{1}{k}\mathbf{c}_k, \quad k = 1, \ldots, n.$$

Here

$$\mathbf{c}_k := \begin{bmatrix} c_1^k \\ c_2^k \\ \vdots \\ c_s^k \end{bmatrix}, \quad k = 0, 1, \ldots.$$

Recall that the condition for symplecticity is $M = O$, where

$$M = BA + A^T B - \mathbf{b}^T \mathbf{b}.$$

Let

$$\mu_{k,\ell} := \mathbf{c}_{k-1}^T M \mathbf{c}_{\ell-1}, \quad k, \ell = 1, 2, \ldots.$$

Clearly,

$$M = O \quad \Leftrightarrow \quad \mu_{k,\ell} = 0, \quad k, \ell = 1, \ldots, s.$$

Let

$$\mathbf{g}_k := \begin{bmatrix} b_1 c_1^k \\ b_2 c_2^k \\ \vdots \\ b_s c_s^k \end{bmatrix}, \quad k = 0, 1, \ldots.$$

It now follows that $M = O$ if and only if

$$\mathbf{g}_{k-1}^T A \mathbf{c}_{\ell-1} + \mathbf{g}_{\ell-1}^T A \mathbf{c}_{k-1} = \left(\mathbf{b}^T \mathbf{c}_{k-1}\right)\left(\mathbf{b}^T \mathbf{c}_{\ell-1}\right), \quad k, \ell = 1, \ldots, s. \tag{3.1}$$

Given that the method is $D(n)$, we have

$$\mathbf{g}_{\ell-1}^T A \mathbf{c}_{k-1} = \frac{1}{k} \mathbf{g}_{\ell-1}^T \mathbf{c}_k = \frac{1}{k} \mathbf{b}^T \mathbf{c}_{k+\ell-1}, \quad k = 1, \ldots, n.$$

Thus, according to (3.1), $\mu_{k,\ell} = 0$, $k, \ell = 1, \ldots, n$, is equivalent to

$$\frac{k+\ell}{k\ell} \mathbf{b}^T \mathbf{c}_{k+\ell-1} = \left(\mathbf{b}^T \mathbf{c}_{k-1}\right)\left(\mathbf{b}^T \mathbf{c}_{\ell-1}\right). \tag{3.2}$$

Let us suppose that the weights are chosen consistently with the underlying quadrature formula, i.e. so as to have $B(s)$. This is the natural way of choosing weights and the loss of generality is miniscule. Then (3.2) yields

$$\mathbf{b}^T \mathbf{c}_{k+\ell-1} = \frac{1}{k+\ell}, \quad k, \ell = 1, \ldots, n.$$

We can deduce straightforwardly:

Proposition 5 If $M = O$, $B(s)$ and $C(n)$ then $B(2n)$. □

Suppose that s distinct numbers $c_1, \ldots, c_s$ and s polynomials

$$p_\ell(t) := \sum_{j=0}^{s-1} p_{j,\ell} t^j, \quad \ell = 1, \ldots, s,$$

are given. We set

$$q_\ell(t) := \frac{1}{\ell!}\left(p_\ell(t) - t^\ell\right), \quad \ell = 1, \ldots, s,$$

and, following Nørsett and Wanner [13], define a *perturbed collocation* scheme for $\mathbf{y}' = \mathbf{f}(t, \mathbf{y})$ as

$$\begin{aligned} &\mathbf{u_n} = \mathbf{y}_n, \quad \mathbf{u}_n \text{ is } s\text{th degree polynomial in } t, \\ &\mathbf{u}_n'(t_n + c_\ell h) = \mathbf{f}(t_n + c_\ell h, \{\mathbf{P}(t_n, \mathbf{u}_n)\}(t_n + c_\ell h)), \quad \ell = 1, \ldots, s, \\ &\mathbf{y}_{n+1} = \mathbf{u}(t_n + h), \end{aligned} \tag{3.3}$$

where the perturbation operator $\mathbf{P}$ is defined by

$$\{\mathbf{P}(\tau,\mathbf{v})\}(t) = \mathbf{v}(t) + \sum_{j=1}^{s} q_j\left(\frac{t-\tau}{h}\right)\mathbf{v}^{(j)}(\tau)h^j.$$

The importance of (3.3) is emphasised in two important theorems of Nørsett and Wanner:

Theorem 6 [13] The perturbed collocation method (3.3) is identical to the Runge–Kutta scheme

$$\begin{array}{c|c} \mathbf{c} & A \\ \hline & \mathbf{b}^T \end{array}, \tag{3.4}$$

where $A = VPJV^{-1}$, $\mathbf{b}^T = \mathbf{1}^T JV^{-1}$ and

$$V = \begin{bmatrix} 1 & c_1 & \cdots & c_1^{s-1} \\ 1 & c_2 & \cdots & c_2^{s-1} \\ \vdots & \vdots & & \vdots \\ 1 & c_s & \cdots & c_s^{s-1} \end{bmatrix}, \quad P = \begin{bmatrix} p_{0,1} & p_{0,2} & \cdots & p_{0,s} \\ p_{1,1} & p_{1,2} & \cdots & p_{1,s} \\ \vdots & \vdots & & \vdots \\ p_{s-1,1} & p_{s-1,2} & \cdots & p_{s-1,s} \end{bmatrix},$$

$$J = \operatorname{diag}\left(1, \frac{1}{2}, \ldots, \frac{1}{s}\right).$$

□

Theorem 7 [13] Suppose that each q_ℓ is of degree at most ℓ for all $\ell = 1,\ldots,s$, that

$$\int_0^1 \tau^j \omega(\tau)\mathrm{d}\tau = 0, \quad j = 0,\ldots,r-1$$

where

$$\omega(t) := \prod_{\ell=1}^{s}(t-c_\ell)$$

and that

$$\int_0^1 \tau^j q_\ell(\tau)\mathrm{d}\tau = 0, \quad j = 0,\ldots,s+r-\ell-1,\ \ell = 1,\ldots,m.$$

Then (3.3) (and hence the corresponding Runge–Kutta method (3.4)) is of order $s+r$. □

Not every Runge–Kutta method corresponds to perturbed collocation, in the sense of Theorem 6. However, almost all methods of interest do and the loss of generality in considering perturbed collocation instead of Runge–Kutta is marginal. More importantly, the order conditions of Theorem 7 are sufficient, but by no means necessary. Note that all the known families (due to Butcher, Ehle and Chipman) of s-stage Runge–Kutta methods of order at least $2s-2$ fit into these conditions. However, there is here a loss of generality, which we acknowledge and willingly trade off for the better analytic tools available in the perturbed collocation formulation.

It is clear from the conditions of Theorem 7 that, for nontrivial r, some q_ℓs must vanish identically. We make this explicit by saying that (3.3) (or – and this should cause no confusion – (3.4)) is $\mathcal{P}_n$ for some $n \in \{0,1,\ldots,s-1\}$ if $q_\ell \equiv 0$ for $\ell = 1,2,\ldots,s-r$. The special case of $\mathcal{P}_1$ where, up to rescaling, $q_s \equiv \omega$, is denoted by $\mathcal{P}_0$ and it corresponds to (unperturbed) collocation (that is, to $\mathbf{P}$ being the identity operator).

Let us examine $\mu_{k,\ell}$ for $\mathcal{P}_n$. Clearly

$$\begin{aligned} A\mathbf{c}_{\ell-1} &= VPJV^{-1}\mathbf{c}_{\ell-1} = VPJ\mathbf{e}_\ell = \frac{1}{\ell}VP\mathbf{e}_\ell \\ &= \frac{1}{\ell}V\begin{bmatrix} p_{0,\ell} \\ p_{1,\ell} \\ \vdots \\ p_{s-1,\ell}\end{bmatrix} = \frac{1}{\ell}\begin{bmatrix} p_\ell(c_1) \\ p_\ell(c_2) \\ \vdots \\ p_\ell(c_s)\end{bmatrix} := \frac{1}{\ell}p_\ell(\mathbf{c}). \end{aligned}$$

Here $\mathbf{e}_\ell$ is the ℓth unit vector. Note that $q_\ell \equiv 0$ implies $p_\ell(t) = t^\ell$, thus $A\mathbf{c}_{\ell-1} = \frac{1}{\ell}\mathbf{c}_\ell$. In particular, $\mathcal{P}_n$ implies $C(s-n)$ for $r = 1,\ldots,s-1$. Moreover, $\mathcal{P}_0$ implies $C(s)$, since then $q_s(\mathbf{c}) = \omega(\mathbf{c}) = \mathbf{0}$.

Theorem 8 The only symplectic s-stage Runge–Kutta method in $\mathcal{P}_0$ is the $(2s)$-order Gauss–Legendre scheme.

Proof. We have already seen that symplecticity in $\mathcal{P}_0$ implies $C(s)$. Thus, by Proposition 5, $B(2s)$ and the underlying quadrature is of the maximal order $2s$. This is attainable if and only if ω is the sth Legendre polynomial, shifted to $[0,1]$, which yields precisely the Gauss-Legendre method. □

$\mathcal{P}_0$ leads us precisely in the direction that we were so reluctant to follow in Section 1. This motivates the analysis of methods in $\mathcal{P}_1$. We commence by explicit evaluation of linear stability functions of $\mathcal{P}_1$ methods (without assuming for the time being that the method is symplectic). Let

$$E_{m,s} := \begin{bmatrix} 0 & 0 & \cdots & 0 & \delta_m \\ \gamma_{m+1} & 0 & \cdots & 0 & \delta_{m+1} \\ 0 & \gamma_{m+2} & \ddots & \vdots & \delta_{m+2} \\ \vdots & \ddots & \ddots & 0 & \vdots \\ 0 & \cdots & 0 & \gamma_{s-1} & \delta_{s-1} \end{bmatrix},$$

where $\{\gamma_j\}$, $\{\delta_j\}$ are arbitrary constants. Set $e_{m,s}(\lambda) := \det(E_{m,s} - \lambda I)$. Expanding in the top row, it follows at once that

$$e_{m,s}(\lambda) = -\lambda e_{m+1,s}(\lambda) + (-1)^{s+m-1}\left(\prod_{j=m+1}^{s-1}\gamma_j\right)\delta_m.$$

Thus, and since

$$e_{s-1,s}(\lambda) = -\lambda + \delta_m,$$

straightforward induction affirms that

$$e_s(\lambda) := e_{0,s}(\lambda) = (-1)^s\left\{\lambda^s - \left(\prod_{j=0}^{s-1}\gamma_j\right)\sum_{\ell=0}^{s-1}\frac{\delta_\ell}{\prod_{j=0}^{\ell}\gamma_j}\lambda^\ell\right\},$$

where $\gamma_0 := 1$.

Let $R(z) := P(z)/Q(z)$ be the stability function of the method. Then $Q(z) = \det(I - zA)$. In our case

$$Q(z) = \det\left(I - zVPJV^{-1}\right) = \det(I - zPJ) = (-1)^s z^s e_s(z^{-1}),$$

where $\gamma_j = 1/j$, $j = 1,\ldots,s-1$, and $\delta_j = p_{s,j}$, $j = 0,\ldots,s-1$. It follows that

$$Q(z) = \frac{1}{s!}\sum_{\ell=0}^{s} q_s^{(s-\ell)}(0)z^\ell. \tag{3.5}$$

The representation (3.5) is what Nørsett called the *C-polynomial form* [13]. In particular, (3.5) and order $\geq s$ imply that

$$P(z) = \frac{1}{s!}\sum_{\ell=0}^{s} q_s^{(s-\ell)}(1)z^\ell.$$

$M = O$ implies that $P(z) = Q(-z)$. Consequently

$$q_s^{(\ell)}(0) = (-1)^\ell q_s^{(\ell)}(1), \quad \ell = 0,\ldots,s,$$

and q_s is necessarily either symmetric or anti-symmetric with respect to $t = \frac{1}{2}$, depending on the parity of s. In other words, the expansion of q_s in Legendre polynomials $\{\hat{P}_n\}$ (shifted to $[0,1]$) features only $\hat{P}_{s-2j}$, $j = 0,1,\ldots,[s/2]$.

Since the method is $\mathcal{P}_1$ and symplectic, we have $C(s-1)$, hence, by virtue of Proposition 5, $B(2s-2)$. Consequently, $\mu_{k,\ell} = 0$ for $k,\ell = 1,\ldots,s-1$

and, by symmetry, it remains to verify that $\mu_{s,\ell} = 0$ for $\ell = 1, \ldots, s$. (3.1) yields

$$\begin{aligned} \mu_{s,\ell} &= \frac{1}{s}\mathbf{g}_{\ell-1}^T p_s(\mathbf{c}) + \frac{1}{\ell}\mathbf{b}^T\mathbf{c}_{s+\ell-1} - \frac{1}{s\ell}, \quad \ell = 1, \ldots, s-1, \\ \mu_{s,s} &= \frac{2}{s}\mathbf{g}_{s-1}^T p_s(\mathbf{c}) - \frac{1}{s^2}, \end{aligned}$$

where we use the convention that

$$v(t) := \sum_{j=0}^{n} v_j t^j \quad \Leftarrow \quad v(\mathbf{c}) = \sum_{j=0}^{n} v_j \mathbf{c}_j.$$

The substitution $p_s(\mathbf{c}) = \mathbf{c}_s + s! q_s(\mathbf{c})$ results in

$$\begin{aligned} \mathbf{g}_{\ell-1}^T q_s(\mathbf{c}) &= \frac{1}{s!\ell}\left(1 - (s+\ell)\mathbf{b}^T\mathbf{c}_{s+\ell-1}\right) := \alpha_\ell, \quad \ell = 1, \ldots, s-1, \\ \mathbf{g}_{s-1}^T q_s(\mathbf{c}) &= \frac{1}{2s!s}\left(1 - 2s\mathbf{b}^T\mathbf{c}_{2s-1}\right) := \alpha_s, \end{aligned}$$

a system of s equations in the unknowns $q_s(c_1), \ldots, q_s(c_s)$. Since we know that the leading coefficient of q_s is $-1/s!$, these s values are sufficient to determine q_s uniquely. Note that $B(2s-2)$ implies that $\alpha_\ell = 0$, $\ell = 1, \ldots, s-2$. Thus, letting $d_j := b_j q_s(c_j)$, $j = 1, \ldots, s$, we have

$$V^T\mathbf{d} = \alpha_{s-1}\mathbf{e}_{s-1} + \alpha_s\mathbf{e}_s. \tag{3.6}$$

We can easily write down the explicit solution of (3.6). Let

$$\omega_\ell(t) := \frac{\omega(t)}{(t - c_\ell)} = \sum_{j=0}^{s-1} \nu_{\ell,j} t^j, \quad \ell = 1, \ldots, s.$$

Then

$$\left(V^{-1}\right)_{k,\ell} = \frac{\nu_{\ell,k-1}}{\omega_\ell(c_\ell)}, \quad k, \ell = 1, \ldots, s$$

(cf. [8]), and it follows that

$$d_\ell = \frac{1}{\omega_\ell(c_\ell)}\left\{\alpha_s - \left(\sum_{j=1}^{s} c_j - c_\ell\right)\alpha_{s-1}\right\}, \quad \ell = 1, \ldots, s$$

and

$$b_\ell = \frac{1}{\omega_\ell(c_\ell)}\int_0^1 \omega_\ell(t)\mathrm{d}t, \quad \ell = 1, \ldots, s.$$

This finally produces

$$q_s(c_\ell) = \frac{1}{\int_0^1 \omega_\ell(t)\mathrm{d}t}\left\{\alpha_s - \left(\sum_{j=1}^{s} c_j - c_\ell\right)\alpha_{s-1}\right\}, \quad \ell = 1, \ldots, s. \tag{3.7}$$

Given $\mathbf{c}$ and subject to $B(2s-2)$, we can use (3.7) to derive the correcponding *unique* symplectic method from $\mathcal{P}_1$. In general, its linear stability function possesses complex conjugate pairs of zeros. However, frequently we can choose our two remaining degrees of freedom so as to force real zeros.

We hope to provide an explicit presentation of all symplectic $\mathcal{P}_1$ methods whose stability function has only real zeros in a future paper. In the remainder of the present article we consider just the two simplest cases, $s=2$ and $s=3$.

When there are just two stages we explicitly evaluate

$$\begin{aligned} d_1 &= \frac{1-2c_2}{8(c_2-c_1)}(2c_2-4c_1c_2+2c_1^2-1), \\ d_2 &= \frac{1-2c_1}{8(c_2-c_1)}(1-2c_1+4c_1c_2-2c_2^2) \end{aligned}$$

and

$$b_1 = \frac{2c_2-1}{2(c_2-c_1)}, \quad b_2 = \frac{1-2c_1}{2(c_2-c_1)}.$$

Consequently

$$\begin{aligned} q_2(c_1) &= \frac{1}{4}(1-2c_2+4c_1c_2-2c_1^2), \\ q_2(c_2) &= \frac{1}{4}(1-2c_1+4c_1c_2-2c_2^2). \end{aligned}$$

and easy calculation affirms that

$$\begin{aligned} q_2(t) &= \frac{1}{4}(1-2c_1)(1-2c_2)+\frac{1}{2}t-\frac{1}{2}t^2 \\ &= -\frac{1}{12}\hat{P}_2(t)+\frac{1}{6}(2-3(c_1+c_2)+6c_1c_2)\hat{P}_0(t). \end{aligned}$$

The denominator of R is

$$Q(z) = 1-\frac{1}{2}z-\frac{1}{4}(1-2c_1)(1-2c_2)z^2$$

and it has real zeros if either $c_1, c_2 < \frac{1}{2}$ or $c_1, c_2 > \frac{1}{2}$. One of the zeros is negative and one positive, as can be deduced from the analysis of Section 2.

The case $s=3$ is more interesting. Since $B(4)$, we need to express one of the c_ℓs in terms of the remaining two. Short calculation verifies that, without loss of generality,

$$c_1 = \frac{3-4(c_2+c_3)+6c_2c_3}{4-6(c_2+c_3)+12c_2c_3}.$$

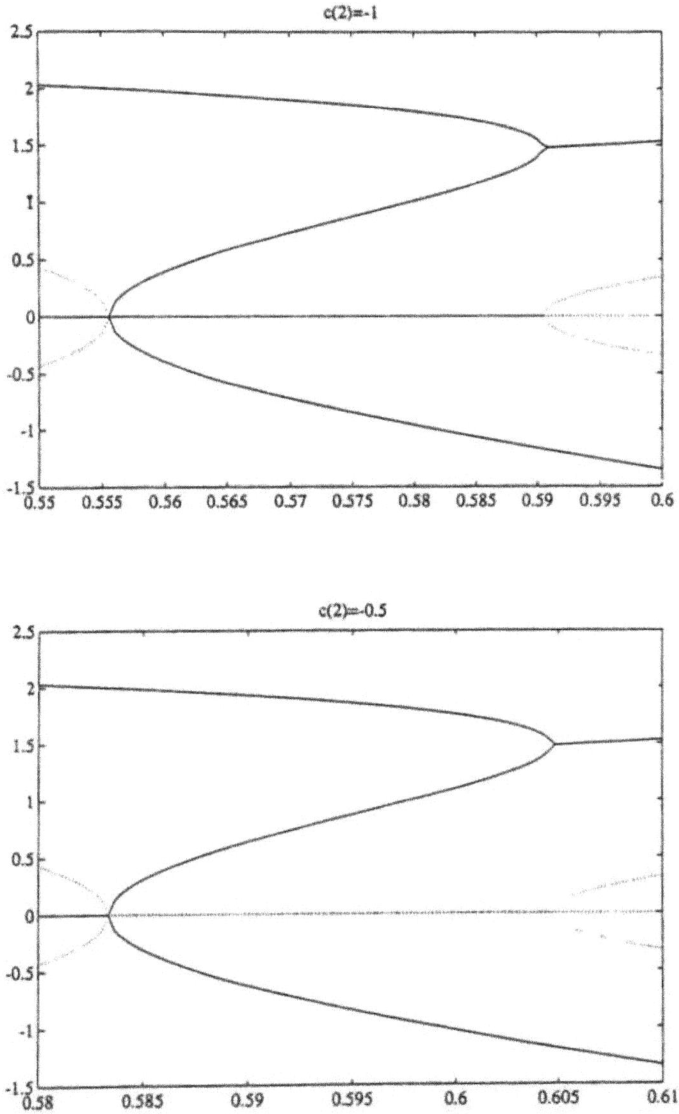

Figure 2 Bifurcation diagrams for c_3.

It is possible subsequently to evaluate α_2, α_3, the weights *etc.*, culminating in an explicit form of q_3, and check whether the zeros of Q are real. We have performed a search for real zeros on a computer: setting c_2, we plot the bifurcation diagram of zeros of Q as a function of c_3. Figure 2 presents two such diagrams, for $c_2 = -\frac{1}{2}$ and $c_2 = -1$ respectively, for a 'relevant' range of c_3. The real part of zeros is denoted by a solid line, whereas the dotted line stands for the imaginary part. Clearly, the interval where the dotted line lies on the real axis corresponds to the regime of real zeros. We selected two methods from the aforementioned regimes, both with relatively 'nice' rational coefficients:

$$\begin{array}{c|ccc}
-3 & -\frac{11}{9300} & -\frac{16279}{2850} & \frac{20815}{7672} \\
-\frac{1}{2} & -\frac{2497}{18600} & \frac{73}{1425} & -\frac{2102}{5041} \\
\frac{26}{43} & \frac{56}{11749} & \frac{536}{3575} & \frac{6460}{14357} \\
\hline
 & -\frac{11}{4650} & \frac{146}{1425} & \frac{12920}{14357}
\end{array},$$

$$Q(z) = 1 - \frac{1}{2}z - \frac{239}{344}z^2 + \frac{803}{2064}z^3,$$

and

$$\begin{array}{c|ccc}
-\frac{7}{4} & -\frac{104}{6993} & -\frac{5497}{3456} & -\frac{1297}{8972} \\
-1 & -\frac{3497}{6993} & \frac{239}{4752} & -\frac{6042}{10981} \\
\frac{49}{83} & -\frac{32}{931} & \frac{1234}{7705} & \frac{2446}{5265} \\
\hline
 & -\frac{208}{6993} & \frac{239}{2376} & \frac{4892}{5265}
\end{array},$$

$$Q(z) = 1 - \frac{1}{2}z - \frac{925}{1328}z^2 + \frac{3107}{7968}z^3.$$

Two abscissae of both methods are negative. This would have been highly detrimental – or, at the very least, unusual – for 'standard' ordinary differential equations, but is quite harmless for Hamiltonian equations. Note that the order star corresponding to the first method has been displayed in Figure 1.

To recap, both methods are fourth-order and symplectic and they can be implemented considerably cheaper than the two-stage fourth-order Gauss-Legendre method. Moreover, we believe that the approach of this paper may well lead to higher-order symplectic methods (for $s \geq 4$) with similarly helpful features.

References:

[1] V.I. Arnold (1978), *Mathematical Methods of Classical Mechanics*, Springer-Verlag, New York.

[2] J.C. Butcher (1964), "Implicit Runge–Kutta processes", *Math. Comp.* **18**, 50–64.

[3] J.C. Butcher (1988), "Towards efficient implementation of singly-implicit methods", *ACM Trans. on Math. Software* **14**, 68–75.

[4] Feng Kang (1986), "Difference schemes for Hamiltonian formalism and symplectic geometry", *J. Comput. Math.* **4**, 279–289.

[5] J. Guckenheimer and P. Holmes (1983), *Nonlinear Oscillations, Dynamical Systems, and Bifurcations of Vector Fields*, Springer-Verlag, New York.

[6] E. Hairer, S.P. Nørsett and G. Wanner (1987), *Solving Ordinary Differential equations I: Nonstiff Problems*, Springer-Verlag, Berlin.

[7] A. Iserles and S.P. Nørsett (1989), "Order stars and rational approximants to exp(z)", *Appl. Num. Math.* **5**, 63–70.

[8] A. Iserles and S.P. Nørsett (1990), "On the theory of parallel Runge–Kutta methods", *IMA J. Num. Anal.* **10**, 463–488.

[9] A. Iserles and S.P. Nørsett (1991), *Order Stars: Theory and Applications*, Chapman and Hall, London, to appear.

[10] J.D. Lambert (1987), "Developments in stability theory for ordinary differential equations", in *The State of the Art in Numerical Analysis* (A. Iserles and M.J.D. Powell, eds), Oxford University Press, Oxford.

[11] F. Lasagni (1988), "Canonical Runge–Kutta methods", *ZAMP* **39**, 952–953.

[12] R.S. MacKay (1991), "Some aspects of the dynamics and numerics of Hamiltonian systems", to appear in *Numerics of Dynamics and Dynamics of Numerics, Bristol 1990* (D.S. Broomhead and. A. Iserles, eds), Oxford University Press, Oxford.

[13] S.P. Nørsett and G. Wanner (1981), "Perturbed collocation", *Numer. Math.* **38**, 193–208.

[14] R. Ruth (1983), "A canonical integration technique", *IEEE Trans. Nucl. Sci.* **30**, 2669–2671.

[15] J.M. Sanz-Serna (1988), "Runge–Kutta schemes for Hamiltonian systems", *BIT* **28**, 877–883.

[16] J.M. Sanz-Serna (1990), "The numerical integration of Hamiltonian systems", to appear in *Computational Differential Equations, London 1989*

(J. Cash and I. Gladwell, eds), Oxford University Press, Oxford.

[17] J.M. Sanz-Serna and L. Abia (1990), "Order conditions for canonical Runge–Kutta schemes", to appear in *SIAM J. Num. Anal.*

[18] Y.B. Suris (1989), "Canonical transformations generated by methods of Runge–Kutta type for the numerical integration of the system $x'' = -\partial U/\partial x$", *Zh. Vychisl. Mat. i Mat. Fiz.* **29**, 202–211 (in Russian)

EUCLID'S ALGORITHM AND ACCELERATED ITERATIVE METHODS

ELIAS A. LIPITAKIS

Department of Informatics
Athens University of Economics
Athens, Greece

Abstract

The well-known Euclid's algorithm is presented as a general iterative method (the E-algorithm) for solving a wide class of linear problems and its relationship to known accelerated implicit and explicit iterative schemes as Richardson, Chebychev and Conjugate Gradient-type methods is given. Some computational work estimates of the E-algorithm are also presented.

1. INTRODUCTION

Many problems of Mathematical Physics and Engineering are related to (special cases of) the second-order linear parabolic Partial Differential Equation (PDE):

$$\varphi(x)\frac{\theta U(x,t)}{\theta t} = \sum_{i=1}^{n}\frac{\theta}{\theta x_i}\left(K_i(x)\frac{\theta U(x,t)}{\theta x_i}\right) + \sum_{i=1}^{n} G_i(x)\frac{\theta U(x,t)}{\theta x_i} -$$

$$- \sigma(x)U(x,t) + S(x,t), \qquad x\in R,\ t>0, \qquad (1.1)$$

where R is a given finite connected region in Euclidean n-dimensional space, with boundary conditions

$$a(x)\cdot U(x,t) + b(x)\cdot\frac{\theta U(x,t)}{\theta n}\quad \gamma(x), \qquad x\in C\ ,\quad t>0\ , \qquad (1.1a)$$

where C is the external boundary of R.

In recent years several iterative methods have been introduced for solving the semi-discrete analogue of equation (1.1), viz.

$$Au = b\ , \qquad (1.2)$$

where A is a large sparse real unsymmetric (nxn)-matrix, b is a given (nx1)-column vector and u is an (nx1)-column vector to be determined. Some of these methods have attractive features of good convergence rates as well as reasonable storage requirements.

The main objective of this article is based on the search for a common origin of a wide class of accelerated iterative methods and the existence of a general accelerated iterative scheme, representing a large category of iterative methods, which can be efficiently used for the numerical solution of initial/boundary-value problems. In Varga's article [21] the author uses Pade approximation, i.e. a particular type of rational approximation of exp (-tA), to approximate the exponential function and then using the fact that the exponential of a matrix is the solution of a semi-discrete approximation to a parabolic PDE, he shows that various forms of Pade approximation are no more than the numerical methods for solving the PDE, i.e. the Crank-Nicholson method. It is also stated that not all known methods for approximating solution of parabolic PDE's can be obtained from Pade approximations, i.e. multilevel approximations.

In this article we develop a generalized form of Euclid's algorithm, i.e. the E-algorithm, and then show that various accelerated iterative methods for solving large sparse linear systems of algebraic equations can be considered as special cases of the E-algorithm for specific choices of the parameters in question. The E-algorithm can be regarded in terms of discrete approximations to PDE's and adapted for use in approximating the solution of initial/boundary-value problems.

2. THE E-ALGORITHM

Euclid's algorithm is considered to be the oldest-known documented algorithm. In its original version [3], it appears as a technique for determining the highest common factor (or greatest common divisor) of two positive integers. It is also known [3] that Euclid's only explicit reasons for introducing the above algorithmic process are to determine the greatest common divisor and to use a criterion for incommensurability which then is never used explicitly. A different approach of heuristic character can be derived using the principles of "anthyphairesis" (operation of reciprocal subtractions) and "anthyphairetic ratios theory" [3]. The algorithm then can be equivalently stated as follows: Let us consider two given positive integer numbers A and B (assuming without any loss of generality that A > B) and the "anthyphairetic" ratios $Q_1, Q_2, \dots, Q_N$, where the ratios Q_i, $i=1(1)N$, are defined from the following division relationships:

$$A = BQ_1 + R_1 \ ; \quad B = R_1 Q_2 + R_2 \tag{2.1}$$

and

$$R_i = R_{i+1} Q_{i+2} + R_{i+2}, \quad i \geq 2 \ . \tag{2.2}$$

The algorithm can be equivalently stated in the following "pseudo-code" form:

Euclid's Algorithm

Purpose: $A/B = (Q_1, Q_2, Q_3, \ldots, Q_N)$,

Input: positive integer numbers A and B

Output: "anthyphairetic" ratios $Q_1, Q_2, Q_3, \ldots, Q_N$.

step 1: read the given positive integer number A,B

step 2: set i=1

step 3: if A>B then set $A = BQ_1 + R_1$; $B = R_1Q_2 + R_2$

step 4: while i≤N repeat steps 5-7

step 5: $R_i = R_{i+1}\, Q_{i+2} + R_{i+2}$

step 6: if $R_{i+2} = 0$ then

OUTPUT $(Q_1, Q_2, \ldots, Q_{N+2})$

STOP

step 7: set i=i+1

step 8: OUTPUT ("Method failed after N-iterations, N=",N)

STOP

The relationships (2.1), (2.2) can then be equivalently written as:

$$R_i = R_{i+1}\, Q_{i+2} + R_{i+2}\,, \qquad i \geq -1\,, \tag{2.3}$$

where

$$R_{-1} \equiv A \quad \text{and} \quad R_0 \equiv B, \tag{2.4}$$

while the original problem can be stated on

$$R_i/R_{i+1} = (Q_{i+2}, Q_{i+3}, Q_{i+4}, \ldots, Q_N),\ i \geq -1, \tag{2.5}$$

with R_i, R_{i+1} the given positive integer numbers and Q_j, $j=i+2, \ldots, N$ the derived "anthyphairetic" ratios. Typical examples are the ratios 60/26 and 118/27, i.e.

60 = 26.2 + 8		118 = 27.4 + 10	
26 = 8.3 + 2	and	27 = 10.2 + 7	
8 = 2.4 + 0		10 7.1 + 3	
		7 = 3.2 + 1	
		3 = 1.3 + 0	(2.6)

which correspond respectively to the following "anthyphairetic" schemes:

$$60/26 = (2,3,4) \quad \text{and} \quad 118/27 = (4,2,1,2,3)\ . \tag{2.6a}$$

The recursive equations (2.3), (2.4) can be written in the following parametrized iterative form (henceforth called the E-algorithm):

$$R_{i+1} = \pi_i R_i + \tau_i R_{i-1} + \delta_i, \quad i \geq 0, \tag{2.7}$$

where π_i, τ_i are sequences of pre-conditioned iterative parameters and δ_i is a "correction"-term. Throughout this paper we will refer to the parameters π_i, τ_i, δ_i of the E-algorithm (2.7) as the "E-parameters".

The purpose in introducing the E-algorithm is twofold. Firstly, it can be clearly seen, cf.(section 3), that the iterative schemes for solving large sparse linear systems of algebraic equations are special cases of the E-algorithm. Secondly, the E-algorithm can generate many other useful implicit and explicit iterative schemes of higher-order (with suitable selection of the E-parameters) for solving a wide class of large sparse linear systems. It should be noted that such higher-order iterative schemes have not as yet been used in practical applications. However, it is envisaged that they may be of use on multiprocessor systems, where higher-order implicit and explicit iterative methods can be efficiently used.

The association of the E-algorithm with the classical analysis topic of iterative methods gives a powerful tool for the convergence analysis of this algorithmic scheme, which might lead to more rapidly convergent iterative methods.

3. E-ALGORITHM AND IMPLICIT PRECONDITIONED METHODS

In this Section we establish a correspondence between the E-algorithm and the best known and most widely used iterative methods, following a natural way of selecting the E-parameters involved. Since the E-algorithm and Pade approximation have common origin, i.e. the Euclid's algorithm, we first present a short review of Pade approximation and discuss its relationship with the E-algorithm.

Pade Rational Approximations

It is known from the theory of rational approximation of analytic functions that approximations of exp(-z) can be generated by means of Pade rational approximations [18].

Assumming that f(z) is any analytic function of the form

$$f(z) = a_0 + a_1 z + a_2 z^2 + \dots, \tag{3.1}$$

then approximations to f(z) can be defined by

$$f(z) \simeq n_{p,q}(z)/d_{p,q}(z), \tag{3.2}$$

where $n_{p,q}(z)$ and $d_{p,q}(z)$ are respectively polynomials of degree q and p in z ($d_{p,q}(0) \neq 0$).

Each pair of non-negative integers p,q corresponds to an entry of the so-called "Pade table" for f(z), cf. [21]. It can be verified [9] that the p,q entry of the Pade-table for exp(-z) is composed of

$$n_{p,q}(z) = \sum_{k=0}^{q} \frac{(p+q-k)!\ q!}{(p+q)!\ k!(q-k)!} (-z)^k \tag{3.3}$$

and

$$d_{p,q}(z) = \sum_{k=0}^{p} \frac{(p+q-k)!\ p!}{(p+q)!\ k!(p-k)!} z^k . \tag{3.3a}$$

For example, the first few entries of the Pade table for f(z)= exp(-z) can be easily computed from the equations (3.3) as

$$E_{0,0} = 1 ; \quad E_{0,1} = 1-z ; \quad E_{0,2} = 1 - z + z^2/2$$

$$E_{1,0} = 1/(1+z) ; \quad E_{1,1} = \frac{2-z}{2+z} ; \quad E_{2,0} = 1/(1+z+z^2/2) . \tag{3.4}$$

It should be noted that the rational approximations for exp(-z) can obviously generate matrix approximation of exp (-H), simply by replacing the variable z by the matrix H, i.e.

$$\exp(-H) \simeq [d_{p,q}(H)]^{-1}[n_{p,q}(H)] \quad E_{p,q}(H) . \tag{3.5}$$

Various forms of Pade approximation can be used for approximation of the exponential function and since the exponential of a matrix is the solution of semi-discrete approximations to parabolic PDEs, it can be shown that these forms of Pade approximation coincide with known numerical methods for solving PDE's [21].

The Point-Jacobi Iterative Method

Let us consider the E-algorithm in the form

$$\mathbf{u}_{i+1} = \pi_i \mathbf{u}_i + \delta_i, \qquad i\geq 0 . \tag{3.6}$$

Let us also consider, the decomposition A=D-E-F, where D is a positive diagonal matrix and E and F are respectively strictly lower and upper triangular matrices. Then, by choosing the E-parameters as

$$\pi_i = D(L+U) ; \quad \delta_i \quad D^{-1}\mathbf{b} , \qquad i\geq 0 , \tag{3.7}$$

where $L=D^{-1}E$ and $U=D^{-1}F$, the E-algorithm can be written as

$$\mathbf{u}_{i+1} \quad D(L+U)\ \mathbf{u}_i + D^{-1}\mathbf{b} , \qquad i\geq 0 . \tag{3.8}$$

The iterative scheme (3.6), by considering the parabolic problem (1.1), can be written in the following generalized form

$$\mathbf{u}_{i+1} = \{(1-\Delta t)I + \Delta t(L+U)\}\mathbf{u}_i + \Delta t.D^{-1}\mathbf{b} , \quad i\geq 0 , \quad \Delta t\geq 0 , \tag{3.9}$$

or equivalently,

$$u_{i+1} = (I-\Delta t.H)u_i + \Delta t.D^{-1}b\ , \qquad i\geq 0\ , \quad \Delta t\geq 0, \tag{3.10}$$

where $H = I \quad D^{-1}E - D^{-1}F = I-L-U$.

Then, the term $(I-\Delta t.H)$ can be considered as the consistent Pade matrix approximation $E_{0,1}$ $(\Delta t.H)$, i.e.

$$\exp(-\Delta t.H) \simeq I-\Delta t.H = I-\Delta t.D^{-1}(D-E-F) = E_{0,1}(\Delta t.H), \quad \Delta t\geq 0, \tag{3.11}$$

assumming that Δt is small and only the first two terms of the expansion have been retained.

Let us consider now several choices of the regular splitting $A = M_i-N_i$, i.e. $M_1=D$, $N_1=E+F$; $M_2 = D-E$, $N_2=F$ and

$$M_3 = \frac{1}{\omega}(D-\omega E), \quad N_3 = \frac{1-\omega}{\omega} D + F, \quad \omega \neq 0\ . \tag{3.12}$$

Then, by considering the parabolic problem (1.1) and the Pade approximation $E_{0,1}(\Delta t.C_i^{-1}A)$ for $\exp(-\Delta t.C_i^{-1}A)$ with $C_i\equiv M_i$, $i=1,2,3$ and $\Delta t=1$, it can be shown [21] that the corresponding Pade approximations exactly correspond to point Jacobi, point Gauss-Seidel and point SOR respectively. Furthermore, it can be shown that various forms of E-algorithm can be obtained by using certain consistent Pade matrix approximations.

The General Iterative Method

Let us consider the E-algorithm in the form

$$u_{i+1} = \pi_i u_i + \tau_i u_{i-1} + \delta_i\ , \qquad i\geq 0\ . \tag{3.13}$$

The choice of E-parameters

$$\pi_i = (I-H^{-1}A)\ ; \qquad \tau_i = 0\ ; \qquad \delta_i = H^{-1}b\ , \tag{3.13a}$$

where the non-singular matrix H is chosen such that the computational work required to solve the linear system $Hu = s$ (where s is given) is small compared to that required to solve the system (1.2), leads to the iterative procedure

$$u_{i+1} = (I-H^{-1}A)\ u_i + H^{-1}b\ , \qquad i\geq 0\ . \tag{3.14}$$

It can be clearly seen that by setting

$$G = I \quad H^{-1}A\ ; \qquad k = H^{-1}b \tag{3.15}$$

the iterative procedure (3.5) can be equivalently written as a linear stationary first-order iterative method (the general iterative method) of the form

$$u_{i+1} = G.u_i + k\ , \qquad i\geq 0\ . \tag{3.16}$$

Note that several choices of matrix H lead to well-known iterative methods, i.e. Richardson, block Jacobi, Gauss-Seidel, SOR, SSOR, etc.

Chebychev and Conjugate-Gradient type Iterative Methods

The iterative methods described in the previous section have been linked with the E-algorithm using stationary first-order iterative schemes. Let us now consider the E-algorithm in the form

$$u_{i+1} = \pi_i u_i + \tau_i u_{i-1} + \delta_i \;, \quad i \geq 0 \;, \qquad (3.17)$$

with the E-parameters chosen as

$$\pi_i = (I - \alpha_i H^{-1} A) \;; \quad \tau_i = 0 \;; \quad \delta_i = \alpha_i H^{-1} b \;, \qquad (3.17a)$$

where α_i is a preconditioned sequence of parameters and H is the matrix defined earlier in this section. The choice (3.17a) then leads to the iterative scheme (known as first-order Richardson acceleration method):

$$u_{i+1} = (I - \alpha_i H^{-1} A) u_i + \alpha_i H^{-1} b \;, \quad i \geq 0 \;, \qquad (3.18)$$

while for $\alpha_i = \omega$ we have the Jacobi Overrelaxation method.

Let us consider now the E-algorithm (3.17) as a non-stationary second-order iterative scheme with the following selection of E-parameters:

$$\pi_i = \rho_{i+1}(1 - \gamma_{i+1} + \gamma_{i+1} G) \;; \quad \tau_i = 1 - \rho_{i+1} \;; \quad \delta_i = \rho_{i+1}\gamma_{i+1} k \;, \; i \geq 0 \;, \qquad (3.19)$$

where the parameters ρ_{i+1}, γ_{i+1} are functions of the smallest and largest eigenvalues of matrix G, defined by

$$\gamma_i = 2/(2 - \xi_1 - \xi_N), \; i \geq 0 \;; \quad \rho_{i+1} = (1 - \sigma^2 \rho_i / 4)^{-1} \;, \quad i \geq 2 \;, \qquad (3.20)$$

with $\rho_1 = 1$; $\rho_2 = (1 - \sigma^2/2)^{-1}$ and $\sigma = (\xi_N - \xi_1)/(2 - \xi_N - \xi_1)$, (3.20a)

where ξ_1 and ξ_N denote respectively the smallest and largest engenvalues of G, cf. (3.16).

Then, the resulting iterative scheme from E-algorithm (3.17) and equations (3.17a), (3.19) yields

$$u_{i+1} = \rho_{i+1}\{\gamma_{i+1}(G u_i + k) + (1 - \gamma_{i+1}) u_i\} + (1 - \rho_{i+1}) u_{i-1} \;, \quad i \geq 1 \;. \qquad (3.21)$$

The iterative scheme (3.21) represents a non-stationary second-order Chebychev and Conjugate-Gradient (CG) type method [11], which can be used to accelerate the convergence of the iterative schemes (3.16).

Furthermore, the equations (3.19), assuming that $\gamma_{i+1} = 1$, $i \geq 1$, lead to the following simplified choice of E-parameters:

$$\pi_i = \rho_{i+1} G \;; \quad \tau_i = 1 - \rho_{i+1} \;; \quad \delta_i = \rho_{i+1} k \;, \quad i \geq 0 \;, \qquad (3.22)$$

which in turn, leads to a simplified form of iterative scheme (3.21), i.e.

$$u_{i+1} = \rho_{i+1}(G u_i + k) + (1 - \rho_{i+1}) u_{i-1} \;, \quad i \geq 1 \;. \qquad (3.23)$$

By choosing the E-parameters in an analogous manner as in (3.19), i.e.

$$\pi_i = \rho_{i+1}(1 - \gamma_{i+1} + \gamma_{i+1} G^2) \;; \quad \tau_i = 1 - \rho_{i+1} \;; \quad \delta_i = \rho_{i+1}\gamma_{i+1}(Gk + k) \;, \; i \geq 0 \;, \qquad (3.24)$$

where ρ_{i+1} and γ_i are corresponding iteration parameters [7], we can obtain the following "accelerated double method":

$$z_{i+1} = \rho_{i+1}\{\gamma_{i+1}(G^2 z_i + Gk + k) + (1-\gamma_{i+1})z_i\} + (1-\rho_{i+1})z_{i-1}, \quad i\geq 1 \; . \qquad (3.25)$$

It can be shown [3] that the iterative scheme (3.25) results from the "double-method":

$$u_{i+1} = G^2 u_i + Gk + k \; , \qquad i\geq 0 \; , \qquad (3.26)$$

which in turn, can be obtained by considering two applications of the general iterative scheme (3.16). Note that the iterative scheme (3.23) can be shown that, under certain conditions, is virtually equivalent to the accelerated double method (3.25), i.e. $z_i = u_{2i}$, $i\geq 0$, cf.[7].

Richardson and Chebychev second-order Iterative Methods

Given the non-symmetric linear system $Au = b$ and a family of ellipses of centre d with foci $d \pm c$, let us assume that the E-parameters are selected as

$$\pi_i = (1+\beta_i - \alpha_i A) \; ; \quad \tau_i = -\beta_i \; ; \quad \delta_i = \alpha_i b \; , \quad i\geq 0 \; , \qquad (3.26)$$

where the parameters α_i, β_i are defined as

$$\alpha_i = [d-(c/2)^2\alpha_{i-1}]^{-1} \; ; \quad \beta_i = d.\alpha_i \quad 1 \; , \quad i\geq 1 \qquad (3.27)$$

with

$$\alpha_1 = 2d/(2d^2-c^2) \; ; \quad \beta_1 = d\alpha_0 - 1 \; ; \quad \alpha_0 = 1/d \; ; \quad \beta_0 = 0 \; . \qquad (3.28)$$

Then, the E-algorithm can be equivalently written as a second-order non-stationary iterative scheme (known as Chebychev iterative method):

$$u_{i+1} = (1+\beta_i - \alpha_i A)u_i - \beta_i u_{i-1} + \alpha_i b \; , \quad i\geq 1 \; , \qquad (3.29)$$

or equivalently,

$$u_{i+1} = u_i + \alpha_i r_i + \beta_i(u_i \quad u_{i-1}) \; , \quad i\geq 1 \; , \qquad (3.30)$$

where r_i is the residual factor $r_i \;\; b-Au_i$, and u_0, u_1 are arbitrary [23].

The following simplified choice of the E-parameters:

$$\pi_i = (1+\beta-\alpha A) \; ; \quad \tau_i = -\beta \; ; \quad \delta_i = \alpha . b \; , \quad i\geq 0, \qquad (3.31)$$

where the parameters α,β are defined as

$$\alpha \quad \frac{2}{d\pm(d^2-c^2)^{1/2}} \; ; \qquad \beta = d\alpha \quad 1, \qquad (3.32)$$

leads respectively to a stationary second-order iterative scheme

$$u_{i+1} = (1+\beta-\alpha A)u_i - \beta u_{i-1} + \alpha b \; , \quad i\geq 1 \qquad (3.33)$$

or, equivalently,

$$u_{i+1} = u_i + \alpha r_i + \beta(u_i - u_{i-1}) \; , \quad i\geq 1 \; , \qquad (3.34)$$

which is known as the second-order Richardson's method.

It can be shown that the Chebychev iterative scheme (3.30) is asymptotically equivalent to the linear second-order stationary method (3.34), i.e. the parameters α,β are the asymptotic values of α_i, β_i, cf.[16].

We consider now a fixed (nxn) matrix G, with $\rho(G) < 1$, where $\rho \equiv \rho(G)$ denotes here the spectral radius of G, and the class of Chebychev polynomials $C_i(x)$ of degree $i \geq 0$. Then, by selecting the E-parameters as

$$\pi_i = \omega_{i+1} G \; ; \quad \tau_i = 1-\omega_{i+1} \; ; \quad \delta_i = \omega_{i+1} b \; , \quad i \geq 1 \; , \tag{3.35}$$

where the parameter ω_i is defined by

$$\omega_{i+1} = 1 + \frac{C_{i-1}\,(1/\rho)}{C_{i+1}\,(1/\rho)} \; , \quad i \geq 1 \quad \text{and} \quad \omega_1 = 1 \; , \tag{3.36}$$

and $C_i(x)$ is the known Chebychev polynomial of degree $i \geq 0$, the E-algorithm leads to the following iterative scheme

$$u_{i+1} = \omega_{i+1} G u_i + (1-\omega_{i+1}) u_{i-1} + \omega_{i+1} b \; , \quad i \geq 1 \; , \tag{3.37}$$

or, equivalently,

$$u_{i+1} = \omega_{i+1} \{ G u_i + b \quad u_{i-1} \} + u_{i-1} \; , \quad i \geq 1 \; . \tag{3.38}$$

The iterative method (3.38), known as the Chebychev semi-iterative method, can be derived from the general iterative method (3.16), cf.[21].

Furthermore, the cyclic nature of matrix G, i.e.

$$G = \left[\begin{array}{c|c} 0 & F \\ \hline F^* & 0 \end{array} \right] ,$$

allows the use of the following "cyclic" Chebychev semi-iterative method:

$$u_{2i+1} = \omega_{2i+1} \{ F u_{2i} + b - u_{2i-1} \} + u_{2i-1} \; , \quad i \geq 1 \; , \tag{3.39}$$

$$u_{2i+2} = \omega_{2i+2} \{ F^* u_{2i+1} + b \quad u_{2i} \} + u_{2i} \; , \quad i \geq 0 \; , \tag{3.40}$$

where $\quad u_1 = F u_0 + b. \tag{3.41}$

Note that for solving $(I-G)u = b$, the vectors u and b have been partitioned such that

$$u = Fu + b \; ; \quad u = F^* u + b \; . \tag{3.42}$$

The Chebychev semi-iterative method (3.38) is closely related to the second-order Richardson iterative method [21].

Preconditioned two-step CG Methods

The Normalized Implicit Conjugate Gradient (NICG) method is an efficient normalized form of preconditioned CG methods using the approximate factorization $A \simeq D_r T_r^T T_r D_r$, where A is a symmetric, positive definite (2v+1)-diagonal (nxn) matrix of semi-band width(s) $m_1, m_2, \ldots, m_l$, $l \in [1, v-1]$ (with v denoting here the dimensionality of the problem under consideration), D_r is a diagonal matrix (with positive elements), T_r is an upper triangular matrix (with unit diagonal elements) retaining a certain number of outermost off-diagonal entries in semi-bandwidth(s) $m_1, m_2, \ldots, m_l$, $l \in [1, v-1]$ and T_r^T denotes the transpose of T_r, cf.[11]. Furthermore, a second-order NICG method for the numerical solution of self-adjoint elliptic PDE's in regular domains, based on the previously defined four-term approximate factorization, has been presented in [11].

In the following we establish a correspondence between the E-algorithm (3.17) and preconditioned two-step CG methods. In this case let us assume that the E-parameters are selected as:

$$\pi_i = \rho_{i+2}(1+\gamma_{i+2}A) \; ; \quad \tau_i = 1-\rho_{i+1} \; ; \quad \delta_i = \rho_{i+1}\gamma_{i+1}b \, , \quad i\geq 0 \, , \tag{3.43}$$

where

$$\rho_{i+1} = 1 + \frac{a^*_i}{a^*_{i-1}} \beta^*_i \; ; \quad \gamma_{i+1} = a^*_i/\rho_{i+1} \, , \quad i=0,1,\ldots,k-1 \tag{3.44}$$

with k the smallest integer such that $r_k = b-Au_k$ 0, $\rho_1=1$ and a^*_i, β^*_i are the known CG scalar quantities, as they are defined in the original paper of Hestenes-Stiefel [8]. With such a selection of E-parameters, the E-algorithm becomes

$$u_{i+1} = \rho_{i+1}(1+\gamma_{i+1}A)u_i + (1-\rho_{i+1})u_{i-1} + \rho_{i+1}\gamma_{i+1}b \, , \quad i\geq 1 \tag{3.45}$$

which can be equivalently written as a two-step CG method [1] of the form:

$$u_{i+1} = u_{i-1} + \rho_{i+1}(\gamma_{i+1}r_i + u_i - u_{i-1}), \quad i\geq 1 \, . \tag{3.46}$$

Note that the two-step CG method (3.46) can be easily obtained from the second-order Richardson iterative method (3.34) by using relationships of the well-known CG method [8].

In order to establish a correspondence between the E-algorithm and the second-order NICG method we now proceed in a similar manner by selecting the following values of the E-parameters:

$$\pi_i = \rho^+_{i+2}(1+\gamma^+_{i+2}G) \; ; \quad \tau_i = 1-\rho^+_{i+1} \; ; \quad \delta_i = \rho^+_{i+1}\gamma^+_{i+1}b^+, \quad i\geq 0, \tag{3.47}$$

where

$$\rho^+_{i+1} = 1 + \frac{a^+_i}{a^+_{i-1}} \beta_i \; ; \quad \gamma^+_{i+1} = a^+_i/\rho^+_{i+1} \, , \quad i=0,1,\ldots,k-1, \tag{3.48}$$

with k the smallest integer such that $r^+_k = 0$,

$$r^+_i = b^+ - Gu_i \; ; \quad b^+ = Gb \; ; \quad G = (T_r^T T_r)^{-1} D_r^{-1} A \, , \tag{3.49}$$

while a^+_i, β^+_i are the NICG scalar quantities [11].

Then, the E-algorithm (3.17) leads to the iterative scheme

$$u_{i+1} = \rho^+_{i+1}(1+\gamma^+_{i+1}G)u_i + (1-\rho^+_{i+1})u_{i-1} + \rho^+_{i+1}\gamma^+_{i+1}b^+ \, , \quad i\geq 1 \, , \tag{3.50}$$

which in turn, can be written as a second-order NICG method of the form:

$$u_{i+1} = u_{i-1} + \rho^+_{i+1}(\gamma^+_{i+1}r^+_i + u_i - u_{i-1}) \, , \quad i\geq 1 \, . \tag{3.51}$$

The second-order NICG method has proved to be an efficient preconditioned solution method which also includes a proposal for the selection of certain values of the iteration parameters ρ_i, γ_i involved in solving 2D and 3D elliptic boundary-value problems leading to substantial savings in the computational work. Further details for the convergence rate of the iterative scheme (3.51) and its required computational work for convergence are given in [11]. The relationship of the E-algorithm and multi-step iterative methods can be derived in an analogous manner.

The Block-Stiefel (Chebychev) Iterative Method

Let us consider the E-algorithm (3.17) with the following selection of E-parameters:

$$\pi_i = 1-\omega_i(I-G) \; ; \quad \tau_i = \gamma\omega_i - 1 \; ; \quad \delta_i = (1-\gamma\omega_i)u_{i-2} + \omega_i b \, , \quad i\geq 1 \, , \tag{3.52}$$

with

$$\omega_i = 1/(\gamma - \frac{1}{4\alpha^2}\,\omega_{i-1}), \quad i\geq 1 \; ; \qquad \gamma = \beta/\alpha \tag{3.52a}$$

$$\alpha = 2(\xi_N^* - \xi_1^*) \; ; \qquad \beta \quad (\xi_1^* + \xi_N^*)/(\xi_N^* - \xi_1^*) \, , \tag{3.52b}$$

$$u_1 = u_0 + \gamma^{-1} r_0 \, , \tag{3.52c}$$

where ξ_1^* is a lower bound of the smallest eigenvalue, ξ_N^* is an upper bound of the largest eigenvalue of the coefficient matrix and $r_i = b-(I-G)u_i$, $i\geq 0$. Then, the equations (3.17), (3.52) yield the iterative scheme

$$u_{i+1} = (1-\omega_i(I-G))u_i + (\gamma\omega_i - 1)u_{i-1} + (1-\gamma\omega_i)u_{i-2} + \omega_i b \, , \quad i\geq 1 \, , \tag{3.53}$$

which can be equivalently written in the form

$$u_{i+1} = u_i + \Delta u_i \, , \quad i\geq 1 \, , \tag{3.54}$$

where

$$\Delta u_i = \omega_i r_i + (\gamma\omega_i - 1)\Delta u_{i-1} \, , \quad i\geq 1 \, . \tag{3.54a}$$

The iterative scheme (3.54) is known as the Block-Stiefel (Chebychev) iterative method [20]. In the following Tables 1 and 2 several cases of the E-algorithm are given, which are no more than various preconditioned methods as indicated previously in this section.

E - Parameters			Implicit Preconditioned method
π_i	ι_i	δ_i	
$D(L+U)$	0	$D^{-1}b$	Point-Jacobi [21,23]
$(I-H^{-1}A)$	0	$H^{-1}b$	General iterative scheme [21,23]
$(I-\alpha_i H^{-1}A)$	0	$\alpha_i H^{-1}b$	Richardson (first-order)
$\rho_{i+1}(1-\gamma_{i+1}+\gamma_{i+1}G)$	$(1-\rho_{i+1})$	$\rho_{i+1}\gamma_{i+1}\kappa$	Conjugate-Gradient type methods (second-order) [8]
$\rho_{i+1}(1-\gamma_{i+1}+\gamma_{i+1}G^2)$	$(1-\rho_{i+1})$	$\rho_{i+1}\gamma_{i+1}(G\kappa+\kappa)$	Accelerated double method [7]
$1+\beta-\alpha A$	$-\beta$	$\alpha.b$	Richardson (Second-order) [21]
$\omega_{i+1}G$	$1-\omega_{i+1}$	$\omega_{i+1}b$	Chebychev semi-iterative [23]
$\rho_{i+2}(1+\gamma_{i+2}A)$	$1-\rho_{i+1}$	$\rho_{i+1}\gamma_{i+1}b$	Preconditioned (two-step) Conjugate-Gradient [11]
$\rho^{+}_{i+2}(1+\gamma^{+}_{i+2}G)$	$1-\rho^{+}_{i+1}$	$\rho^{+}_{i+1}\gamma^{+}_{i+1}b^{+}$	Normalized Implicit Conjugate-Gradient (NICG) [11]
$1-\omega_i(I-G)$	$\gamma\omega_{i-1}$	$(1-\gamma\omega_i)u_{i-2}+\omega_i b$	Block-Stiefel [20]

Table 1: Certain cases of the E-algorithm, which are no more than various implicit preconditioned methods.

4. E-ALGORITHM AND EXPLICIT PRECONDITIONED METHODS

Apart of the implicit preconditioned methods described so far in the text, it can be shown that analogous behavior show the counterpart explicit preconditioned methods, which are particularly applicable for solving large linear systems on parallel and vector processors, cf. [4,10,15]. In the following we will examine the correspondence of the E-algorithm with such explicit iterative methods.

Explicit AIM Methods

It is known that approximate factorization methods can be obtained by considering the simple replacement of the coefficient matrix A by a matrix $(A+\Pi_r)$ such that

$$A+\Pi_r = L_r U_r , \tag{4.1}$$

where L_r and U_r are sparse (strictly) lower and upper triangular matrices respectively [15]. The choice of Π_r, assuming that the matrix $(A + \Pi_r)$ can be factored in sparse triangular matrices leading to "easily-solvable" systems $(L_r U_r)u = b$, proves to be of great importance

[14] for the convergence rate of accelerated semi-direct methods used for solving the linear system (1.2).

Let us assume that M_r, a non-singular real (nxn) matrix, be the inverse of $(A+\Pi_r)$, i.e. $M_r \equiv (L_r U_r)^{-1}$, $r \in [1,m-1]$, (4.2) where r is "fill-in" parameter, i.e. the number of the outermost-off diagonal entries which are retained in semi-bandwidth m.

The explicit iterative methods, based on the generalized Approximate Inverse Matrix (AIM) techniques [11], are originated from known algorithmic procedures of inverting a real (nxn) matrix A, which can be factorized as $A \doteq L_r U_r$, by determining the elements of A^{-1} without inverting the decomposition factors L_r and U_r, cf.[5]. The effectiveness of the explicit iterative methods for solving a certain class of initial/boundary-value problems on regular domains is related to the fact that the exact inverse of the sparse coefficient matrix A (although is full) exhibits a similar "fuzzy" structure as A, i.e. the largest elements are clustered around the principal diagonal and main diagonals, cf. [15].

Let us consider now the E-algorithm in the form (3.17) and assume that the E-parameters are respectively selected as:

$$\pi_i \quad 1 + \beta_i - \alpha_i M_r A \; ; \quad \tau_i \quad -\beta_i \; ; \quad \delta_i = \alpha_i M_r b \, , \quad i \geq 0 \, , \qquad (4.3)$$

and

$$\pi_i = 1 + \beta \quad \alpha M_r A \; ; \quad \tau_i \quad -\beta \; ; \quad \delta_i = \alpha M_r b \, , \quad i \geq 0 \, , \qquad (4.4)$$

where M_r is an approximate form of the inverse of A and α_i, β_i and α, β are preconditioned sequences of parameters and parameters respectively. Then, the E-algorithm (3.17) yields the following explicit iterative schemes:

$$u_{i+1} = (1 + \beta_i - \alpha_i M_r . A) u_i - \beta_i u_{i-1} + \alpha_i M_r b \, , \quad i \geq 1 \, , \qquad (4.5)$$

and

$$u_{i+1} = (1 + \beta \quad \alpha M_r . A) u_i - \beta u_{i-1} + \alpha M_r b \, , \quad i \geq 1 \, , \qquad (4.6)$$

which are known as the Explicit Chebychev (ECh) and Explicit second-order Richardson (ER2) methods respectively [15].

It should be noted that alternative approximate forms of the inverse M_r (in banded form) can be effectively used, by retaining only δ_U and δ_L ($\delta_U \in [2,n-1]$, $\delta_L \in [1,n]$) diagonals in the upper and lower triangular parts of M_r, the remaining elements being just not computed at all, cf.[15]. The main motive for the derivation of generalized AIM techniques lies on the fact that they can be efficiently used in conjuction with explicit iterative schemes leading to effective composite semi-direct solution methods, which posses a high-degree of explicitness and therefore are particularly suitable for solving large linear systems of algebraic equations on multiprocessor systems.

Furthermore, we state that analogous relationships to those derived in (3.46), (3.51) can be derived for the Explicit CG-type methods. The establishment of correspondence between the E-algorithm and other explicit preconditioned methods can be achieved in a similar manner as it is shown in the following presentation.

Generalized AGE Iterative Methods

The Generalized Alternating Group Explicit (AGE) iterative method, based on the known ADI method [19], has been recently developed for solving multidimensional bowndary-value problems by improving the convergence rate and attaining higher order accuracy [4]. The AGE iterative method can be derived as follows. Let us consider the regular splitting

$$A = D + G_1 + G_2 \tag{4.7}$$

where D is a non-negative diagonal matrix and G_1, G_2, D satisfy the following conditions:

i) the matrices $(G_i+\theta.D+\rho.I)$, i=1,2, are non-singular for any $\theta\geq 0$, $\rho>0$,

ii) the systems $x = G_1^{-1} v$ and $y - G_2^{-1} z$ (4.8)

can be easily solved explicitly, for any vectors v, z.

Note that G_1 and G_2 are either small (2x2) block systems or can be made so by a suitable permutation of their rows and corresponding columns. The systems of case (ii) can be easily solved in the sense that the work required is much less than would be required to solve the original system (1.2) directly.

The linear system (1.2) can be written in the form

$$(G_1+G_2+D)\ u = b \tag{4.9}$$

or, equivalently,

$$(G_1+\theta.D+\rho I)\ u = b - (G_2+(1-\theta)D - \rho I)\ u, \tag{4.10}$$

$$(G_2+\theta.D+\rho^* I)\ u = b - (G_2+(1-\theta)D - \rho^* I)\ u\ . \tag{4.10a}$$

Proceeding in an analogous manner as in ADI method [19] the system (4.10) can be written as

$$(G_1 + \theta D + \rho I)\ u_{i+1/2} = b - (G_2 + (1-\theta)D \quad \rho I)\ u_i\ , \tag{4.11}$$

$$(G_1 + \theta^* D + \rho^* I)\ u_{i+1} \quad b - (G_1 + (1-\theta^*)D - \rho^* I)\ u_{i+1/2}\ . \tag{4.11a}$$

Assuming that $\theta = \theta^* = 1/2$ and $\rho = \rho^*$ the following simplified version of (4.11) (known as the AGE iterative scheme) can be obtained:

$$(G_1^* + \rho I)\ u_{i+1/2} = b \quad (G_2^* \quad \rho I)\ u_i\ , \tag{4.12}$$

$$(G_2^* + \rho I)\ u_{i+1} = b - (G_1^* - \rho I)\ u_{i+1/2}\ , \tag{4.12a}$$

where

$$G_1^* = G_1 + D/2\ ; \qquad G_2^* = G_2 + D/2\ . \tag{4.12b}$$

By modifying the basic iteration (4.12) we obtain the following generalized AGE scheme:

$$(G_1 + \rho_{i+1} I)\, u_{i+1/2} = (\rho_{i+1} I - G_2)\, u_i + b\,, \tag{4.13}$$

$$(G_2 + \rho_{i+1} I)\, u_{i+1} = (G_2 - (1-\omega)\, \rho_{i+1} I)\, u_i + (2-\omega)\, \rho_{i+1}\, u_{i+1/2}\,, \tag{4.13a}$$

where ω is a non-negative acceleration parameter.

The choice $\omega=0$ leads to the AGE scheme (4.12), after using the first stage to express $G_1 u_{i+1/2}$ in terms of $G_2 u_i$ and thereby economising on the evaluation of the right hand side vectors. For $\omega=1$ the iteration (4.13) is analogous to the Douglas-Rachford iterative methods for lines [1] and can efficiently used for solving multidimensional boundary-value problems. Note that for G_1, G_2 symmetric and positive-definite and for fixed acceleration parameter $\rho>0$, the generalized AGE scheme converges for $0\le\omega\le2$.

Let us now consider the E-algorithm (3.17) with the following choice of E-parameters:

$$\pi_i = (\rho_i I + G_2)^{-1} (\rho_i I + G_1)^{-1} (\rho_i I - G_1)\, (\rho_i I - G_2)\;; \qquad \tau_i=0,$$

$$\delta_i = 2\rho_i\, (\rho_i I + G_2)^{-1} (\rho_i I + G_1)^{-1}\, b\,, \quad i\ge0\,, \tag{4.14}$$

where G_1, G_2 are real matrices, cf. (2.11), as defined earlier in this section and $\rho_i, i\ge0$ is a preconditioned sequence of positive parameters.

Then, the E-algorithm (3.17) with such a selection of E-parameters yields the following iterative scheme:

$$u_{i+1} = (\rho_i I+G_2)^{-1}(\rho_i I+G_1)^{-1}\, [(\rho_i I-G_1)(\rho_i I-G_2)u_i + 2\rho_i b\,]\,, \quad i\ge0\,, \tag{4.15}$$

which can be equivalently written in the form

$$(\rho_i I+G_1)u_{i+1/2} = (\rho_i I-G_2)u_i + b\,, \quad i\ge0\,, \tag{4.16}$$

$$(\rho_i I+G_2)u_{i+1} = (\rho_i I-G_1)u_{i+1/2} + b\,, \quad i\ge0\,, \tag{4.16a}$$

which is known as the AGE scheme, cf. [4].

Let us now select the E-parameters as follows

$$\pi_i = 1+\omega_{i+1}(\rho_i I+G_2)^{-1}\, (\rho_i I+G_1)^{-1} A\;; \qquad \tau_i=0\,,$$

$$\delta_i = -\omega_{i+1}\, (\rho_i I+G_2)^{-1}\, (\rho_i I+G_1)^{-1}\, b\,, \quad i\ge0\,, \tag{4.17}$$

where the sequence ω_i, $i\ge0$, are determined by

$$\omega_i=2[\,(\lambda_1+\lambda_n) - (\lambda_1-\lambda_n)t_i\,]^{-1}\,, \quad i=1,2,\dots,v\,, \tag{4.18}$$

where $$t_i = \cos[(2i-1)\pi/2i]\,, \quad i=1,2,\dots,v \tag{4.18a}$$

are zeros of the i-th Clebychev polynomial.

Then, the E-algorithm (3.17) can be re-written as

$$u_{i+1} = u_i+\omega_{i+1}(\rho_i I+G_2)^{-1}(\rho_i I+G_1)^{-1}(Au_i-b)\,, \quad i\ge0\,, \tag{4.19}$$

which is the Richardson method applied to the AGE algorithm.

Let us now consider the following selection of E-parameters:

$$\pi_i = \omega_i H_i \ ; \quad \tau_i = (1-\omega_i) \ ; \quad \delta_i = 2\omega_i\rho_i(\rho_i I+G_2)^{-1}(\rho_i I+G_1)^{-1}b \ , \quad i\ge 0 \ , \tag{4.20}$$

where H_i is the iteration matrix defined by

$$H_i \equiv (\rho_i I+G_2)^{-1}(\rho_i I+G_1)^{-1}(\rho_i I-G_1)(\rho_i I-G_2) \ , \quad i\ge 0 \ . \tag{4.21}$$

Then, the E-algorithm (3.17) leads to the iterative scheme

$$u_{i+1} = \omega_i(H_i u_i + 2\rho_i(\rho_i I+G_2)^{-1}(\rho_i I+G_1)^{-1}b)+(1-\omega_i)\ u_{i-1} \ , \quad i\ge 0 \ , \tag{4.22}$$

which is the Clebychev semi-iterative method applied to the AGE algorithm. It has been shown that both iterative schemes (4.19) and (4.22) converge faster than the original Richardson and semi-direct Clebychev methods respectively, cf.[6].

The Method of Fractional Steps

The AGE method, which is based on the combination of certain elementary first order difference processes permitting the reduction of a given complicated problem into a sequence of simpler problems,can be considered as a fractional (splitting-up) method [17,22].

Given the linear system $Au=b$, where $A=G_1+G_2$, with $G_1>0$, $G_2>0$, an approximate solution can be obtained by the iterative scheme

$$B(u_{i+1}-u_i)/\eta + Au_i = b \ , \quad i\ge 0 \ , \tag{4.23}$$

where the operator B is defined as

$$B = (I + \frac{\eta}{2} G_1)\ (I + \frac{\eta}{2} G_2) \tag{4.24}$$

with η a preconditioned parameter chosen to maximize the asymptotic convergence rate of the iterative scheme (4.23).

Let us now consider the following selection of E-parameters:

$$\pi_i = (I + \frac{\eta}{2} G_2)^{-1}\ (I + \frac{\eta}{2} G_1)^{-1}\ (I - \frac{\eta}{2} G_1)(I \quad \frac{\eta}{2} G_2) \ ; \qquad \tau_i=0 \ ;$$

$$\delta_i = \eta\ (I + \frac{\eta}{2} G_2)^{-1}\ (I + \frac{\eta}{2} G_1)^{-1}\ b \ , \qquad i\ge 0 \ . \tag{4.25}$$

Then, the E-algorithm (3.17) with such a selection of E-parameters yields the following iterative scheme

$$u_{i+1} = (I + \frac{\eta}{2}G_2)^{-1}(I + \frac{\eta}{2}G_1)^{-1}\ [(I - \frac{\eta}{2}G_1)(I - \frac{\eta}{2}G_2)\ u_i+\eta b] \ , \quad i\ge 0 \ , \tag{4.26}$$

which is known as the fractional-step iterative method.

It can be easily seen that the iterative method can be splitted-up in the following sequence of simpler operations:

$$z_i = Au_i - b \ ; \qquad (I + \frac{\eta}{2} G_1)\ z_{i+1/3} = z_i \ ; \tag{4.27}$$

$$(I + \frac{\eta}{2} G_2)\ z_{i+2/3} = z_{i+1/3} \ ; \qquad u_{i+1} = u_i - \eta z_{i+2/3} \ , \tag{4.27a}$$

which by the trivial nature of G_1 and G_2 matrices can all be simplified by expressing them as explicit processes.

By selecting the E-parameters as

$$\pi_i=1 \;;\quad \tau_i=0 \;;\quad \delta_i = [\frac{1}{\eta}(I + \frac{\eta^2}{2} G_1G_2) + \frac{1}{2} A]^{-1} \cdot b \;,\quad i\geq 0 \;, \tag{4.28}$$

the E-algorithm (3.17) becomes

$$\frac{1}{\eta}(I + \frac{\eta^2}{4} G_1G_2)(u_{i+1}-u_i) + \frac{1}{2} A (u_{i+1}-u_i) = b \;,\quad i\geq 0 \;, \tag{4.29}$$

which, in turn, coincides in approximation order with the Grank-Nicolson scheme

$$\frac{1}{\eta}(u_{i+1}-u_i) + \frac{1}{2} A (u_{i+1}-u_i) = b \;,\quad i\geq 0 \;, \tag{4.30}$$

provided that the solution is "smooth".

Finally, let us consider the following selection of E-parameters:

$$\pi_i = (I + \theta G_1)^{-1} (I + \theta G_2)^{-1} \{(I + \theta G_1)(I + \theta G_2) - \eta A\}$$
$$\tau_i=0 \;;\quad \delta_i = \eta b \;,\quad i\geq 0 \;, \tag{4.31}$$

where G_1, G_2 are commutative matrices and θ, η are preconditioned parameters defined as $\theta = 1/\sqrt{\xi_1 . \xi_2}$, where ξ_1, ξ_n are the extreme eigenvalue of $B^{-1} A$, cf.(4.23), and η is selected according to Chebychev acceleration given by (4.18a). Then, using the selection (4.31) the E-algorithm (3.17) yields the general fractional-step iterative scheme

$$\frac{1}{\eta}(I + \theta G_1)(I + \theta G_2)(u_{i+1}-u_i) + A u_i = b \;,\quad i\geq 0 \;, \tag{4.32}$$

which can be made to achieve the asymptotic rate of convergence under certain conditions for G_1, G_2, θ, η described in the previous paragraph.

Multiple Explicit Jacobi Iteration and its Parametric Forms

It is known that the Jacobi method, since the solution of the lth iteration is totally dependent on the solution obtained from the (l-1)th iteration, possesses a high level of inherent parallel implementation and its synchronous (and asynchronous) version is well suited for parallel implementation on machines of SIMD and MIMD type. The disadvantage due to its slow convergence rate can be tackled by incorporating the method within first or second-order acceleration schemes. The AIM semi-direct methods based on the multiple explicit Jacobi iteration have been proved very effective for solving numerically large sparse linear systems on vector or parallel processors [15]. In the following we establish a correspondence between the E-algorithm and Multiple explicit Jacobi method.

We consider, for the sake of analysis, the regular splitting:

$$A = D-L-U \quad D(I-B), \tag{4.33}$$

where $D = \mathrm{diag}(A)$, L and U are strictly lower and upper triangular parts of A and $B = D^{-1}(L+U)$. It is known that a "good" approximation M_μ of the inverse $A^{-1} = (\sum_{k=0}^{+\infty} B^k)D^{-1}$ can be obtained by considering only the first μ-terms, i.e.

$$M_\mu = (\sum_{k=0}^{\mu-1} B^k)D^{-1}, \quad \mu>0 . \tag{4.34}$$

Let us consider now the E-algorithm in the form

$$u_{i+1} = \pi_i u_i + \tau_i u_{i-1} + \delta_i , \quad i\geq 0 , \tag{4.35}$$

and let us assume that the E-parameters can be selected as follows:

$$\pi_i = \prod_{i=0}^{\nu-1} (I \quad \alpha_i M_\mu .A) \; ; \; \tau_i = 0 \; ; \; \delta_i = \alpha_i M_\mu .b , \quad i\geq 0 , \tag{4.36}$$

where M_μ is the pseudoinverse (4.34) and α_i a preconditioned sequence of parameters. Then, the E-algorithm (4.35) yields the following first-order non-stationary iterative scheme:

$$u_{i+1} = G^\nu_{\alpha_{i,\mu}} u_i + g_{\alpha_{i,\mu}} , \quad i\geq 0 , \tag{4.37}$$

where

$$G^\nu_{\alpha_{i,\mu}} = \prod_{i=0}^{\nu-1} (I - \alpha_i H_\mu); \quad H_\mu = (\sum_{k=0}^{\mu-1} B^k)D^{-1}A , \quad i\geq 0 , \quad \mu>0 , \tag{4.37a}$$

and

$$g_{\alpha_{i,\mu}} = \alpha_i .M_\mu b = \alpha_i \; (\sum_{k=0}^{\mu-1} B^k)D^{-1}b , \quad i\geq 0 . \tag{4.37b}$$

The iterative scheme (4.37) is known as the multiple explicit Jacobi (μ-EJ) method. This method,provided that its spectral radius ρ satisfies the convergence condition $\rho(G^\nu_{\alpha_{i,\mu}})<1$, in combination with the (Lanczos) economized Chebychev polynomials of degree μ, has proved to be very effective for solving elliptic boundary-value problems on parallel processors [10].

Let us consider now the E-algorithm (4.37) and the following selection of E-parameters:

$$\pi_i = -\omega^*_{i+1} .\alpha_i \; (\sum_{k=0}^{\mu-1} \omega^+_k B^k)D^{-1}A \; ; \quad \tau_i = 1-\omega^*_{i+1} ,$$

$$\delta_i = \omega^*_{i+1} .\alpha_i \; (\sum_{k=0}^{\mu-1} \omega^+_k B^k)D^{-1}b , \quad i\geq 1 , \tag{4.38}$$

where ω^*_i are the Chebychev parameters defined by

$$\omega^*_{i+1} = 1/(1-\beta^2\omega^*_i/4) \; ; \quad \omega^*_2 = 1/(1-\beta^2/2) \; ; \quad \omega^*_1=1 , \quad i\geq 2 , \tag{4.39}$$

and ω^+_i are the coefficients of a preconditioned Chebychev polynomial $f_{\mu-1}(x) \equiv \sum_{i=0}^{\mu-1} w_i x^i$, (i.e. $f^+_{\mu-1}(x) = \sum_{i=0}^{\mu-1} w^+_i x^i$, with $w^+_i = w_i/w_0$, $i\in[1,\mu-1]$; $w^+_0=1$).

By $\beta \equiv \rho(\bar{G}^{\nu}_{\alpha_{i,\mu}})$ we denote here the spectral radius of $\bar{G}^{\nu}_{\alpha_{i,\mu}}$, satisfying the convergence condition $\beta<1$, where

$$\bar{G}^{\nu}_{\alpha_{i,\mu}} = \prod_{i=0}^{\nu-1}(I-\alpha_i\bar{H}_\mu) \;;\quad \bar{H}_\mu = (\sum_{k=0}^{\mu-1}\omega^+_k B^k)D^{-1}A \,,\quad i\geq 0 \,,\quad \mu>0 \,, \tag{4.40}$$

With this choice of E-parameters, the E-algorithm (4.35) yields the following iterative scheme,

$$u_{i+1} = \omega^*_{i+1}\left\{[I-\alpha_i(\sum_{k=0}^{\mu-1}\omega^+_k B^k)D^{-1}Au_i+\alpha_i(\sum_{k=0}^{\mu-1}\omega^+_k B^k)D^{-1}b-u_{i-1}]\right\} + u_{i-1} \,,\quad i\geq 1 \,, \tag{4.41}$$

or equivalently,

$$u_{i+1} = [\omega^*_{i+1}u_i + (1-\omega^*_{i+1})u_{i-1}] + \omega^*_{i+1}\alpha_i(\sum_{k=0}^{\mu-1}\omega^+_k B^k)r^+_i \,,\quad i\geq 1 \,, \tag{4.42}$$

where
$$r^+_i = D^{-1}r_i = D^{-1}(b - Au_i) \,. \tag{4.42a}$$

The non-stationary iterative scheme (4.42), known as the second-order parametric Chebychev multiple explicit Jacobi (ω_μ-EJ) method, can be used, in combination with (Lanczos) economized Chebychev polynomial and Neumann series of certain degree, very effectively for solving explicitly large sparse linear systems resulting from the discretization elliptic and parabolic PDE's [10].

E - Parameters			Explicit Preconditioned method
Π_i	τ_i	δ_i	
$1+\beta-\alpha M_r A$	$-\beta$	$\alpha M_r . b$	Explicit Richardson [15]
$1+\beta_i-\alpha_i M_r A$	$-\beta_i$	$\alpha_i M_r . b$	Explicit Chebychev [15]
$(\rho_i I+G_2)^{-1}(\rho_i I+G_1)^{-1}$ $(\rho_i I-G_1)(\rho_i I-G_2)$	0	$2\rho_i(\rho_i I+G_2)^{-1}$ $(\rho_i I+G_1)^{-1}.b$	Generalized AGE Method [4]
$1+\omega_{i+1}(\rho_i I+G_2)^{-1}.$ $(\rho_i I+G_1)^{-1} A$	0	$-\omega_{i+1}(\rho_i I+G_2)^{-1}.$ $(\rho_i I+G_1)^{-1}.b$	Richardson + AGE method [4]
$\omega_i H_i$	$1-\omega_i$	$2\omega_i\rho_i(\rho_i I+G_2)^{-1}.$ $(\rho_i I+G_1)^{-1}\, b$	Chebychev + AGE method [4]
$(I+\frac{\eta}{2}G_2)^{-1}(I+\frac{\eta}{2}G_1)^{-1}.$ $(I-\frac{\eta}{2}G_1)(I-\frac{\eta}{2}G_2)$	0	$n(I+\frac{\eta}{2}G_2)^{-1}.$ $(I+\frac{\eta}{2}G_1)^{-1}.b$	Fractional-step iterative method [4,17,22]
1	0	$[\frac{\eta}{2}(I+\frac{\eta^2}{2}G_1G_2)$ $+\frac{1}{2}A]^{-1}.b$	Grank-Nickolson scheme [4,17,22]
$(I+\theta G_1)^{-1}(I+\theta G_2)^{-1}.$ $\{(I+\theta G_1)(I+\theta G_2)-\eta A\}$	0	$\eta.b$	General fractional-step iteration [4,17,22]
$\prod_{i=0}^{\nu-1}(I-\alpha_i M_\mu .A)$	0	$\alpha_i M_\mu .b$	Multiple explicit Jacobi (μ-EJ) [10]
$-\omega_{i+1}\alpha_i(\sum_{\kappa=0}^{\mu-1}\omega_\kappa^+ B^\kappa).$	$1-\omega_{i+1}$	$\omega_{i+1}\alpha_i(\sum_{\kappa=0}^{\mu-1}\omega_\kappa^+ B^\kappa).$	Chebychev + multiple explicit Jacobi [10]
$D^{-1}A$		$D^{-1}\, b$	

Table 2: Certain cases of the E-algorithm, which are no more than various explicit preconditioned methods.

5. CONVERGENCE ANALYSIS

Approximate LU-factorization procedures for solving parabolic and elliptic difference equations in three-space variables in regular domains have been presented in [13]. The solution of the resulting linear system $Au = b$ by approximate factorization techniques [i.e.

$A - M_r$, where A is a large sparse seven-diagonal (nxn) matrix of semibandwidths m and p and M is an approximate factorization of the form $L_r U_r$, $L_r D_r L_r^T$ or $D_r T_r^T T_r D_r$] can be obtained as a convergence limit of a sequence of vectors $\{u_{i+1}\}$ generated by the E-algorithm, i.e.

$$u_{i+1} = \pi_i u_i + \tau_i u_{i-1} + \delta_i \ , \quad i \geq 0 \ . \tag{5.1}$$

In this section we present some theoretical and practical results for the implementation of the E-algorithm and relate its rate of convergence to the amount of "fill-in" terms occuring in the decomposition factors in general. It is known that approximate factorization algorithms are tediously complicated, especially in the three-dimensional case. In the forthcoming presentation we derive some estimates of the computational work required by the E-algorithm for solving efficiently 3D-elliptic boundary-value problems.

Let us consider the E-algorithm in the form (5.1) with the following selection of the E-parameters:

$$\pi_i = I - \alpha_i M_r^{-1} A \ ; \qquad \iota_i = 0 \ ; \qquad \delta_i = \alpha_i M_r^{-1} b \ , \tag{5.2}$$

where α_i is a pre-conditioned sequence of parameters. Then, the E-algorithm (5.1) can be equivalently written as

$$M_r (u_{i+1} - u_i) = \alpha_i r_i \ , \quad i \geq 0 \ , \tag{5.3}$$

where r_i is the residual factor r_i b Au_i. The iteration matrix H_r^v of the iterative scheme (5.2) is defined by

$$H_r^v = \prod_{i=0}^{v-1} (I - \alpha_i \Psi_r) \ ; \qquad \Psi_r \equiv M_r^{-1} A \ . \tag{5.4}$$

In the following, we establish work estimates for the number of computations of E-algorithm (5.1) required to reduce the L_2-norm of the error by a factor $\epsilon > 0$, using an optimal parameter sequence when the coefficient matrix A is positive definite. It should be noted that if the $r_1 = m-1$ and $r_2 = p-1$ outermost off-diagonal entries have been retained in the decomposition factors L_r, U_r, then $L_r \equiv L$ and $U_r \equiv U$ (the "limit" case), where L and U are strictly lower and upper triangular matrices retaining the (m-1) and (p-1) outermost off-diagonal entries respectively. The coefficient matrix A is then factorized exactly, i.e. A=LU, to yield a direct algorithmic procedure for the FD solution, cf.[13].

Let $\lambda_j(\Psi_r)$, $j \in [1,n]$, be the eigenvalues of Ψ_r, where $\lambda_{min}(\Psi_r) \equiv \lambda_1(\Psi_r)$ and $\lambda_{max}(\Psi_r) \equiv \lambda_n(\Psi_r)$ denote the eigenvalues of smallest and largest algebraic value of Ψ_r respectively. Let us also assume throughout this section that

$$\|A\| \equiv \|A\|_1 = \max_{j \in [1,n]} \{ \sum_{i=1}^{n} |A_{i,j}| \} \ . \tag{5.5}$$

Lemma 1: Let H be a real $(n \times n)$ non-singular matrix and M_H be the M-condition number of H. Then, it holds

$$\|H^{-1}\| \le \frac{M_H^2}{n^2} \frac{1}{\|H\|} . \tag{5.6}$$

Lemma 2: Let $\Psi_r \equiv M_r^{-1}A$ be the preconditioning matrix of the iterative scheme (5.3), with M_r as defined earlier in this Section. Let m and p $(p>m)$ be positive integer numbers denoting the semi-bandwidths of the $(n \times n)$ coefficient-matrix A, and $r_1 \in [1,m-1]$, $r_2 \in [1,p-1]$ be the fill-in parameters of the LU factors. Then, we have

$$\frac{1}{\Lambda_0} \frac{1}{1 + C_1^{(r)}} \le \|\Psi_r\| \le -1 + 2\, C_2^{(m,p,r)} , \tag{5.7}$$

where $C_2^{(m,p,r)}$ is a constant (strictly greater than 1) depending on m,p,r_1,r_2; $C_1^{(r)}$ is a constant depending on r_1, r_2 (independent of the mesh size h); and $\Lambda_0 \equiv M_0^2/n^2$ with M_0 the M-condition number of $A \equiv LU$.

Let us consider now the positive numbers ξ_1, ξ_n such that

$$\xi_n \le -1 + 2\, C_2^{(m,p,r)} \quad \text{and} \quad \xi_1 \ge \frac{1}{\Lambda_\Psi \Lambda_0} \frac{1}{1+C_1} , \tag{5.8}$$

where C_1 $(C_1 \ge C_1^{(r)})$ is a constant independent of h, r_1, r_2 and $\Lambda_\Psi \equiv M^2/n$, with M the M-condition number of Ψ_r. Then, it can be readily seen that

$$\xi_n = O(h^{-1}(r_1^{-1} + h^{-1}r_2^{-1})) , \tag{5.9}$$

while the behavior of the values of $C_1^{(r)}$, $C_2^{(m,p,r)}$ is closely related to the eigenvalue-bound distribution for the 3D-model problem (i.e. the Laplace equation in the unit cube with zero boundary values) as $r_1 \to m-1$ and $r_2 \to p-1$, cf.[12].

The following theorem governs the rate of convergence of the E-algorithm (4.1):

Theorem 1: Let $A=LU$ and M_r be positive definite matrices, as defined earlier in this Section. Suppose that there exist positive numbers ξ_1, ξ_n such that

$$(LUu, u)/(M_r u, u) \in [\xi_1^*, \xi_n^*] , \quad \forall\, u \neq 0 ,$$

where ξ_1 is independent of h and $\xi_n = O(h^{-1}r_1^{-1} + h^{-2}r_2^{-1})$. Then, the number of iterations of the iterative scheme (5.1) required to reduce the L_2-norm of the error by a factor $\varepsilon>0$ for proper values of $\{\alpha_i\}$, $i=1,2,\ldots,v-1$ is given by

$$O(h^{-1/2}r_1^{-1/2}r_2^{-1/2}\,(r_2 + h^{-1}r_1)^{1/2} \log\varepsilon^{-1}) .$$

Furthermore, the total number of arithmetic operations required for the calculation of the solution u_v is given by

$$O(h^{-7/2}(r_1+r_2)(r_2+h^{-1}r_1)^{1/2}r_1^{-1/2}r_2^{-1/2} \log\varepsilon^{-1}) .$$

Proof: From the hypothesis of the theorem the convergence factor becomes

$$\frac{\xi_1+\xi_n}{\xi_n-\xi_1} = 1 + \frac{2}{(\xi_n/\xi_1)-1} = 1 + 0\ (hr_1r_2(r_2+h^{-1}r_1)^{-1}) \ . \qquad (5.10)$$

The maximum absolute value of the polynomial $F_v(x)$ of degree v on $[\xi_1, \xi_n]$, is given by

$$\max_{x\in[\xi_1,\xi_n]} [F_v(x)] = \left[C_v(\frac{\xi_1+\xi_n}{\xi_n-\xi_1})\right]^{-1} , \qquad (5.11)$$

where C_v is the known Chebychev polynomial of degree v, i.e.

$$C_v(x) \equiv \cosh(v\cosh^{-1}x),\ \cosh^{-1}x \equiv \log[x+(x^2-1)^{1/2}] \ , \quad x\geq 1 \ .$$

Then, the number of iterations of the iterative scheme (5.1) required to make (5.11) less than $\varepsilon > 0$ is given by

$$v = 0(h^{-1/2}r_1^{-1/2}r_2^{-1/2}\ (r_2 + h^{-1}r_1)^{1/2}\ \log\varepsilon^{-1}) \ .$$

Since each iteration requires $0(h^{-3}(r_1+r_2))$ arithmetic operations to be carried out, we conclude that the total number of arithmetic operations involved in the determination of u_v, is given by

$$0(h^{7/2}(r_1 + r_2)(r_2+h^{-1}r_1)^{1/2}\ r_1^{-1/2}\ r_2^{-1/2}\ \log\varepsilon^{-1}) \ .$$

Finally, we state that analogous results for the E-algorithm in the case of the Explicit Iterative Methods [10] can be obtained in a similar manner.

REFERENCES

[1] DOUGLAS J., RACHFORD H.H. (1956): On the numerical solution of heat conduction problems in two or three space variables, Trans. Amer. Math. Soc. 82, 421-439.

[2] ENGELI, M., GINSBURG T., RUTISHAUSER H., STIEFEL E. (1959): Refined iterative methods for computation of the solution and the eigenvalues of self-adjoint B.V. problems, in: Mitt. Inst. Angew. Math. 8 (Birkhauser Zurich).

[3] EUCLID (348 B.C.): Elements, Book VII: 1 & 2, Book X:2 & 3, Proc. Platonic Academy 4009.

[4] EVANS D.J. (1985): Group explicit iterative methods for solving large linear systems, Intern. J. Computer Math. 17, 81-108.

[5] FADEEVA V.M.: Computational methods of Linear Algebra, transl. by C.D. Benster (1959), Dover publ. Inc., N.Y. 1950.

[6] GOURLAY A.R.: The acceleration of the Peaceman-Rachford method by Chebychev polynomials, Comp. J. 10, 378-382.

[7] HAGEMAN L.A., LUK F.T., YOUNG D.M. (1980): On the equivalence of certain iterative acceleration methods, SIAM J. Numer. Anal. 17, 852-873.

[8] HESTENES M.R., STIEFEL E. (1952): Methods of Conjugate Gradients for solving linear systems, J.Res.Nat.Bur. Standards 49, 409-436.

[9] HUMMEL P.M., SEEBECK C.L. (1949): A generalization of Taylor's Theorem, Amer. Math. Monthly 56, 243-247.

[10] LIPITAKIS E.A. (1984): Solving elliptic boundary-value problems on parallel processors by Approximate Inverse Matrix semi-direct methods based on multiple Jacobi iteration, Intern. J. of Comp. and Maths with Applics 10, 171-184.

[11] LIPITAKIS E.A. (1984): The use of second-degree normalized implicit Conjugate-Gradient methods for solving large sparse systems of linear equations, J. of Comp. and Applied Maths 11, 39-47.

[12] LIPITAKIS E.A. (1986): Approximate root-free factorization techniques for solving elliptic difference equations in three-space variables, Linear Algebra and its Applics 76, 247-269.

[13] LIPITAKIS E.A., EVANS D.J. (1980): Solving non-linear elliptic difference equations by extendable sparse factorization procedures, Computing 24, 325-339.

[14] LIPITAKIS, E.A., EVANS D.J. (1981): The rate of convergence of an approximate matrix factorization semi-direct method, Numer. Math. 36, 237-251.

[15] LIPITAKIS E.A., EVANS D.J. (1987): Explicit semi-direct methods based on approximate inverse matrix techniques for solving BV

problems on parallel processors, Math. & Comp. in Simulation 29, 1-17.

[16] MANTEUFFEL, T.A. (1982): Optimal parameters for linear second-degree stationary iterative methods, SIAM J. Numer. Anal. 19, 833-839.

[17] MARCHUK G.I.: Methods of Numerical Mathematics, Springer-Verlag, 1975.

[18] PADE H. (1892): Sur la representation approchee d'une function par des fractions rationnelles, these, Ann. de l'Ec. Nor. 3, 93.

[19] PEACEMAN D.W., RACHFORD H.H. (1955): The numerical solution of parabolic and elliptic differential equations, J.Soc. Ind. Appl. Math. 3, 28-41.

[20] STIEFEL E. (1958): Kernel polynomials in linear algebra and their numerical applications, National Bureau of Standards, Applied Math. Series 49, 1-22.

[21] VARGA R.: Matrix Iterative Analysis, Prentice-Hall, 1962.

[22] YANENKO N.N. (1971): The method of fractional steps, Springer-Verlag.

[23] YOUNG D. (1971): Iterative solution of large linear systems, Academic Press.

Prof. E.A. Lipitakis
Department of Informatics
Athens University of Economics
76 Patission Street
Athens 104 34 - Greece.

AN EFFICIENT FINITE ELEMENT METHOD FOR NONLINEAR DIFFUSION PROBLEMS

OWE AXELSSON
Department of Mathematics, University of Nijmegen
NL-6525 ED Nijmegen, The Netherlands

IVAR GUSTAFSSON
Department of Computer Science, Chalmers University of Technology
S-412 96 Göteborg, Sweden

Abstract. A mixed variable finite element method is used for the derivation of an efficient iterative method for nonlinear diffusion problems. The formulation has two advantages as compared to classical finite element methods:
(i) updating of the material coefficients is simplified
(ii) the discrete approximation is much more accurate for problems with (almost) discontinuous coefficients where the discontinuity occurs in the interior of the elements.

Numerical tests with various iterative methods such as Picard's method and Newton's method and with various preconditioning techniques are performed and compared.

Key Words. Nonlinear diffusion problems, mixed variable finite element method, preconditioned iterative methods, incomplete factorization

1. Introduction. Consider the finite element method based on the primal variational formulation for a diffusion problem

$$-\nabla\cdot(a\nabla u)=f \quad , x\in\Omega\subset R^n \tag{1.1a}$$

$$u=g_1 \quad , x\in\Gamma_D \tag{1.1b}$$

$$a\nabla u\cdot\hat{n}=g_2 \quad , x\in\Gamma_N=\Gamma\setminus\Gamma_D \,, \tag{1.1c}$$

where $a=a(u\,,\nabla u)>0$. Let $V_N=SPAN\{\phi_1,\phi_2,\cdots,\phi_N\}$, where ϕ_i are continuous piecewise polynomials of degree k on a finite element mesh on Ω and let $u_h=\sum_{j=1}^{N}\alpha_j\phi_j$, $u_h=g_1$ on Γ_D be the corresponding finite element approximation of u.

We then have

$$\int_\Omega a\nabla u_h\cdot\nabla\phi_i\,d\Omega=\int_\Omega f\,\phi_i\,d\Omega+\int_{\Gamma_N}g_2\,\phi_i\,d\Omega$$

or

$$K(u_h\ .\ \nabla u_h)\,\alpha = f^{(1)} + f^{(2)} \tag{1.2}$$

where $K_{ij} = \int\limits_{\Omega} a\nabla\phi_j \cdot \nabla\phi_i\, d\Omega$, $f_i^{(1)} = \int\limits_{\Omega} f\phi_i\, d\Omega + \int\limits_{\Gamma_N} g_2\phi_i\, d\Omega$ and $f^{(2)}$ occurs because of the Dirichlet boundary condition on Γ_D.

In practice we have to use numerical integration on each element and an assembling process for the evaluation of K and $f^{(1)}$. This method suffers from two disadvantages:

(i) If a is a function of u (and of ∇u), updating of K means that the numerical integration and assembling must be repeated.

(ii) Unless lines of discontinuities of a happens to be meshlines, the approximation u_h is less accurate near such lines. Similar considerations are valid in subregions where a has sharp gradients.

We propose a mixed variable method which does not suffer from the above disadvantages (see Section 2). The method leads to a system of nonlinear equations of the form

$$BM^{-1}(u_h\ .\ \nabla u_h)B^T\alpha = f\,, \tag{1.3}$$

where B only depends on the mesh but not on the material property a. The coefficients in M depend on a, and hence on u and/or ∇u in general but M is often a diagonal matrix.

We prove that for linear and quadratic finite elements and appropriate quadrature rules the method is identical to the usual method and hence (for smooth problems) shares the same advantages as this as far as asymptotic accuracy is consemed, i.e. for a , f , Ω , g_1 , g_2 smooth enough

$$\|u - u_h\|_0 = O(h^{s+1})|u|_{s+1}\,,\ h \to 0$$

for some s , $0 \le s \le k$. Here, $|\cdot|$ is the Sobolev seminorm and h is a meshsize parameter.

As we shall see, however, for problems with discontinuous coefficients for which the line of discontinuity is not a boundary line of elements, it is much more accurate than the classical method if we perform the integration within the "discontinuous" elements with greater care. Such an observation for a similar method for a one-dimensional problem was already made by Babuska and Osborn [8]. The same property is valid for problems with material coefficients with steep gradients. The present paper is an extention of an earlier report [5]. For a more general discussion of the $BM^{-1}B^T$-form for stiffness matrices, see [6].

In Section 2, we present the mixed variable finite element method. In Section 3, we discuss the discretization errors of the new method and compare them with those of the classical formulation. In Section 4, various iterative methods for the solution of (1.3) is discussed. In particular, for reasons of efficiency, we advocate preconditioned iterative methods where a preconditioner for $BM^{-1}B^T$ is constructed from the lowest order finite elements. Such a method has been used earlier in [4].

The presented method is of interest for nonlinear diffusion problems and electromagnetic field problems for instance, but should also simplify the solution of the unknown coefficient problem i.e. the problem to fit an unknown function a to some measurements of the solution u. It has also clear advantages in connection with free boundary value problems where one rarely can fit a mesh to the free boundary.

2. The mixed variable formulation. Let $z = a\nabla u$. Then (1.1) takes the form

$$\frac{1}{a}z - \nabla u = 0 \, , \; x \in \Omega \tag{2.1a}$$

$$-\nabla \cdot z = f \, , \; x \in \Omega \tag{2.1b}$$

$$u = g_1 \, , \; x \in \Gamma_D \tag{2.1c}$$

$$z \cdot \hat{n} = g_2 \, , \; x \in \Gamma_N \tag{2.1d}$$

The variational formulation of (2.1) is

$$\int_\Omega \frac{1}{a} z \cdot \hat{z} \, d\Omega - \int_\Omega \nabla u \cdot \hat{z} \, d\Omega = 0 \, , \; \forall \, \hat{z} \in [L^2(\Omega)]^n \tag{2.2a}$$

$$\int_\Omega z \cdot \nabla v \, d\Omega = \int_\Omega f v \, d\Omega + \int_{\Gamma_N} g_2 v \, d\Gamma \, , \; \forall \, v \in H_0^1(\Omega) = \{v \in H^1(\Omega) \, , \; v = 0 \text{ on } \Gamma_D\} \tag{2.2b}$$

As usual, if the solution of (2.2) is sufficiently regular, it is also a solution of (2.1) and (1.1).

For the discretization of (2.2), we construct finite element subspaces $V \subset H_0^1(\Omega)$ and $W \subset [L^2(\Omega)]^n$. The finite element variational formulation then reads:
Find $u_h \in V_1 \subset H^1(\Omega)$, $u_h = g_1$ on Γ_D and $z_h \in W$ such that

$$\int_\Omega \frac{1}{a} z_h \cdot \hat{z}_h \, d\Omega - \int_\Omega \nabla u_h \cdot \hat{z}_h \, d\Omega = 0 \, , \; \forall \, \hat{z}_h \in W \tag{2.3a}$$

$$\int_\Omega z_h \cdot \nabla v_h \, d\Omega = \int_\Omega f v_h \, d\Omega + \int_{\Gamma_N} g_2 v_h \, d\Gamma \, , \; \forall \, v_h \in V \tag{2.3b}$$

By $u_h = g_1$ we mean that $u_h = g_1$ at the nodepoints on Γ_D, but is in general an approximation of g_1 on Γ_D, derived for instance by using isoparametric elements.

We consider at first a finite element triangulation of Ω, i.e. $\bigcup_{l=1}^{T} e_l = \Omega$ with two choices of finite element basis functions corresponding to $k = 1$ and $k = 2$, respectively:

(a) the piecewise linear - piecewice constant combination

(b) the piecewise quadratic - piecewise linear combination

In case (a), we let $\phi_i^{(l)}$, $i = 1, 2, 3$ be the standard linear Lagrangean basis functions on the element e_l and $\psi_4^{(l)} = 1$ be the constant function with support just on e_l. See Figure 2.1 for the location of nodepoints. With $N_j^{(l)}$ being the nodepoints on e_l we thus have

$$u_h|_{e_l} = \sum_{i=1}^{3} u_h(N_i^{(l)})\phi_i^{(l)}$$

$$z_h|_{e_l} = z_h(N_4^{(l)})\psi_4^{(l)}$$

Similarly, in case (b), we let $\psi_i^{(l)}$, $i = 1, \cdots, 6$ be the standard quadratic Lagrangean basis functions on e_l and $\psi_i^{(l)}$, $i = 4, 5, 6$ be linear Lagrangean basis functions associated to the mid-edge points of e_l, with support just on e_l, see Figure 2.1 for the location of nodepoints. For instance,

$$\psi_4^{(l)}(N_j^{(l)}) = \begin{cases} 1, & j = 2, 4, 3 \\ 0, & j = 5, 6 \\ -1, & j = 1 \end{cases}$$

Observe that the global basis functions $\{\phi_i\}_{i=1}^{N}$ have support on several elements while the global basis functions $\{\psi_i\}_{i=1}^{N'}$ have support on just one element. As a consequence, in case (b) there are two sets of unknowns z_h at each mid-edge point (except on the boundary).

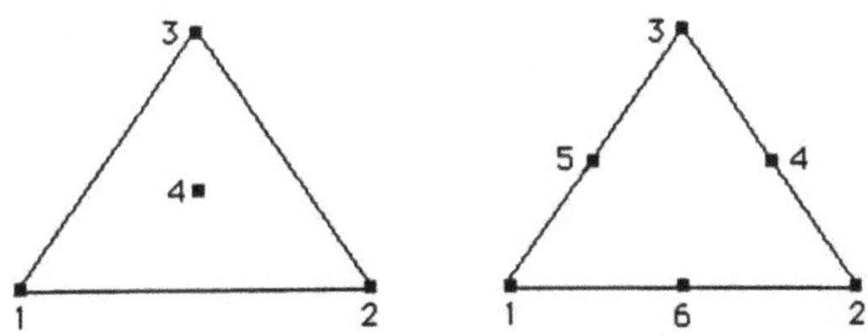

Figure 2.1 Location of nodepoints for finite element basis functions, case (a) to the left and case (b) to the right.

In two space dimensions we use $\{(\psi_i\,,0)^T\}_{i=1}^{N'}\cup\{(0\,,\psi_i)^T\}_{i=1}^{N'}$ as a basis for W and similarly in three space dimensions. However, for simplicity, in the following we confine our study to two space dimensions. We express the approximations as $z_h=\sum_{i=1}^{N'}\beta_i^{(1)}(\psi_i\,,0)^T+\sum_{i=1}^{N'}\beta_i^{(2)}(0\,,\psi_i)^T$ and $u_h=\sum_{i=1}^{N}\alpha_i\phi_i$. The system of nonlinear equations corresponding to (2.3) then takes the form

$$\begin{bmatrix} M^{(1)} & 0 & -B^{(1)T} \\ 0 & M^{(2)} & -B^{(2)T} \\ B^{(1)} & B^{(2)} & 0 \end{bmatrix}\begin{bmatrix}\beta^{(1)}\\ \beta^{(2)}\\ \alpha\end{bmatrix}=\begin{bmatrix} g^{(1)}\\ g^{(2)}\\ f\end{bmatrix} \tag{2.4}$$

where, with e_j and Ω_i being the support of ψ_j and ϕ_i, respectively,

$$M_{ij}^{(1)}=M_{ij}^{(2)}=\int_{e_j}a^{-1}\phi_i\psi_j\,d\Omega \tag{2.5a}$$

$$B_{ij}^{(1)}=\int_{e_j}\psi_j\frac{\partial}{\partial x}\phi_i\,d\Omega \tag{2.5b}$$

$$B_{ij}^{(2)}=\int_{e_j}\psi_j\frac{\partial}{\partial y}\phi_i\,d\Omega \tag{2.5c}$$

$$f_i=\int_{\Omega_i}f\,\phi_i\,d\Omega+\int_{\Omega_i\cap\Gamma_N}g_2\,\phi_i\,d\Gamma \tag{2.5d}$$

and $g^{(1)}$ and $g^{(2)}$ occurs because of $u_h=g_1$ on Γ_D.

Observe that for an unisotropic problem e.g. modelling the flow about an airfoil for instance, $M_{ij}^{(1)}\neq M_{ij}^{(2)}$ in (2.5a) due to different material coefficients $a^{(1)}$ and $a^{(2)}$, say. The method presented here, however, works well also for this more general type of problems.

We note that the nonlinearity appears only in the material property a and that the matrices $B^{(1)}$ and $B^{(2)}$ do not depend on a and can hence be calculated once and for all before starting an iterative process for solving the nonlinear system.

The equations (2.4) may be written

$$\begin{bmatrix} M & -B^T \\ B & 0\end{bmatrix}\begin{bmatrix}\beta\\ \alpha\end{bmatrix}=\begin{bmatrix}g\\ f\end{bmatrix}$$

and by elimination of β we get

$$BM^{-1}B^T\alpha=f-BM^{-1}g \tag{2.6a}$$

or

$$[B^{(1)}M^{(1)-1}B^{(1)T}+B^{(2)}M^{(2)-1}B^{(2)T}]\alpha=f-B^{(1)}M^{(1)-1}g^{(1)}-B^{(2)}M^{(2)-1}g^{(2)} \quad (2.6b)$$

For the evaluation of $M_{ij}^{(k)}$, $k=1,2$, in general we have to use numerical integration. In case (b) we use the rule (exact for quadratic polynomials)

$$\int_{e_l} f\,d\Omega=\frac{1}{3}area(e_l)\sum_{j=4}^{6}f(N_j^{(l)}) \quad (2.7)$$

where $N_j^{(l)}$, $j=4,5,6$ are the mid-edge points, and then we get

$$M_{ij}^{(k)}=\frac{1}{3}area(e_j)\,a^{-1}(N^{(j)})\,\delta_{ij}\,,\ k=1,2$$

where δ_{ij} is the Kronecker delta and $N^{(j)}$ is the point associated to ψ_j i.e. the point where $\psi_j=1$. We observe that $M^{(k)}$, $k=1,2$ are diagonal matrices which simplifies the updating and thus the iterative solution method.

In case (a) we may also use the simple one-point quadrature rule

$$\int_{e_l} f\,d\Omega=area(e_l)\,f(N_0^{(l)}) \quad (2.8)$$

where $N_0^{(l)}$ is the centre of gravity of the element.

In general, for triangles and p-point quadrature rules, where $p=\frac{k(k+1)}{2}$ for a k-degree finite element approximation, $M^{(1)}$ and $M^{(2)}$ (and hence M) become diagonal if Lagrangean basis functions $\{\psi_i\}$ are associated to the quadrature-points.

A similar method may be developed for quadri-laterals. For rectangular elements we let $V=SPAN\{\{\phi_i^{(l)}\}_{i=1}^{m^2},\ l=1,2,\cdots,R\}$, where R is the number of rectangles and $\phi_i^{(l)}$ are bi-polynomial Lagrangean basis functions of degree $m-1$ in the $m\times m$ Lobatto quadrature points. Further, we let $W=SPAN\{\psi_i^{(l)}\}_{i=1}^{(m-1)^2}$, where $\psi_i^{(l)}$ are the Lagrangean basis functions of degree $m-2$ in the $(m-1)\times(m-1)$ Gaussian quadrature points. For the location of nodepoints in the case $m=3$ see Figure 2.2. It is left to the interested reader to establish that also in this case the matrices $M^{(k)}$, $k=1,2$ are diagonal.

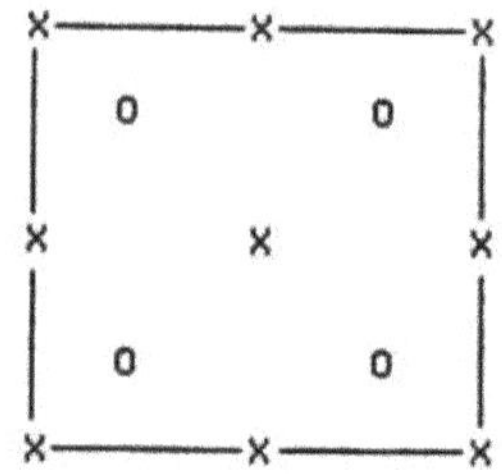

Figure 2.2 Location of Lobatto (x) and Gauss (o) quadrature points (m = 3).

3. Discretization errors. At first we prove that the mixed variable and classical formulations are identical if basis functions and quadrature rules are chosen accordingly. Since the generalization is straightforward, we confine the proof to a triangulation in two space dimensions.

Let $\{N_m^{(l)}\}_{m=1}^p$ be the quadrature points, where $p = \frac{k(k+1)}{2}$ for a k-degree finite element approximation, and let the Lagrangean basis functions $\{\psi_i\}_{i=1}^p$ be associated to these points. For the classical formulation (1.2) we then have

$$K_{ij} = \int_\Omega a \nabla\phi_j \cdot \nabla\phi_i \, d\Omega = K_{ij}^{(1)} + K_{ij}^{(2)}$$

where $K_{ij}^{(1)} = \int_\Omega a \frac{\partial\phi_i}{\partial x}\frac{\partial\phi_j}{\partial x} \, d\Omega$ and $K_{ij}^{(2)} = \int_\Omega a \frac{\partial\phi_i}{\partial y}\frac{\partial\phi_j}{\partial y} \, d\Omega$. Using the actual quadrature rule we get

$$K_{ij}^{(1)} = \sum_{l=1}^{T} \sum_{m=1}^{p} \gamma_m^{(l)} a(N_m^{(l)}) \frac{\partial\phi_i}{\partial x}(N_m^{(l)}) \frac{\partial\phi_j}{\partial x}(N_m^{(l)}) \tag{3.1}$$

where T is the number of triangles and $\gamma_m^{(l)}$ are quadrature-weights.

For the mixed formulation we have, see (2.5) and (2.6),

$$B^{(1)}M^{(1)-1}B^{(1)T} = \sum_{k=1}^{N'} b_{ik}^{(1)}(m_{kk}^{(1)})^{-1} b_{jk}^{(1)} = \sum_{k=1}^{N'} \int_\Omega \psi_k \frac{\partial\phi_i}{\partial x} \, d\Omega \, / \left[\int_\Omega a^{-1}\psi_k^2 \, d\Omega\right]^{-1} \int_\Omega \psi_k \frac{\partial\phi_j}{\partial x} \, d\Omega.$$

Using the quadrature rule and observing that ψ_k has support on just one element, say e_k, and that $\psi_k = 0$ in all points but one, say N_k with quadrature-weight γ_k, where $\psi_k = 1$, we then have

$$B^{(1)}M^{(1)-1}B^{(1)T} =$$

$$= \sum_k^{N'} \sum_{m=1}^{p} \gamma_m^{(k)} \psi_k (N_m^{(k)}) \frac{\partial \phi_i}{\partial x}(N_m^{(k)}) \left(\sum_{m=1}^{p} \gamma_m^{(k)} a^{-1}(N_m^{(k)}) \psi(N_m^{(k)})^2 \right)^{-1} \sum_{m=1}^{p} \gamma_m^{(k)} \psi_k (N_m^{(k)}) \frac{\partial \psi_j}{\partial x}(N_m^{(k)})$$

$$= \sum_k^{N'} \gamma_k \frac{\partial \phi_i}{\partial x}(N_k) \{\gamma_k a^{-1}(N_k)\}^{-1} \gamma_k \frac{\partial \phi_j}{\partial x} N_k = \sum_k^{N'} \gamma_k a(N_k) \frac{\partial \phi_i}{\partial x}(N_k) \frac{\partial \phi_j}{\partial x}(N_k) \tag{3.2}$$

We recall that if a point (node) belongs to more than one element, different indices, say k_1 and k_2, refer to this same node i.e. $N_{k_1} = N_{k_2}$. Rewriting (3.2) as a sum over elements then gives (3.1). In a similar way, one proves that $K_{ij}^{(2)} = B^{(2)}M^{(2)-1}B^{(2)T}$ and that the right-hand sides of (1.2) and (2.6b) are the same. Hence, we get the same discretization $u_h = \sum_{j=1}^{N} \alpha_j \phi_j$ with well known error bounds

$$\|u - u_h\|_0 \le C\, h^{s+1} |u|_{s+1} \;, \quad 0 \le s \le k. \tag{3.3}$$

For this kind of error estimates for nonlinear problems, see further [9], [11] and [1].

We remark that the formulations are not in general identical. Indeed, it depends on the numerical integration. For instance, the formulas (2.8) and (2.7) are appropriate for case (a) and (b), respectively. However, with case (a) i.e. linear finite element discretization and the three-point quadrature rule (2.7) the approximations are not the same. Actually, one of the aims of this paper is to show that the mixed variable formulation is much more accurate for discontinuous and rapidly varying material coefficients. To achieve this, we apparently have to perform more accurate integration than what is given by these quadrature rules. A consequence of using more accurate integration is that the nice property of $M^{(1)}$ and $M^{(2)}$ being diagonal matrices in general gets lost. However they will be block-diagonal.

In the following, we present first some numerical tests for the nonlinear problem (1.1) with Ω being the unit square $0 \le x \le 1$, $0 \le y \le 1$ and $\Gamma_D = \Gamma$. Linear ($k = 1$) or quadratic ($k = 2$) finite elements on a uniform rightangled triangulation and the quadrature rules (2.8) and (2.7) have been used, respectively. Hence, the classical and mixed variational methods are identical.

(P1) $a(u) = 1 + \gamma u$ with $\gamma = 2$ and $k = 2$

(P2) $a(u) = 6 + \frac{10}{\pi} \arctan \gamma(u_0 - u)$ with $u_0 = 1/32$, $\gamma = 1$ and $k = 2$

(P3) the same as (P2) but with $\gamma = 100$

(P4) $a(u) = \gamma^{-1} + \arctan \gamma u$ with $\gamma = 10$ and $k = 1$

(P5) $a(u) = \gamma + u$ with $\gamma = 10^{-2}$ and $k = 1$

The problem data, f and g_1, were chosen such that the true solution of (P1), (P2) and (P3) is $u = x(1-x)y(1-y)$. For (P4) and (P5), the true solution is chosen according to

$$v(u) = \int_0^u a(s)ds \tag{3.4}$$

where $v(x,y) = x^3y - xy^3$ and $v = xy$, respectively. We note that (in both cases), $\Delta v = 0$ i.e. v is a harmonic function. For (P5) we readily find $v = \gamma u + \frac{u^2}{2}$ and hence

$$u = -\gamma + \sqrt{\gamma^2 + 2v} = \frac{2xy}{\gamma + \sqrt{\gamma^2 + 2xy}}$$

where the last equality avoids cancellation of digits when $v = xy$ is small.

For (P4), the integral equation (3.4) is solved to determine u on (parts of) the boundary (and in the interior of Ω too, if the discretization error has to be calculated).

The following discretization errors in averaged l_2-norm i.e. $[\frac{1}{N}\sum_{i=1}^{N}(u_h(N_i) - u(N_i))^2]^{1/2}$, where N_i, $i = 1, \cdots, N$ are the nodes, were achieved for some values of the meshsize h.

h	1/6	1/12	1/24
P1	$.12\ 10^{-3}$	$.74\ 10^{-5}$	$.47\ 10^{-6}$
P2	$.34\ 10^{-4}$	$.21\ 10^{-5}$	$.13\ 10^{-6}$
P3	$.79\ 10^{-1}$	$.27\ 10^{-2}$	$.13\ 10^{-3}$
P4	$.16\ 10^{-2}$	$.54\ 10^{-3}$	$.19\ 10^{-3}$
P5	$.82\ 10^{-1}$	$.23\ 10^{-2}$	$.65\ 10^{-3}$

For (P4) and (P5) the result is in agreement with (3.3) while for (P1), (P2) and (P3) we notice an $O(h^4)$ asymptotic behaviour which is one order more than can be expected and is due to a cancellation effect in the pointwise error.

More interesting is the behaviour of the discretization error for a problem with discontinuous material coefficient and where the line of discontinuity cuts through the interior of some elements. Such a case appears for moving boundary value problems, for instance. We shall consider our testproblem above with

$$\text{(P6)} \quad a = \begin{cases} 1\ , & x \le ½ \\ a_0\ , & x > ½ \end{cases} ,$$

i.e. a linear, essentially one-dimensional problem.

The problem data were chosen such that the exact solution is

$$u = \begin{cases} x(\gamma - x) & ,\ 0 < x \le ½ \\ (1-x)(x+1-\gamma)a_0^{-1} & ,\ ½ \le x < 1 \end{cases}$$

where $\gamma = (a_0+3)/(2a_0+2)$ so that u is continuous and $a\nabla u$ is continuously differentiable. In this example the integrals involved are calculated exactly for the two methods, the primal (classical) and the mixed formulations, which therefore are not identical.

We notice, as is also pointed out in [8] for a one-dimensional and for the piecewise linear - piecewise constant combination i.e. $k = 1$, that in the mixed formulation the material coefficients appear as harmonic averages.

The following discretization errors in the averaged l_2-norm were achieved for different values of the meshsize h and different values of the discontinuity-parameter a_0.

a_0	10		100		1000	
h	Primal	Mixed	Primal	Mixed	Primal	Mixed
$\frac{1}{5}$	$.81\ 10^{-2}$	$.50\ 10^{-3}$	$.17\ 10^{-1}$	$.75\ 10^{-3}$	$.18\ 10^{-1}$	$.94\ 10^{-3}$
$\frac{1}{9}$	$.49\ 10^{-2}$	$.46\ 10^{-3}$	$.94\ 10^{-2}$	$.13\ 10^{-2}$	$.10\ 10^{-1}$	$.14\ 10^{-2}$
$\frac{1}{17}$	$.27\ 10^{-2}$	$.46\ 10^{-3}$	$.49\ 10^{-2}$	$.92\ 10^{-3}$	$.52\ 10^{-2}$	$.99\ 10^{-3}$

We see that already for these fairly coarse meshes the mixed method gives excellent results and much better than for the standard finite element method. This latter method seems to converge with a rate $O(h)$ in the l_2-norm. When $h \to 0$ the error in the mixed method must converge with the same rate as the standard method. Since the former method is already so accurate for coarse meshes, the price to be paid is that the rate of convergence for intermediate values of h is slow or even absent. This plateau-effect was also observed for certain finite element approximations of singular perturbation problems, see [3]. Needless to say, we are happy with the already accurate approximation for the coarse meshes.

Accurate computations of gradients. We finish this section by a remark on the possibility to calculate the gradient ∇u in certain directions and at certain points with good accuracy. At first we note that the variable $z = a\nabla u$ is approximated by z_h to a lower accuracy than u_h approximates u. For $k = 1$ (the linear-constant combination) and sufficiently regular meshes, u_h satisfies (3.3) with $s = 1$ i.e. at a point P, the approximation u_P fulfills

$$u_P = u(P) + c(P)h^2 + O(h^3) \tag{3.5}$$

If u is sufficiently smooth, $c(P)$ is also smooth. In order to explain our idea we consider the following meshpoints in a uniform mesh

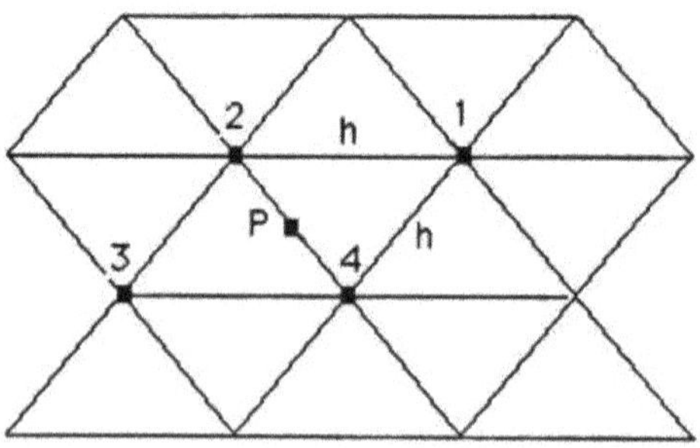

Let $\nabla u \cdot e_{ij}$ be the gradient in the direction $i \to j$. Then (3.5) gives, since $c(2) - c(4) = O(h)$ and $c(1) - c(3) = O(h)$,

$$\nabla u \cdot e_{42} = \frac{u_2 - u_4}{h} + O(h^2)$$

$$\nabla u \cdot e_{31} = \frac{u_1 - u_3}{h\sqrt{3}} + O(h^2) ,$$

for instance. In a similar way one can calculate ∇u in any direction and at any point to an accuracy of $O(h^2)$.

We leave to the interested reader to show that, using the same technique together with Richardson-extrapolation, it is possible for higher order finite elements as well to calculate ∇u with accuracy $O(h^{k+1})$, if u is sufficiently smooth.

4. Iterative solution methods. For the nonlinear system of equations (2.6), we shall consider iterative solution methods. In order to simplify notations, we let $A = BM^{-1}B^T$ and

$b = f - BM^{-1}g$. We recall that the nonlinearity appears only in $M = M(u_h, \nabla u_h) = M(\alpha)$ so that the system to be solved may be written

$$A(\alpha) \equiv BM^{-1}(\alpha)B^T\alpha = b \equiv f - BM^{-1}(\alpha)g \tag{4.1}$$

In particular, we shall advocate preconditioned iterative methods where the preconditioning is based on spectral equivalence. Let $A^{(k)}$ denote the matrix of (4.1) when finite elements of order k is used. For the system (4.1) with $k = 2$ we here use $A^{(1)}$ as preconditioning matrix. A simple preconditioned Picard's method then reads

$$A^{(1)}(\alpha^{(m)})(\alpha^{(m+1)} - \alpha^{(m)}) = -\tau_m(A^{(2)}(\alpha^{(m)})\alpha^{(m)} - b), \quad m = 0, \cdots \tag{4.2}$$

where τ_m is a so called damping parameter, present in order to enlarging the domain of convergence.

In each iteration of (4.2) we have to solve a linear system of equations with matrix $A^{(1)}(\alpha^{(m)})$. This could be done by a direct method of elimination type but we consider iterative methods for this system too and we speak about inner iterations while the iterations in (4.2) are called outer iterations. This permits us to keep the factored form of the matrix $A^{(1)} = BM^{-1}B^T$. In a direct method we would have to update and form the matrix after every outer iteration while in an iterative method we just have to update M.

It also becomes clear why we choose this type of preconditioning. The matrix $B^{(1)}$ is more sparse than $B^{(2)}$ which simplifies the iterations. Furthermore it is well known, see [4] and [10], that the matrices $A^{(1)}$ and $A^{(k)}$ are spectrally equivalent i.e. there exists positive constants c_1 and c_2, independent of the meshsize h, such that

$$c_1 x^T A^{(1)} x \le x^T A^{(k)} x \le c_2 x^T A^{(1)} x, \quad \forall x \in R^N \tag{4.3}$$

where N is the order of the matrices (the number of degrees of freedom). For instance, for a linear problem with constant coefficient a, $k = 2$ and isosceles triangles, $c_1 = 2/3$ and $c_2 = 4/3$. These values are also valid asymptotically as $h \to 0$ for a nonlinear problem with smoothly varying material coefficients.

Another advantage with the iterative approach is that the residuals can be calculated more accurately i.e. with less round-off errors, see [4]. The idea in this reference of using differences of first order is already present in the actual factored form of $A^{(k)} = BM^{-1}B^T$.

We present numerical results for the problem (P1) in Section 3 with $a(u) = 1 + \gamma u$, where γ is a parameter used to vary the degree of nonlinearity. The matrix M is diagonal since the recommended quadrature rule in Section 3 is used. Further we let τ_m in (4.2) be constant. It follows from (4.3) and the comments on the spectral equivalence bounds below this formula that the optimal parameter for a linear problem, for h sufficiently small, should be

$$\tau_m \equiv \tau = \frac{2}{4/3 + 2/3} = 1$$

In our nonlinear problem a value $0.9 \le \tau \le 1$ has turned out to be optimal.

As starting approximation and stopping criterion we have used $\alpha^{(0)} = 0$ and a relative residual error $\varepsilon_o = 10^{-4}$, respectively. The number of outer iterations which was required is denoted by p_o.

For the solution of the linear system with matrix $A^{(1)}$ we have used the preconditioned conjugate gradient method with the MIC(2)-method for the preconditioning, see [10]. For the inner iterations we use the zero-vector as starting approximation and a relative residual error $\varepsilon_i = 10^{-2}$ as stopping criterion. The modified incomplete factorization was performed in the starting point only and was not updated.

	p_o			p_i		
h	$\gamma=2$	$\gamma=10$	$\gamma=50$	$\gamma=2$	$\gamma=10$	$\gamma=50$
1/6	6	7	11	12	19	42
1/12	6	7	10	18	20	49
1/24	6	6	8	25	26	51

Table 4.1 The number of outer and inner iterations for preconditioned Picard's method applied to the problem (P1) with $a(u) = 1 + \gamma u$ for different values of γ and h.

We observe that the number of outer itertions is fairly insensitive to the degree of nonlinearity of the problem in particular for small values of h. The number of inner iterations is in agreement with the theoretical result $p_i = O(h^{-1/2})$ (for solving one problem). If the MIC-factorization is updated, the number of inner iterations can be reduced. For instance, with $\gamma = 50$ and $h = 1/24$ the total number of inner iterations reduces to 38 if the factorization is updated every 2'nd outer iteration and to 35 if it is updated every outer iteration. Since updating costs little compared to other work involved, we recommend a frequent updating of the MIC-factorization unless the problem is almost linear.

Similar results were obtained for the test problem (P2) with different values of the parameter γ, if the stepzise h was sufficiently small. For large values of γ, say $\gamma \ge 100$ and h large, say $h \ge 1/8$, the convergence sometimes turned out to be slow or even absent. We realize that in general one has to choose a sufficiently fine mesh.

For a mildly nonlinear problem it is possible to use a preconditioned nonlinear variant of the conjugate gradient method, see for instance [2], for the outer iterations as well. For the present test problem (P1), this method converges for $0 \le \gamma \le 9$ and for $\gamma = 2$ (for instance) it is much faster than preconditioned Picard's method. However, we are interested in robust methods i.e methods which are not only reliable for mildly nonlinear problems and hence we exclude this method from the following discussion.

If an efficient solver for linear systems with BB^T (a black-box routine) is present and if the coefficient a is complicated (requires many operations to be calculated), the following variant of preconditioned Picard's method could be recommended:

$$BB^T(\alpha^{(m+1)} - \alpha^{(m)}) = -\tau_m BMB^T(BB^T)^{-1}(BM^{-1}B^T\alpha^{(m)} - b) \ , \quad m = 0, \cdots \tag{4.4}$$

In this method each iteration involves two solutions of systems with BB^T We also observe that the nonlinearity is not present at all in the solution of the linear systems; a is just needed for the calculation of the righthand side.

For the test problem (P1) with $\gamma = 50$ and $h = 1/24$, still only 6 outer iterations were required for the method (4.4) compared to 8 iterations for the method (4.2). The choice of which one is preferable depends on a and the efficiency of the BB^T-solver.

One advantage with the method (4.4) is that it requires few iterations for problems with almost constant coefficient a, i.e. when $M = c_1 I$. By a proper scaling this can be made true also for unisotropic (almost constant) problems i.e. when $M^{(1)} \approx c_1 I$ and $M^{(2)} \approx c_2 I$.

For completeness we also present and study the faster convergent, but somewhat more complicated, Newton's method. For the equation $-\nabla\cdot(a(u)\nabla u) = f$, we derive this method by introducing corrections δu and δz of u and z, respectively. This gives, omitting higher order terms in the corrections,

$$-\nabla\cdot(a(u+\delta u)\nabla(u+\delta u)) = f$$

$$-\nabla\cdot(a(u)\nabla(\delta u)) - \nabla\cdot(\frac{\partial a}{\partial u}(u)\delta u\,\nabla u) = f + \nabla\cdot(a(u)\nabla u)$$

$$-\nabla\cdot(a(u)(\nabla\delta u + \frac{1}{a(u)}\frac{\partial a}{\partial u}(u)\delta u\,\nabla u)) = f + \nabla\cdot z$$

By $\delta z = a(u)(\nabla\delta u + \frac{1}{a(u)}\frac{\partial a}{\partial u}(u)\delta u \nabla u) = a(u)(\nabla\delta u - \frac{\partial}{\partial u}(\frac{1}{a})(u))z$ we may formulate a mixed variable Newton's method:

$$\begin{cases} \frac{1}{a(u^{(m)})}\delta z - \nabla\delta u + \delta u \frac{\partial}{\partial u}(\frac{1}{a})(u^{(m)})z^{(m)} = 0 \\ -\nabla\cdot\delta z = f + \nabla\cdot z^{(m)} \end{cases}, \quad m = 0, \cdots \tag{4.5}$$

where $\delta u = u^{(m+1)} - u^{(m)}$ and $\delta z = z^{(m+1)} - z^{(m)}$.

A discrete version of (4.5), based on finite elements, can be written in matrix-form

$$\begin{bmatrix} M & -C^T \\ B & 0 \end{bmatrix} \begin{bmatrix} \beta^{(m+1)} - \beta^{(m)} \\ \alpha^{(m+1)} - \alpha^{(m)} \end{bmatrix} = \begin{bmatrix} g \\ f - BM^{-1}B^T\alpha^{(m)} \end{bmatrix} \tag{4.6}$$

Here, M, B, f and g are defined as in (2.4) - (2.5) and C is given by

$$c_{ij} = b_{ij} - \int_\Omega \frac{\partial}{\partial u}(\frac{1}{a})(u^{(m)})z^{(m)}\psi_j \phi_i \, d\Omega = b_{ij} + \int_\Omega \frac{1}{a(u^{(m)})}\frac{\partial a}{\partial u}(u^{(m)})\nabla u^{(m)}\psi_j \phi_i \, d\Omega$$

where ψ_j and ϕ_i are basis functions defined as in Section 2.

Eliminating β and introducing a damping parameter τ_m then gives the following (damped) Newton's method:

$$BM^{-1}(\alpha^{(m)})C^T(\alpha^{(m)})(\alpha^{(m+1)} - \alpha^{(m)}) = -\tau_m(BM^{-1}(\alpha^{(m)})B^T\alpha^{(m)} - b) \tag{4.7}$$

where $b = f - BM^{-1}g$ and we explicitly indicates the dependence on $\alpha^{(m)}$. Compare this method with preconditioned damped Picard's method (4.2).

We observe that C depends on $\alpha^{(m)}$ so we have lost the nice property of having to update just a diagonal matrix. We conclude that Newton's method (4.7) is competitive only if the convergence is much faster than for preconditioned Picard's method (4.2) or (4.4). For another implementation of Newton's method where the matrix C does not occur, see [6].

In Table 4.2 we present the number of outer iterations for these methods with the zero-vector as starting approximation and a relative residual error $\varepsilon_o = 10^{-4}$ as stopping criterion for the problems (P4) and (P5) for different values of γ and h. For all methods, the result is presented for the (experimentally obtained) optimal value of the constant damping parameter $\tau_m \equiv \tau$.

	P4, $\gamma = 1$			P4, $\gamma = 10$			P4, $\gamma = 100$			P5, $\gamma = 1$			P5, $\gamma = 10^{-2}$		
h	4.2	4.4	4.7	4.2	4.4	4.7	4.2	4.4	4.7	4.2	4.4	4.7	4.2	4.4	4.7
1/4	5	5	3	6	6	4	5	6	4	4	3	3	6	9	5
1/8	5	9	3	6	16	4	5	15	4	4	3	3	7	10	5
1/16	5	7	3	6	14	4	6	22	4	4	3	3	7	9	5

Table 4.2 The number of iterations for preconditioned Picard's method (4.2), the variant of this method (4.4) and Newton's method (4.7) for problems (P4) and (P5) (see Section 2) with different values of the parameter γ and the stepsize h.

It was mentioned above that the method (4.4) is preferable for almost constant coefficient problems. It becomes clear from these results that it is less efficient for problems where the degree of nonlinearity is larger. The two other methods, preconditioned Picard's method (4.2) and Newton's method (4.7) seam to be robust. The difference in convergence rate between these two methods is minor so it does not pay off to use the more complicated and more expensive Newton's method. This is the reason why we do not give any details regarding the solution of the linear systems involved in each iteration of this method.

5. Conclusions. It has been demonstrated how much simpler and efficient a mixed variable finite element method for a nonlinear diffusion problem can be as compared to the standard finite element method. The iterative solution method and the updating of the stiffness matrix is simplified in the new method.

For problems with composite materials or rough material coefficients, a much coarser mesh can be used for the mixed variable method for the same order of accuracy. For a model problem it has been pointed out how the material coefficients appear as harmonic averages in this method.

Numerical experiments show that, for solving the arising nonlinear system of equations, a simple preconditioned Picard's method is highly competitive as compared to the somewhat more complicated methods of Newton's type. A preconditioning based on spectral equivalence is used for higher order finite elements in order to simplify the solution of the linear system of equations to be solved in each iteration.

The method is well suited for implementation on parallel architectures because most of the work in each iteration is matrix-vector products. For an example of parallel implementation of preconditioned methods of this type, see [7].

Finally, we comment that the presented method is more stable for rounding errors than the standard method because the residual is calculated by first order differences.

REFERENCES

[1] O. AXELSSON, *Error estimates for Galerkin Methods for Quasilinear Parabolic and Elliptic Differential Equations in Divergence form*, Numer. Math. 28 (1977), 1-14.

[2] O. AXELSSON and U. NÄVERT, *On a Graphical Package for Nonlinear Partial Differential Equation Problems*, Information Processing 77, ed: B. Gilchrist, North-Holland Publishing Company, Amsterdam, 1977, 103-108.

[3] O. AXELSSON, *Stability and Error Estimates of Galerkin Finite Element Approximations for Convection-diffusion Equations*, IMA J. Numer. Anal. 1 (1981), 329-345.

[4] O. AXELSSON and I. GUSTAFSSON, *A Preconditioned Conjugate Gradient Method for Finite Element Equations, which is Stable for Rounding Errors*, Information Processing 80, ed: S. H. Lavington, North-Holland Publishing Company, Amsterdam, 1980, 723-728.

[5] O. AXELSSON and I. GUSTAFSSON, *An efficient Finite Element Method for Nonlinear Diffusion Problems*, Research Report 84.06 R, Department of Computer Science, Chalmers University of Technology, Göteborg, Sweden (1984).

[6] O. AXELSSON and J. MAUBACH, *On the Updating and Assembly of the Hessian Matrix in Finite Element Methods*, Comp. Meth. Appl. Mech. Eng. 71 (1988), 41-67.

[7] O. AXELSSON, G. CAREY and G. LINDSKOG, *On a Class of Preconditioned Iterative Methods for Parallel Computers*, Int. J. Numer. Methods Eng. 27 (1989), 637-654.

[8] I. BABUSKA and J. OSBORN, *Generalized Finite Element Methods: Their Performance and their Relation to Mixed Methods*, SIAM J. Numer. Anal. 20 (1983), 510-536.

[9] P. CIARLET, *The Finite Element Method for Elliptic Problems*, North-Holland Publishing Company, Amsterdam, 1978.

[10] I. GUSTAFSSON, *Modified Incomplete Cholesky (MIC) Methods*, Preconditioning Methods - Theory and Applications, ed: D. Evans, Gordon and Breach Publishers, New York - London - Paris (1983), 265-293.

[11] M. F. WHEELER, *A Priori L^2-estimates for Galerkin Approximations to Parabolic Partial Differential Equations*, SIAM J. Numer. Anal. 10 (1973), 713-759.

Generalized Stability Of Difference Method For Nonlinear Initial Value Problems And Its Application

GUO BEN-YU

Shanghai University of Science and Technology, Shanghai, China

Abstract. Generalized stability of difference method for nonlinear initial value problems is proposed. Such stability implies the convergence with some reasonable conditions. As an example, a conservative scheme is given for RLW equation. The generalized stability is analyzed, from which the convergence follows. The relation between the stability and the conservation is also discussed.
Keywords. Nonlinear initial value problems, generalized stability, RLW equation.

1. Introduction

Richardson [1] and Courant, Friedrichs, Lewy [2] proposed the stability of difference schemes, which plays an important role in numerical solutions of partial differential equations. Lax, Richtmeyer [3] developed the theory of stability and convergence (Also see [4]). But this framework is applicable only for linear problems. Stetter [5] and Keller [6] studied nonlinear stability and local stability respectively. They have been used successfully for nonlinear ordinary differential equations. On the other hand, the author [7-9] proposed generalized stability which has been applied widely to numerical solutions of nonlinear partial differential equations. Since the above theory was given in general version, it might be difficult to be used for nonlinear initial value problems. In this paper, we present weak generalized stability of difference method for nonlinear initial value problems. Such stability implies the convergence provided that the approximation error satisfies some reasonable conditions. In order to show how to use the above framework, we construct a difference scheme for nonlinear RLW equation. If we approximate the nonlinear term in this equation suitably, then the numerical solution keeps the discrete conservation. We use energy method to analyze the generalized stability and convergence, and find out that the suitable approximation of nonlinear term also makes the scheme stabler in the sense of generalized stability.

2. Generalized Stability

Let B_1 and B_2 be Banach spaces with the norms $\|\cdot\|_1$ and $\|\cdot\|_2$, $B_1 \subseteq B_2$. P and L are differential operators with respect to the variables in the spaces, mapping $U \in B_1$ to B_2. For simplicity, assume that P is linear, and the inverse P^{-1} exists. Let B_3 be a Banach space with the norm $\|\cdot\|_3$, whose element f is a map from the interval $(0, T]$ to B_2. Let U_0 be a given element of B_1 describing the initial state. We consider the nonlinear operator equation

$$\begin{cases} \frac{d}{dt}(PU(t)) = LU(t) + f(t), & 0 < t \leq T, \\ U(0) = U_0. \end{cases} \tag{2.1}$$

We define a genuine solution of (2.1) as a one parameter family $U(t)$ such that $U(t)$ is in the domains of P and L for each $t \in [0, T]$ and

$$\|\frac{1}{s}[PU(t+s) - PU(t)] - LU(t) - f(t)\|_2 \to 0, \qquad \text{as } s \to 0.$$

Let D_1 and D_3 be such subsets of B_1 and B_3 that for any $U_0 \in D_1$ and $f \in D_3$, (2.1) has a unique genuine solution. We suppose that D_1 and D_3 are dense in B_1 and B_3.

Now, let h be a positive parameter. $B_{q,h}$ is a finite dimensional Banch space with the norm $\| \quad \|_{q,h}$, approximating the space B_q. Assume that $\mathrm{Dim}(B_{1,h}) = \mathrm{Dim}(B_{2,h})$. Let $r_{q,h}$ be a continuous restriction operator from B_q to $B_{q,h}$, and satisfies the following conditions

(i) for all $V \in B_1, r_{1,h}V = r_{2,h}V$,

(ii) for all $V \in B_q, \|r_{q,h}V\|_{q,h} \to \|V\|_q$, as $h \to 0$, $q = 1, 2, 3$.

Next, let $d = d(h) > 0$ be the mesh spacing of the variable t, and

$$u_{h,t}(t) = \frac{1}{d}[u_h(t+d) - u_h(t)].$$

Let $P_h u_h(t)$ and $L_h(u_h(t), u_h(t+d))$ be the approximations to $PU(t)$ and $LU(t)$. Similarly, $f_h(t)$ and $u_{0,h}$ are the approximations to $f(t)$ and U_0 respectively. We consider the approximate equation

$$\begin{cases} (P_h u_h(t))_t = L_h(u_h(t), u_h(t+d)) + f_h(t), \\ u_h(0) = u_{0,h}. \end{cases} \tag{2.2}$$

Suppose that for each t, $u_h(t)$ is determined uniquely by (2.2). The simplest case is that P_h^{-1} exists and

$$L_h(u_h(t), u_h(t+d)) = \overline{L}_h(u_h(t)).$$

The approximation error for $t > 0$ is defined as

$$\begin{aligned} R_h(U(t)) = &- r_{2,h}\frac{d}{dt}(PU(t)) + (P_h r_{1,h}U(t))_t + r_{2,h}LU(t) \\ &- L_h(r_{1,h}U(t), r_{1,h}U(t+d)) + r_{2,h}f(t) - f_h(t). \end{aligned}$$

If $U(t)$ is the solution of (2.1), then

$$R_h(U(t)) = (P_h r_{1,h}U(t))_t - L_h(r_{1,h}U(t), r_{1,h}U(t+d)) - f_h(t).$$

If the following conditions are fulfilled

(i) $\|r_{1,h}U_0 - u_{0,h}\|_{1,h} \to 0$, as $h \to 0$,

(ii) for all $U(t) \in D_1, t \leq T$ and $f \in D_3$,

$$\|R_h U(t)\|_{2,h} \to 0, \qquad \text{as } h \to 0,$$

then we say that the approximation (2.2) is consistent with (2.1).

If for all $U(t)$, the solution of (2.1) and $u_h(t)$, the corresponding solution of (2.2),

$$\|r_{1,h}U(t) - u_h(t)\|_{1,h} \to 0, \quad \text{as } h \to 0, \qquad 0 \leq t \leq T,$$

then we say that (2.2) is convergent.

We now turn to the stability. The usual stability is as follows.

Definition 1. If there are positive constants h_0 and $C(T)$ such that for all $h \leq h_0$ and $t \leq T$,

$$\|u_h^{(1)}(t) - u_h^{(2)}(t)\|_{1,h} \leq C(T)(\|u_{0,h}^{(1)} - u_{0,h}^{(2)}\|_{1,h} + \|f_h^{(1)} - f_h^{(2)}\|_{3,h}) \tag{2.3}$$

where $u_h^{(q)}$ is the unique solution of (2.2) with the data $u_{0,h}^{(q)}$ and $f_h^{(q)}$, then we say that (2.2) is stable.

According to Lax's theory (see[4]), we conclude that if (2.1) is well posed with linear operators P and L, and (2.2) is consistent with (2.1), then the stability given by Definition 1 is equivalent to the convergence. But nonlinear problems do not satisfy (2.3) usually. Indeed, such stability is too strong for most of nonlinear problems. Firstly $C(T)$ depends only on T. Secondly there is no restriction on $\|u_{0,h}^{(1)} - u_{0,h}^{(2)}\|_{1,h}$ and $\|f_h^{(1)} - f_h^{(2)}\|_{3,h}$. Thus we propose the following stability.

Definition 2. If there are non-negative constant $M(u_h, h, T)$ and positive constants h_0 and $N(u_h, h, T)$ such that for all $h \leq h_0$,

$$\|u_{0,h}^{(1)} - u_{0,h}^{(2)}\|_{1,h} + \|f_h^{(1)} - f_h^{(2)}\|_{3,h} \leq N(u_h^{(1)}, h, T)$$

implies

$$\|u_h^{(1)}(t) - u_h^{(2)}(t)\|_{1,h} \leq M(u_h^{(1)}, h, T)(\|u_{0,h}^{(1)} - u_{0,h}^{(2)}\|_{1,h} + \|f_h^{(1)} - f_h^{(2)}\|_{3,h}), \quad 0 \leq t \leq T, \tag{2.4}$$

then we say that (2.2) is of weak generalized stability at $u_h^{(1)}$.

There is relation between the above stability and the convergence.

Theorem 1. *Let $U(t)$ and $u_h(t)$ be the unique solutions of (2.1) and (2.2) respectively. If (2.2) is of weak generalized stability at $r_{1,h}U(t)$, and*

$$\|r_{1,h}U_0 - u_{0,h}\|_{1,h} + \|R_h(U)\|_{3,h} \leq N(r_{1,h}U, h, T),$$

then for all $t \leq T$,

$$\|r_{1,h}U(t) - u_h(t)\|_{1,h} \leq M(r_{1,h}U, h, T)(\|r_{1,h}U_0 - u_{0,h}\|_{1,h} + \|R_h(U)\|_{3,h}).$$

Proof. We have

$$\begin{cases} (P_h r_{1,h} U(t))_t = L_h(r_{1,h}U(t), r_{1,h}U(t+d)) + f_h(t) + R_h(t), \\ r_{1,h}U(0) = r_{1,h}U_0. \end{cases} \tag{2.5}$$

By comparing (2.2) with (2.5), the conclusion follows from (2.4).

The simplest case is called generalized stability or g-stability (see [10]). It means that

$$M(u_h, h, T) = M(u_h, T)h^{\bar{s}}, \qquad N(u_h, h, T) = N(u_h, T).$$

The infimum of such values $\bar{s}$ is called index of generalised stability, denoted by s. But we can only get upper bound of s usually. Clearly if $u_{0,h} = r_{1,h}U_0$, $\|R_h(U)\|_{3,h} = O(h^q)$ and $s < q$, then (2.2) is convergent with the accuracy $O(h^q)$. Furthermore, if $s \le 0$, then (2.2) is convergent provided that (2.2) is consistent with (2.1). It can be seen that the smaller the index s, the stabler the scheme. In particular, if $s = -\infty$, then there is no restriction on the errors of data and so generalized stability turns to usual stability.

Remark 1. It is not necessary that P is a linear operator.

Remark 2. It is not necessary that (2.1) is a differential equation of first order with respect to t.

3. Application

In this section, we take RLW equation as an example to show how to use the framework in section 2. This equation is also called Benjamin, Bona, Mahony equation (see [11]), and is of the form

$$\begin{cases} \dfrac{\partial U}{\partial t} + U\dfrac{\partial U}{\partial x} - \dfrac{\partial^3 U}{\partial t \partial x^2} = f(x,t), & -\infty < x < \infty, \qquad 0 < t \le T, \\ U(x,0) = U_0(x), & -\infty < x < \infty. \end{cases} \tag{3.1}$$

Assume that $f(x,t)$ and $U_0(x)$ have the period 1 for the variable x and thus $U(x+1,t) = U(x,t)$. (3.1) possesses three conservations (see [12]). For instance, if $f \equiv 0$, then

$$\int_0^1 [U^2(x,t) + (\frac{\partial U}{\partial x}(x,t))^2]dx = \text{constant}. \tag{3.2}$$

Now let $h = \frac{1}{J}$ be the mesh spacing of the variable x, J being a positive integer. Define

$$I_h = \{x | x = jh, 1 \le j \le J\}.$$

We shall use the following notations

$$
\begin{aligned}
v_x(x,t) &= \frac{1}{h}[v(x+h,t) - v(x,t)], \\
v_{\bar{x}}(x,t) &= v_x(x-h,t), \\
v_{\hat{x}}(x,t) &= \frac{1}{2}v_x(x,t) + \frac{1}{2}v_{\bar{x}}(x,t), \\
v_{x\bar{x}}(x,t) &= (v_x(x,t))_{\bar{x}}.
\end{aligned}
$$

We introduce the following discrete scalar product and norms

$$
\begin{aligned}
(v(t), w(t)) &= h\sum_{j=1}^{J} v(jh,t)w(jh,t), \\
\|v(t)\|^2 &= (v(t), v(t)), \\
|v(t)|_1^2 &= \frac{1}{2}\|v_x(t)\|^2 + \frac{1}{2}\|v_{\bar{x}}(t)\|^2, \\
\|v(t)\|_1^2 &= \|v(t)\|^2 + |v(t)|_1^2.
\end{aligned}
$$

Let v, w and y have the period 1 for the variable x. By Abel's formula,

$$(v_x(t), w(t)) + (w_{\bar{x}}(t), v(t)) = 0 \tag{3.3}$$

and so

$$2(v_{x\bar{x}}(t), w(t)) + (v_x(t), w_x(t)) + (v_{\bar{x}}(t), w_{\bar{x}}(t)) = 0. \tag{3.4}$$

It is easy to verify that

$$2(v_t(t), v(t)) = \|v(t)\|_t^2 - d\|v_t(t)\|^2, \tag{3.5}$$

$$(v_t(t), v(t) + v(t+d)) = \|v(t)\|_t^2. \tag{3.6}$$

The combination of (3.4) with (3.5) leads to

$$2(v_t(t), v_{x\bar{x}}(t)) + (|v(t)|_1^2)_t - d|v(t)|_1^2 = 0. \tag{3.7}$$

We next construct a difference scheme for solving (3.1). In order to approximate the nonlinear term $U\frac{\partial U}{\partial x}$, we define

$$J_a(v(x,t), w(x,t)) = aw(x,t)v_{\hat{x}}(x,t) + \frac{1}{2}(1-a)(w(x,t)v(x,t))_{\hat{x}}, \quad 0 \le a \le 1.$$

In particular, (3.3) implies

$$(J_{\frac{1}{3}}(v(t), w(t)), y(t)) + (J_{\frac{1}{3}}(y(t), w(t)), v(t)) = 0 \tag{3.8}$$

and so $J_{\frac{1}{3}}(v(x,t), w(x,t))$ is skew symmetric. Let $u(x,t)$ be the approximation to $U(x,t)$. The difference scheme for (3.1) is

$$\begin{cases} u_t(x,t) + J_a(u(x,t) + bdu_t(x,t), u(x,t)) - u_{tx\bar{x}}(x,t) = f(x,t), & x \in I_h, \quad t \geq 0, \\ u(x+1,t) = u(x,t), & x \in I_h, \quad t \geq 0, \\ u(x,0) = u_0(x), & x \in I_h \end{cases} \tag{3.9}$$

where b is a parameter and $0 \leq b \leq 1$. If $b = 0$, then we can solve $u(x,t)$ for each t, by the same recursion formula. Otherwise we should use different recursion formulas at different time t.

We are going to check the conservation. Let $a = \frac{1}{3}, b = \frac{1}{2}$ and $f \equiv 0$. By multiplying the first formula of (3.9) by $u(x,t) + u(x,t+d)$ and summing the product for $x \in I_h$, we have from (3.3) and (3.6) that

$$\|u(t)\|_1^2 = \|u(0)\|_1^2 \tag{3.10}$$

which is a reasonable discrete analogy of (3.2). Thus we expect that scheme (3.9) with $a = \frac{1}{3}$ gives better numerical result.

We turn to analyze generalized stability. Let $\tilde{f}(x,t)$ and $\tilde{u}_0(x)$ be the errors of f and u_0, which induce the error of $u(x,t)$, denoted by $\tilde{u}(x,t)$. For simplicity, assume that $b = 0$. Then

$$\begin{cases} \tilde{u}_t(x,t) + J_a(\tilde{u}(x,t), u(x,t) + \tilde{u}(x,t)) & \\ \qquad + J_a(u(x,t), \tilde{u}(x,t)) - \tilde{u}_{tx\bar{x}}(x,t) = \tilde{f}(x,t), & x \in I_h, \quad t \geq 0, \\ \tilde{u}(x+1,t) = \tilde{u}(x,t), & x \in I_h, \quad t \geq 0, \\ \tilde{u}(x,0) = \tilde{u}_0(x), & x \in I_h. \end{cases} \tag{3.11}$$

We first take the scalar product of the first formula of (3.11) with $2\tilde{u}(x,t)$ to obtain from (3.3), (3.5), (3.7) and (3.8) that

$$\|\tilde{u}(t)\|_t^2 - d\|\tilde{u}_t(t)\|^2 + |\tilde{u}(t)|_{1,t}^2 - d|\tilde{u}_t(t)|_1^2 + F_1(t) + F_2(t) = 2(\tilde{u}(t), \tilde{f}(t)) \tag{3.12}$$

where

$$\begin{aligned} F_1(t) &= 2(\tilde{u}(t), J_a(\tilde{u}(t), u(t) + \tilde{u}(t))) \\ &= 3(1-a)(\tilde{u}(t), J_{\frac{1}{3}}(\tilde{u}(t), u(t) + \tilde{u}(t))) \\ &\qquad + (3a-1)(\tilde{u}(t)\tilde{u}_{\hat{x}}(t), u(t) \dot{+} \tilde{u}(t)) \\ &= (3a-1)(\tilde{u}(t)\tilde{u}_{\hat{x}}(t), u(t) + \tilde{u}(t)), \\ F_2(t) &= 2(\tilde{u}(t), J_a(u(t), \tilde{u}(t))). \end{aligned} \tag{3.13}$$

Next, let m be an undetermined constant. We take the scalar product of the first formual of (3.11) with $md\tilde{u}_t(x,t)$ to get that

$$md\|\tilde{u}_t(t)\|^2 + md|\tilde{u}_t(t)|_1^2 + F_3(t) + F_4(t) = md(\tilde{u}_t(t), f(t)) \tag{3.14}$$

where

$$F_3(t) = md(\tilde{u}_t(t), J_a(\tilde{u}(t), \tilde{u}(t))),$$
$$F_4(t) = md(\tilde{u}_t(t), J_a(u(t), \tilde{u}(t)) + J_a(\tilde{u}(t), u(t))).$$

Let $\varepsilon > 0$. By putting (3.12) and (3.14) together, we obtain

$$(\|\tilde{u}(t)\|_1^2)_t + d(m-1-\varepsilon)\|\tilde{u}_t(t)\|_1^2 + \sum_{q=1}^{4} F_q(t) \le \|\tilde{u}(t)\|^2 + (1+\frac{dm^2}{\varepsilon})\|\tilde{f}(t)\|^2. \tag{3.15}$$

Let C be a positive constant depending only on $u(x,t)$, which may be different in different cases. We have the following discrete embedding inequality

$$\|v^2(t)\|^2 \le \tilde{C}\|v(t)\|_1^4$$

where $\tilde{C} > 0$ is an absolute constant. Thus

$$\begin{aligned} |F_1(t)| &\le |3a-1|(|\tilde{u}(t)|_1^2 + \|\tilde{u}^2(t)\|^2 + C\|\tilde{u}(t)\|^2) \\ &\le C|3a-1|(\|\tilde{u}(t)\|_1^2 + \|\tilde{u}(t)\|_1^4). \end{aligned} \tag{3.16}$$

Moreover

$$|F_2(t)| \le C(\|\tilde{u}(t)\|^2 + |\tilde{u}(t)|_1^2). \tag{3.17}$$

Furthermore, by Jessen's formula (see [13]), we have

$$\|v(t)w(t)\|^2 \le \frac{1}{h}\|v(t)\|^2\|w(t)\|^2$$

and so

$$\begin{aligned} |F_3(t)| &\le \varepsilon d\|\tilde{u}_t(t)\|^2 + \frac{Cdm^2}{\varepsilon}\|\tilde{u}(t)\tilde{u}_{\hat{x}}(t)\|^2 \\ &\le \varepsilon d\|\tilde{u}_t(t)\|^2 + \frac{Cdm^2}{\varepsilon h}\|\tilde{u}(t)\|^2|\tilde{u}(t)|_1^2. \end{aligned} \tag{3.18}$$

Finally

$$|F_4(t)| \le \varepsilon d\|\tilde{u}_t(t)\|^2 + \frac{Cdm^2}{\varepsilon}(\|\tilde{u}(t)\|^2 + |\tilde{u}(t)|_1^2). \tag{3.19}$$

By substituting (3.16)-(3.19) into (3.15), we obtain

$$\begin{aligned} &(\|\tilde{u}(t)\|_1^2)_t + d(m-1-3\varepsilon)\|\tilde{u}_t(t)\|_1^2 \\ &\le C(1+\frac{dm^2}{\varepsilon})\|\tilde{u}(t)\|_1^2 + C(|3a-1| + \frac{dm^2}{\varepsilon h})\|\tilde{u}(t)\|_1^4 + (1+\frac{dm^2}{\varepsilon})\|\tilde{f}(t)\|^2. \end{aligned} \tag{3.20}$$

Let $m \geq 1 + 3\varepsilon$, and

$$E(t) = \|\tilde{u}(0)\|^2 + d(1 + \frac{dm^2}{\varepsilon})\sum_{s=0}^{t-d}\|\tilde{f}(s)\|^2.$$

By summing (3.20) for $t = 0, d, 2d, \cdots$, we obtain

$$\|\tilde{u}(t)\|_1^2 \leq E(t) + Cd\sum_{s=0}^{t-d}[(1 + \frac{dm^2}{\varepsilon})\|\tilde{u}(s)\|_1^2 + |3a - 1| + \frac{dm^2}{\varepsilon h})\|\tilde{u}(s)\|_1^4]. \quad (3.21)$$

We shall use the following lemma.

Lemma (see [14]). Suppose that the following conditions are fulfilled

(i) $z(t)$ is a non-negative function for $t = 0, d, 2d, \cdots$,

(ii) G and C_q are non-negative constants, p is a constant,

(iii) for all $t \leq t_0$,

$$z(t) \leq G + d\sum_{s=0}^{t-d}(C_1 z(s) + C_2 h^{-p} z^2(s)),$$

(iv) $Ge^{(C_1+C_2)t_0} \leq h^p$ and $z(0) \leq G$.

Then for all $t \leq t_0$,

$$z(t) \leq Ge^{(C_1+C_2)t}.$$

By the above lemma and (3.21), we conclude that

Theorem 2. *If $d = O(h^p), p \geq 1$ and*

$$E(T)e^{C_3 T} \leq C_4, \qquad \textit{for} \quad a \neq \frac{1}{3},$$

$$E(T)e^{C_3 T} \leq C_4 h^{1-p}, \qquad \textit{for} \quad a = \frac{1}{3},$$

then for all $t \leq T$,

$$\|\tilde{u}(t)\|_1^2 \leq E(t)e^{C_3 t}$$

where C_4 is a positive constant depending only on $u(x,t)$, and

$$C_3 = C(1 + \frac{dm^2}{\varepsilon} + \frac{dm^2}{\varepsilon h} + |3a - 1|).$$

Remark 3. Theorem 2 shows the upper bound of index of generalized stability of scheme (3.9), i.e.,

$$s \le 0, \quad \text{for} \quad a \ne \frac{1}{3}, \quad p \ge 1,$$
$$s \le 1 - p, \quad \text{for} \quad a = \frac{1}{3}, \quad p \ge 1.$$

Thus if we use skew symmetric approximation $J_{\frac{1}{3}}(u(t), u(t))$, then not only the discrete conservation (3.10) holds, but also the main nonlinear error term $F_1(t)$ vanishes (see (3.13)), which leads to smaller index of generalised stability. This fact shows the relation between the conservation and the generalised stability.

Finally we consider the convergence. If $U \in C^4(I) \times C^2[0,T]$, then the approximation error for $t > 0$ is $O(d + h^2)$. Thus the following conclusion follows from Theorem 1 and Theorem 2.

Theorem 3. *Let U and u be the solutions of (3.1) and (3.9) respectively, and $u_0 = U_0, d = O(h^p), p \ge 1$. Then for any fixed value T and all $t \le T$,*

$$\|U(t) - u(t)\|_1 = O(h^p + h^2).$$

Remark 4. The technique in this section can be used for many other nonlinear initial value problems, for instance, the vorticity equation, Navier-Stokes equation, atmospheric equation, Dirac equation, Clein-Gordon equation and Schrōdinger equation, ect.(see [14]).

REFERENCES

[1] Richardson,L.F., Phil. Trans. Soc. Lond., 210A(1910), 307-357.

[2] Courant,R., Friedrichs,F. and Lewy,H., Math. Ann., 100(1924), 32-74.

[3] Lax,P.D. and Richtmeyer,R. D., Comm. Pure and Appl. Math., 9(1956), 267-293.

[4] Richtmeyer,R.D. and Morton,K.W., Difference Methods for Initial Value Problems, Interscience, New York, 1967.

[5] Stetter,H.J., Stability of nonlinear discretization algorithms, in ≪ Numerical Solution of Differential Equations ≫, 111-123, Ed. by Bramble, J.H., Academic Press, New York, 1966.

[6] Keller,H.B., Math. Comp., 29(1975), 464-476.

[7] Guo Ben-yu, Tech. Report of SUST(1965), Also see Acta Math. Sinica, 17(1974), 242-258.

[8] Guo Ben-yu, Scientia Sinica, 25A(1982), 702-715.

[9] Guo Ben-yu, Scientia Sinica, 32A(1989), 897-907.

[10] Griffiths,D.F., Bulletin of IMA, 18(1982), December, 210-215.

[11] Benjamin,T.B., Bona,J.L. and Mahony,J.J., Phil. Trans. Soc. Lond., 272A(1972), 47-48.

[12] Olver,P.J., Math. Proc. Camb. Phil. Soc., 85(1979), 143-160.

[13] Hardy,G.H., Littlewood,J.E. and Polya, G., Inequality, 2'nd ed., Cambridge University Press, 1952.
[14] Guo Ben-yu, Difference Methods for Partial Differential Equations, Science Press, Beijing, 1988.

Professor Guo Ben-yu
Shanghai University of Science and Technology
Shanghai City, 201800
People's Republic of China

AN ε-UNIFORMLY CONVERGENT NON-CONFORMING FINITE ELEMENT METHOD FOR A SINGULARLY PERTURBED ELLIPTIC PROBLEM IN TWO DIMENSIONS

J.J.H. MILLER AND S. WANG

Numerical Analysis Group, Trinity College, Dublin 2, Ireland

Abstract: An exponentially fitted Petrov-Galerkin, non-conforming finite element method for a stationary singularly perturbed problem in two dimensions is presented. This method is based on the formulation of a discrete bilinear form. Existence, uniqueness and stability of the approximate solution are proved with respect to a discrete energy norm and an error bound is given for the approximate solution which is independent of the small parameter ε. The associated linear system is structurally symmetric and can be transformed into a symmetric and positive-definite M–matrix. Numerical results are given which show that the method is superior to classical finite element methods.

1. Introduction

In this paper we present an ε-uniformly convergent Petrov-Galerkin finite element method for the numerical solution of convection-dominated problems of the form:

$$-\nabla \cdot (\varepsilon \nabla u - \mathbf{a} u) + G u = F \quad \text{in } \Omega \tag{1.1}$$

$$u|_{\partial\Omega} = 0 \tag{1.2}$$

where $\Omega = (0,1)^2 \in I\!R^2$, $\partial\Omega$ denotes the boundary of Ω, $\mathbf{n}$ denotes the unit outward normal vector on $\partial\Omega$, ε is a small positive parameter, $G \in C^0(\overline{\Omega}) \cap H^1(\Omega)$ and $F \in L^2(\Omega)$. We assume that $\mathbf{a} = (a^{(1)}, a^{(2)})$ is such that

$$G + \frac{\nabla \cdot \mathbf{a}}{2} \geq 0 \tag{1.3}$$

The Petrov-Galerkin finite element method is presented in the next section. In Section 3 the existence, uniqueness and stability of the approximate solution are established. An error bound for the approximate solution is given in Section 4 which is independent of ε. Numerical results are described in Section 5. For brevity the proofs, which will be given elsewhere, are omitted.

In what follows $L^2(\Omega)$ and $W^{m,p}(\Omega)$ denote the usual Sobolev spaces with norms $\|\cdot\|_0$ and $\|\cdot\|_{m,p}$ respectively (cf. for example [1]). The inner product on $L^2(\Omega)$ and $(L^2(\Omega))^2$ is denoted by $(\cdot,\cdot)$ and the kth order seminorm on $W^{m,p}(\Omega)$ is denoted $|\cdot|_{k,p}$. The Sobolev spaces $W^{m,2}(\Omega)$ are denoted by $H^m(\Omega)$ and the corresponding norms and seminorms are denoted by respectively $\|\cdot\|_m$ and $|\cdot|_k$. We put $\mathbf{L}^2(\Omega) = (L^2(\Omega))^2$ and $H_0^1(\Omega) = \{v \in H^1(\Omega) : v|_{\partial\Omega} = 0\}$. We use $|\cdot|$ to denote absolute value, Euclidean length or area, depending on the context.

2. The Finite Element Formulation

In this section we present the Petrov-Galerkin finite element method for the numerical solution of the following variational problem corresponding to (1.1–2).

Problem 2.1: *Find* $u \in H_0^1(\Omega)$ *such that for all* $v \in H_0^1(\Omega)$

$$(\varepsilon \nabla u - \mathbf{a}u, \nabla v) + (Gu, v) = (F, v) \tag{2.1}$$

It is easy to verify that if (1.3) holds then Problem 2.1 has a unique solution.

Before discussing the finite element approximation to Problem 2.1, we make the following assumption

Assumption 2.1: *The coefficient* $\mathbf{a} = (a^{(1)}, a^{(2)})$ *is a constant vector on* Ω.

To discuss the method we first define some meshes on Ω. Let $\mathcal{Q}$ denote the following family of meshes on Ω

$$\mathcal{Q} = \{Q_{h^{(1)},h^{(2)}} : 0 < h^{(1)}, h^{(2)} \leq 1\}$$

where $Q_{h^{(1)},h^{(2)}}$ denotes a rectangular mesh with uniform step sizes $h^{(1)}$ in x-direction and $h^{(2)}$ in y-direction.

Definition 2.1: *The family of meshes* $\mathcal{Q}$ *is regular if there exist constants* $\gamma_1 > 0$ *and* $\gamma_2 > 0$, *independent of* $h^{(1)}$ *and* $h^{(2)}$, *such that for all* $h^{(1)}, h^{(2)} \in (0,1]$

$$\gamma_1 \leq \frac{h^{(1)}}{h^{(2)}} \leq \gamma_2 \tag{2.2}$$

We assume hereafter that $\mathcal{Q}$ is regular. In this case $Q_{h^{(1)},h^{(2)}}$ can be denoted simply by Q_h, where $h = \max\{h^{(1)}, h^{(2)}\}$ denotes the mesh parameter of Q_h.

For each $Q_h \in \mathcal{Q}$, let $X_h = \{x_i\}_1^{N_V}$ be the set of all vertices of Q_h and $E_h = \{e_i\}_1^{N_E}$ the set of all edges of Q_h.

Definition 2.2: *The Dirichlet tessellation* D_h *corresponding to the mesh* Q_h *is defined by* $D_h = \{d_i\}_1^{N_X}$ *where*

$$d_i = \{x : |x - x_i| < |x - x_j|, x_j \in X_h, j \neq i\} \tag{2.3}$$

for all $x_i \in X_h$ *(cf. [2])*

We remark that for each $x_i \in X_h$, the boundary ∂d_i of the tile d_i is also a rectangle (see Fig. 2.1)

The Dirichlet tessellation D_h is a rectanglular mesh dual to Q_h. The number of its elements is equal to the number of vertices in V_h. We define $\mathcal{D} = \{D_h : 0 < h \leq 1\}$ to be the family of all such meshes.

A second mesh dual to Q_h is defined as follows. With each edge $e_k \in E_h$ we associate an open quadrilateral b_k having as its vertices the two end-points of e_k and the circumcentres of the rectangles having e_k as a common edge. From Fig.2.1 it is clear that b_k consists of two triangles and that $|b_k| > 0$. The set of all these b_k forms a quadrilateral mesh on Ω which is denoted by B_h. This is also a dual mesh to Q_h and the number of its elements is equal to the number of edges in E_h. We put $\mathcal{B} = \{B_h : 0 < h \leq h_0\}$.

Corresponding to the two families of meshes Q_h and D_h, we now construct two families of finite-dimensional spaces $S_h \subset L^2(\Omega)$ and $T_h \subset L^2(\Omega)$ as follows. Without loss of generality, we assume that the edges and vertices are numbered so that $\{e_i\}_1^M$ is the set of all edges in E_h not on $\partial\Omega_D$ and $\{x_i\}_1^N$ is the set of all nodes in X_h not on $\partial\Omega_D$.

For each $e_i \in E_h$ we define two exponential functions $\phi_i^{(j)}$ $(j = 1, 2)$ on e_i by

$$\begin{aligned} &\frac{d}{d\mathbf{e}_i}\left(\varepsilon\frac{d\phi_i^{(j)}}{d\mathbf{e}_i} - a_i\phi_i^{(j)}\right) = 0 \\ &\phi_i^{(j)}(x_{i,1}) = \delta_{j1}, \quad \phi_i^{(j)}(x_{i,2}) = \delta_{j2} \end{aligned} \tag{2.4}$$

where $\mathbf{e}_i$ is the unit tangential vector along the edge e_i, $a_i = \mathbf{a} \cdot \mathbf{e}_i$, δ_{ij} denotes the Kronecker delta and $x_{i,1}$ and $x_{i,2}$ denote the two end-points of e_i, as shown in Fig.2.2(a). We note that a_i is either equal to $a^{(1)}$ or to $a^{(2)}$ since all edges are parallel to one of the axes. We then extend $\phi_i^{(j)}$ to b_i by taking $\phi_i^{(j)}$ to be constant along directions perpendicular to e_i. These can be further extended to basis functions on Ω as follows

$$\phi_i^{(j)} = \begin{cases} \phi_i^{(j)} & \text{on } b_i \\ 0 & \text{otherwise} \end{cases}$$

Letting

$$I_i = \{j : e_{j,k} \in E_h\} \tag{2.5}$$

be the index set of all neighbour nodes of x_i, where $e_{j,k}$ denotes the edge connecting x_i and x_j (see Fig.2.2(b)) and $\phi_{i,j}^{(k)}$ $(k = 1, 2)$ be the corresponding basis functions, we define a basis function for Q_h by $\phi_i(x, y) = \sum_{j \in I_i} \phi_{i,j}^{(1)}(x, y)$. We choose the solution space to be $S_h = \text{span}\{\phi_i\}_1^N$. Obviously $S_h \subset L^2(\Omega)$. For any sufficiently smooth function u we can easily show that the S_h-interpolant u_I of u satisfies

$$\frac{d}{de_i}(\varepsilon\frac{du_I}{de_i} - a_i u_I) = 0 \quad \text{on } e_i$$
$$u_I(x_{i,j}) = u(x_{i,j}),\ j = 1,2$$

for $i = 1,2,\ldots,N$. Solving the above problem we get

$$f_i = \varepsilon\frac{du_I}{de_i} - a_i u_I = \frac{\varepsilon}{|e_i|}(B(\frac{a_i|e_i|}{\varepsilon})u(x_{i,2}) - B(-\frac{a_i|e_i|}{\varepsilon})u(x_{i,1})) \tag{2.6}$$

where $B(x)$ denotes the Bernoulli function defined by

$$B(x) = \begin{cases} \frac{x}{e^x-1} & x \neq 0 \\ 1 & x = 0 \end{cases} \tag{2.7}$$

This implies that on each b_i we use a constant approximation to $\mathbf{f}\cdot\mathbf{e}_i$, where $\mathbf{f} = \varepsilon\nabla u - \mathbf{a}u$ is the exact flux. We thus have

$$\sup_{x\in\Omega(e_i)} |\mathbf{f}\cdot\mathbf{e}_i - f_i| \leq Ch|\mathbf{f}|_{1,\infty,\Omega(e_i)} \tag{2.8}$$

where C is a positive constant, independent of h, u and ε.

To construct the test space T_h we define a set of piecewise constant basis functions ξ_i, $(i = 1,2,\ldots,N)$ corresponding to the mesh D_h as follows:

$$\xi_i = \begin{cases} 1 & \text{on } d_i \\ 0 & \text{otherwise} \end{cases}$$

We then define $L_h = \text{span}\{\xi_i\}_1^N$.

We introduce the mass lumping operator L such that

$$L(u)(x) = \sum_{x_i\in X_h} u(x_i)\xi_i(x) \qquad \forall x \in \overline{\Omega} \tag{2.9}$$

Using the two finite dimensional spaces S_h and T_h we now define the following discrete Petrov–Galerkin problem

Problem 2.2: *Find $u_h \in S_h$ such that for all $v_h \in T_h$*

$$A(u_h, v_h) + (L(Gu_h), v_h) = (\hat{F}, v_h) \tag{2.10}$$

where $\hat{F}$ is a (quadrature) approximation to F and $A(\cdot,\cdot)$ denotes the bilinear form on $S_h \times T_h$ defined by

$$A(w_h, v_h) = -\sum_{d\in D_h}\int_{\partial d}(\varepsilon\nabla w_h - \mathbf{a}w_h)\cdot\mathbf{n}\gamma_0(v_h|_d)ds \tag{2.11}$$

Here $v_h|_d$ denotes the restriction of v_h to d, $\gamma_0(v|_d)$ denotes the continuous extension of $v|_d$ to ∂d and $\mathbf{n}$ is the unit outward normal vector on ∂d.

Let $u_h = \sum_{i=1}^N u_i\phi_i$, where $\{u_i\}_1^N$ is a set of constants. Substituting this into (2.13) and taking $v_h = \xi_j$, we get, for $j = 1, 2, \ldots, N$,

$$-\sum_{i=1}^{N}\int_{\partial d_j} u_i(\varepsilon\nabla\phi_i - \mathbf{a}\phi_i)\cdot\mathbf{n}ds + G_j u_j|d_j| = \int_{d_j}\hat{F}dx \tag{2.12}$$

where $G_j = G(x_j)$. Notice that ∂d_j consists of a finite number of segments, each of which is the bisector of an edge having x_j as an end-point. We thus have for $j = 1, 2, \ldots, N$

$$-\sum_{k\in I_j}\int_{\partial d_{j,k}}\left(\varepsilon\frac{du_h}{d\mathbf{n}} - \mathbf{a}u_h\right)\Big|_{\Omega(e_{j,k})}ds + G_j u_j|d_j| = \int_{d_j}\hat{F}dx \tag{2.13}$$

where I_j is the index set defined in (2.5). In (2.13) we have used the notation shown in Fig.2.2(b). Using (2.6) we finally obtain from the above

$$\sum_{k\in I_j}\varepsilon\Big(B(\frac{a_{j,k}|e_{j,k}|}{\varepsilon})u_j - B(-\frac{a_{j,k}|e_{j,k}|}{\varepsilon})u_k\Big)\frac{|\partial d_{j,k}|}{|e_{j,k}|} + G_j u_j|d_j| = \int_{d_j}\hat{F}dx \tag{2.14}$$

where $a_{j,k} = \mathbf{a}\cdot\mathbf{e}_{j,k}$ and

$$|\partial d_{j,k}| = \frac{2|b_{j,k}|}{|e_{j,k}|} = \begin{cases} h^{(1)}/2 \text{ or } h^{(2)}/2 & \text{if } e_{jk}\subset\partial\Omega \\ h^{(1)} \text{ or } h^{(2)} & \text{otherwise}\end{cases}$$

The coefficient matrix of (2.14) is only structurally symmetric, *i.e.* a nonzero element of the upper triangular part of the matrix implies that the corresponding element in the lower triangular part is nonzero and *vice versa*. In practice (2.4) can be solved by the CGS methods (cf. [5]) or by other methods for solving unsymmetric linear systems.

Letting $\psi(x,y) = a^{(1)}x + a^{(2)}y$, we have from (2.14)

$$\sum_{k\in B_j}\varepsilon\Big(B(\frac{\psi_j - \psi_k}{\varepsilon})u_j - B(\frac{\psi_k - \psi_j}{\varepsilon})u_k\Big)\frac{|\partial d_{j,k}|}{|e_{j,k}|} + G_j u_j|d_j| = \int_{d_j}\hat{F}dx \tag{2.15}$$

where $\psi_i = \psi(x_i)$. Under the non-degenerate linear transformation

$$u_i = e^{\psi_i/\varepsilon}w_i \quad i = 1, 2, \ldots, N$$

(2.15) becomes

$$\sum_{k\in B_j}\varepsilon\Big(e^{\psi_j/\varepsilon}B(\frac{\psi_j - \psi_k}{\varepsilon})w_j - e^{\psi_k/\varepsilon}B(\frac{\psi_k - \psi_j}{\varepsilon})w_k\Big)\frac{|\partial d_{j,k}|}{|e_{j,k}|} + G_j e^{\psi_j/\varepsilon}w_j|d_j| = \int_{d_j}\hat{F}dx \tag{2.16}$$

The coefficient matrix of (2.17) is a symmetric and positive definite M-matrix (cf.[7]) since $e^{\psi_j/\epsilon}B(\frac{\psi_j-\psi_k}{\epsilon}) = e^{\psi_k/\epsilon}B(\frac{\psi_k-\psi_j}{\epsilon})$. A more general symmetrisation method for (2.14) is discussed in [8].

3. Existence, Uniqueness and Stability

In the previous section we obtained a structurally symmetric matrix which can be transformed to a symmetric and positive-definite M-matrix by pre-multiplication by diagonal matrices with positive entries. The latter implies that the solution to Problem 2.2 exists and is unique. In this section we show that this solution is stable with respect to a discrete energy norm by verifying that the bilinear form in (2.11) is coercive. We start with the following lemma.

Lemma 3.1. *The mass lumping operator defined in (2.9) is surjective from S_h to T_h.*

Proof. The proof is trivial and is omitted here. □

Let $B(\cdot,\cdot)$ be the bilinear form on $S_h \times S_h$ defined by

$$B(u_h, v_h) = A(u_h, L(v_h)) + (L(Gu_h), L(v_h)) \tag{3.1}$$

We consider the following Galerkin problem:

Problem 3.1. *Find a $u_h \in S_h$ such that for all $v_h \in S_h$*

$$B(u_h, v_h) = (\hat{F}, L(v_h)) \tag{3.2}$$

We say that Problem 3.1 is equivalent to Problem 2.2 if any solution to Problem 3.1 is also a solution to Problem 2.2, and *vice versa*.

Lemma 3.2. *Problem 3.1 is equivalent to Problem 2.2.*

Proof. The result is obvious since the operator L is surjective from S_h to T_h by Lemma 3.1. □

For any $u_h \in S_h$, we introduce the functional

$$|u_h|_{1,h} = \sum_{i=1}^{M} (\frac{u_{i,2} - u_{i,1}}{|e_i|})^2 |b_i| \tag{3.3}$$

We have

Lemma 3.3. *The functional $|\cdot|_{1,h}$ defined in (3.3) is a norm on S_h.*

Proof. The result is obvious since for any $u_h \in S_h$, u_h vanishes on $\partial\Omega$. □

Letting $u_h = \sum_{i=1}^N u_i\phi_i \in S_h$, we define the discrete energy norm $|||\cdot|||_h$ on S_h by

$$|||u_h|||_h^2 = h|u_h|_{1,h}^2 + \sum_{i=1}^{N} G_i u_i^2 |d_i| \tag{3.4}$$

We showed in the previous section that the system matrix of (2.17) is a symmetric and positive definite M-matrix. This implies that the solution to Problem 3.1 exists uniquely. The following theorem establishes the stability of the solution to Problem 3.1.

Theorem 3.4. *There is a positive constant C, independent of h, ε and u_h such that*

$$B(u_h, u_h) \geq C|||u_h|||_h^2 \quad \forall u_h \in S_h \tag{3.5}$$

From Lemma 3.2 and Theorem 3.4 we have

Corollary 3.5. *Problem 2.2 has a unique solution and the solution is stable.*

4. ε-Uniform Convergence

In this section we show that the solution to Problem 2.2 converges to the exact solution uniformly in ϵ. We start with the following lemma

Lemma 4.1. *For any $u_h \in S_h$, there is a constant $C > 0$, independent of h and u_h, such that*

$$||L(u_h)||_0 \leq C \sum_{i=1}^{N} \left(\frac{u_{i,2} - u_{i,1}}{|e_i|}\right)^2 |b_i| \tag{4.1}$$

Proof. The lemma has been proved in [3] for the case of triangular meshes. The mesh used here can be regarded as a triangular mesh by dividing each rectangle into two triangles. In this case the Delaunay tessellation remains unchanged. □

For any sufficiently smooth vector-valued function $\mathbf{g}$, we define the discrete seminorm by

$$|\mathbf{g}|_{1,\infty,h} = \Big(\sum_{i=1}^{M} |\mathbf{g}|_{1,\infty,b_i}^2 |b_i|\Big)^{1/2} \tag{4.2}$$

Using Lemma 4.1 we can prove the following theorem which establishes the convergence of the approximate solution.

Theorem 4.2. *Let u and u_h respectively be the solutions to Problems 2.2 and 3.1. Let* $\mathbf{f} = \varepsilon\nabla u - \mathbf{a}u$ *be the exact flux. Then there is a constant $C > 0$, independent of h, ε and u, such that*

$$|||u - u_h|||_h \leq C\left[h^{1/2}|\mathbf{f}|_{1,\infty,h} + h^{-1/2}(\|Gu - L(Gu)\|_0 + \|F - \hat{F}\|_0)\right] \qquad (4.3)$$

We remark that in Thorem 4.2 we showed that the finite element solution converges ε-uniformly to the exact solution in the discrete energy norm which is also independent of ε. We remark that classical results are based on energy norms depending on ε (cf. for example [6]).

5. Numerical Results

In this section we present some numerical results for the validation of the method presented above. The test problem is chosen to be

$$-\varepsilon\nabla^2 u + u_x + u_y + 2u = f \ \text{ in } \ \Omega = (0,1)^2$$
$$u = 0 \ \text{ on } \ \partial\Omega$$

with the exact solution

$$u_{\text{exact}} = xy(1 - e^{\frac{x-1}{\varepsilon}})(1 - e^{\frac{y-1}{\varepsilon}})$$

which is used in [4]. We choose a uniform mesh with 36×36 mesh points and solve the problem for $\varepsilon = 10^{-s}$ for $s = -1, -2, -3, -4$. The results for different values of ε are plotted in Fig.5.1. From the figure we see that our method gives useful results even for $\varepsilon < O(h^2)$, while classical finite element methods with piecewise linear elements often give results with spurious oscillations.

6. Conclusion

In this paper we presented an exponentially fitted Petrov-Galerkin finite element method based on uniform meshes for a stationary advection-diffusion problem. This method is based on a non-conforming finite element formulation. Existence, uniqueness and stability were proved and an error estimate was given. The latter shows that the approximate solution converges to the exact solution ε-uniformly in the discrete energy norm which is also independent of ε. The coefficient matrix of the associated linear system is structurally symmetric and can be transformed into a symmetric and positive-definite M-matrix. Numerical results were given which show that the method is superior to classical finite element methods.

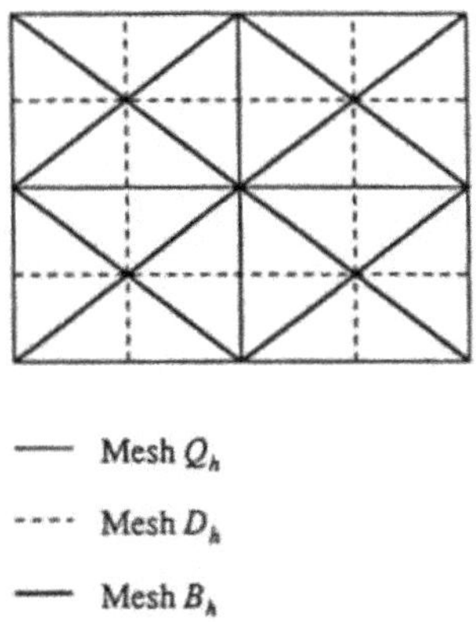

Fig.2.1: Three dual meshes Q_h, D_h and B_h

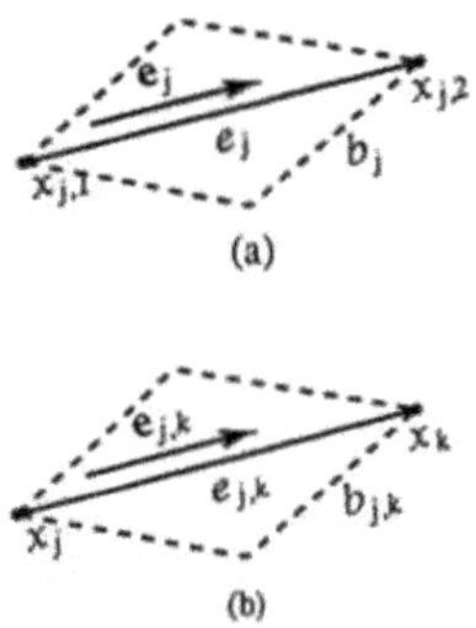

Fig.2.2: Alternative notation for edges, end-points and boxes

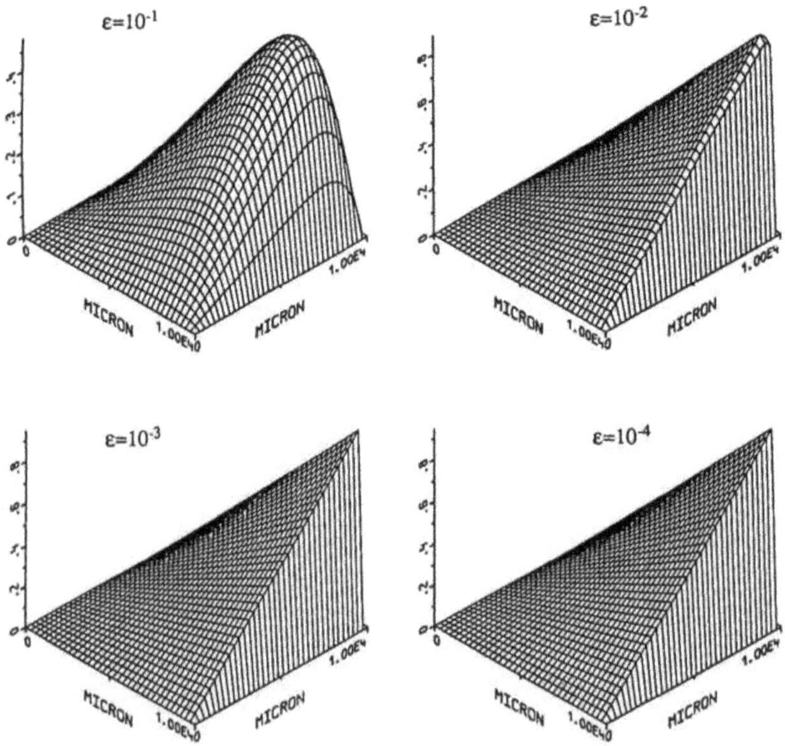

Fig.5.1: Solutions for different values of ε

References

[1] **R.A. Adams** *Sobolev Spaces*, Academic Press, New York (1975)

[2] **G.L. Dirichlet** "Uber die Reduction der positiven quadratischen Formen mit drei unbestimmten ganzen Zahlen", *J. Reine Angew. Math.*, **40**, No.3 (1850) 209–227

[3] **J.J.H. Miller, S. Wang** "A Triangular Mixed Finite Element Method for the Stationary Semiconductor Device Equations", to appear in M^2AN

[4] **U. Risch** "An Upwind Finite Element Method for Singularly Perturbed Elliptic Problems and Local Estimates in the L^∞-norm", M^2AN,**24**, No.2 (1990) 235 – 264

[5] **P. Sonneveld** "CGS, A Fast Lanczos-Type Solver for Nonsymmetric Linear Systems", Rept. of the Dept. of Math. & Informatics, No.84-16, Delft (1984)

[6] **M. Stynes, E. O'Riordan** "Finite Element Methods for Elliptic Convection-Diffusion Problems" in *BAIL V, Proceedings of Boundary and Interior Layers* (eds. B.Y. Guo, J.J.H. Miller, Z.C. Shi), Boole Press, Dublin (1988) 65 – 76

[7] **R.S. Varga** *Matrix Iterative Analysis*, Prentice–Hall, Englewood Cliffs, New Jersey (1962)

[8] **S. Wang** *Numerical Methods for the Solution of the Stationary Semiconductor Device Equations in Two and Three Dimensions*, Ph.D. Thesis, University of Dublin (1989)

HIGH RESOLUTION SCHEMES FOR HYPERBOLIC CONSERVATION LAWS

CHRISTOS D. ANGELOPOULOS

Department of Mechanical Engineering
Imperial College, University of London
London SW7 2BX, U.K.
and
Technology Research Center
Greek Air Force
Athens 16501, Greece

ABSTRACT

In this paper we describe the technique of obtaining high resolution, second order, TVD scalar difference schemes. We propose an approach of extending this technique to systems of equations. A class of limiters is presented which includes a discontinuous limiter due to the author, and various limiters are compared numerically.

1. INTRODUCTION

In this paper we consider numerical solutions of the initial-value problem for hyperbolic systems of conservation laws

$$\begin{aligned} &u_t + f(u)_x = 0, \qquad t > 0, x \in \Re \\ &u(x,0) = u_0(x). \end{aligned} \tag{1.1}$$

Here $u(x,t)$ is a column vector of m unknowns and $f(u)$, the flux, is a vector-valued function of m components. Equation (1.1) is called hyperbolic if all eigenvalues $a^1(u), ..., a^m(u)$ of the Jacobian matrix $A(u)$

$$A(u) = f_u \tag{1.2}$$

are real and the set of right eigenvectors $R^1(u), ..., R^m(u)$ is complete. We assume that the eigenvalues $\{a^i(u)\}$ are arranged in a nondecreasing order

$$a^1(u) \leq a^2(u) \leq \ldots \leq a^m(u).$$

It is well known that even if the initial value u_0 is smooth, the solution to (1.1) typically develops discontinuities as t increases to some $t_0 > 0$. To allow for discontinuous solutions we admit weak solutions that satisfy (1.1) in the sense of distribution theory, i.e.,

$$\int\int_{\Re^+ \times \Re} (w_t u + w_x f(u)) dx dt = 0, \qquad \forall w \in C_0^\infty(\Re^+ \times \Re). \tag{1.3}$$

A piecewise-smooth weak solution of (1.1) satisfies (1.1) pointwise in each smooth region; across each curve of discontinuity the Rankine-Hugoniot relation

$$f(u_R) - f(u_L) = s(u_R - u_L) \tag{1.4}$$

holds, where s is the speed of propagation of the discontinuity, and u_L and u_R are the states on the left and the right, respectively.

However, there can be an infinite number of weak solutions of (1.1) with the same initial data u_0; and an additional principle, the entropy condition, is needed to select the unique physical weak solution [1], [2], [3], [4].

In [5] Van Leer derived a scheme using a flux limiter in his research for the ultimate conservative difference scheme. More recently Chakravarthy and Osher [6] have used limiters, as has Harten [7] who also introduced the notion of TVD (Total Variation Diminishing) to characterize oscillation free schemes. In [8] C. Angelopoulos proposed a discontinuous limiter.

The organization of the rest of this paper is as follows: In §2 we discuss a)Basic theory of TVD schemes; b)Second order accurate TVD schemes using flux limiters; c)Extension to systems of conservation laws. In §3 a numerical comparison of some of the limiters is given.

2. TVD SCHEME FOR HYPERBOLIC CONSERVATION LAWS

We consider explicit finite difference schemes in conservation form which approximate the scalar conservation law

$$\begin{aligned} &u_t + f(u)_x = 0, \qquad t > 0, x \in \Re \\ &u(x,0) = u_0(x) \end{aligned} \tag{2.1}$$

and which we denote by

$$v^{n+1} = L \cdot v^n. \tag{2.2}$$

A general 3-point explicit difference scheme in conservation form can be written as

$$v_i^{n+1} = v_i^n - \lambda(\bar{f}_{i+1/2}^n - \bar{f}_{i-1/2}^n) \tag{2.3}$$

where $\bar{f}_{i+1/2}^n = \bar{f}(v_i^n, v_{i+1}^n), \lambda = \Delta t/\Delta x$, with Δt the time step, and Δx the mesh size. Here $\bar{f}$ is a numerical flux function. We require the numerical flux function $\bar{f}$ to be consistent with the conservation law in the following sense

$$\bar{f}(u,u) = f(u). \tag{2.4}$$

We state now a useful theorem of Lax and Wendroff.

Theorem 2.1: Suppose the difference scheme (2.3) is consistent with the conservation law (2.1). Let v_i^n be a solution of (2.3), with initial values $v_i^0 = \phi(i\Delta x)$. Extend the lattice function v_i^n to continuous values of x, t by setting

$$v(x,t) = v_i^n \qquad i = \left[\frac{x}{\Delta x}\right], \qquad n = \left[\frac{t}{\Delta t}\right]. \tag{2.5}$$

Suppose that for some sequence $(\Delta x)_k \to 0$, $\Delta t/\Delta x = \lambda$, the limit

$$\lim_{(\Delta x)_k \to 0} v(x,t) = u(x,t)$$

exists in the sense of bounded, L_1^{loc} convergence. Then the limit $u(x,t)$ satisfies the weak form of the conservation law.

Proof: See [9].

The total variation of a mesh function is defined by the formula

$$TV(v^n) = \sum_i |v_{i+1}^n - v_i^n| \tag{2.6}$$

and a scheme is called total variation diminishing (TVD) if

$$TV(v^{n+1}) = TV(L \cdot v^n) \leq TV(v^n). \tag{2.7}$$

To determine whether a scheme is TVD, we rewrite the scheme in the form

$$v_i^{n+1} = v_i^n + D_{i+1/2}\Delta v_{i+1/2}^n - C_{i-1/2}\Delta v_{i-1/2}^n \tag{2.8}$$

where

$$\Delta v_{i+1/2}^n = v_{i+1}^n - v_i^n$$

and $C_{i-1/2}$, $D_{i+1/2}$ are functions of v^n.

Lemma 2.2: If the coefficients C and D of equation (2.8) satisfy the inequalities

$$\begin{aligned} &C_{i+1/2} \geq 0, \qquad D_{i+1/2} \geq 0, \\ &0 \leq C_{i+1/2} + D_{i+1/2} \leq 1 \end{aligned} \tag{2.9}$$

then the scheme (2.8) is TVD.

Proof:

$$
\begin{aligned}
TV(v^{n+1}) &= \sum_i |v_{i+1}^{n+1} - v_i^{n+1}| \\
&= \sum_i |v_{i+1}^n + D_{i+3/2}(v_{i+2}^n - v_{i+1}^n) - C_{i+1/2}(v_{i+1}^n - v_i^n) \\
&\quad - v_i^n - D_{i+1/2}(v_{i+1}^n - v_i^n) + C_{i-1/2}(v_i^n - v_{i-1}^n)| \\
&= \sum_i |(1 - C_{i+1/2} - D_{i+1/2})(v_{i+1}^n - v_i^n) \\
&\quad + D_{i+3/2}(v_{i+2}^n - v_{i+1}^n) + C_{i-1/2}(v_i^n - v_{i-1}^n)| \\
&\leq \sum_i (1 - C_{i+1/2} - D_{i+1/2})|v_{i+1}^n - v_i^n| \\
&\quad + \sum_i D_{i+3/2}|v_{i+2}^n - v_{i+1}^n| \\
&\quad + \sum_i C_{i-1/2}|v_i^n - v_{i-1}^n| \\
&= \sum_i |v_{i+1}^n - v_i^n| \\
&= TV(v^n).
\end{aligned}
$$

As an example consider a numerical flux $\bar{f}$ of the form

$$
\bar{f}(v_i, v_{i+1}) = \frac{1}{2}\left[f(v_i) + f(v_{i+1}) - \frac{1}{\lambda}Q(\lambda \bar{a}_{i+1/2})\Delta v_{i+1/2}\right] \tag{2.10}
$$

where

$$
\bar{a}_{i+1/2} = \begin{cases} (f(v_{i+1}) - f(v_i))/\Delta v_{i+1/2}, & \text{when } \Delta v_{i+1/2} \neq 0 \\ \\ a(v_i), & \text{when } \Delta v_{i+1/2} = 0 \end{cases} \tag{2.11}
$$

$$
a(u) = \frac{df(u)}{du}.
$$

Here $Q(x)$ is some function, which is often referred to as the coefficient of numerical viscosity. Using Lemma 2.2 we shall prove the following theorem.

Theorem 2.3: Let $Q(x)$ in (2.10) satisfy the inequalities

$$|x| \le Q(x) \le 1, \qquad 0 \le |x| \le \mu \le 1; \tag{2.12}$$

then the 3-point finite-difference scheme is TVD under the CFL like restriction

$$\lambda \max_i |\bar{a}^n_{i+1/2}| \le \mu. \tag{2.13}$$

Proof: Using the notation

$$\bar{\nu}_{i+1/2} = \lambda \bar{a}_{i+1/2}$$

we rewrite (2.10) as

$$\lambda \bar{f}_{i+1/2} = \lambda f(v_i) - \frac{1}{2}\Big[-\bar{\nu}_{i+1/2} + Q(\bar{\nu}_{i+1/2})\Big]\Delta v_{i+1/2}.$$

So we find

$$\begin{aligned} v_i^{n+1} &= v_i^n - \lambda(\bar{f}_{i+1/2} - \bar{f}_{i-1/2}) \\ &= v_i^n + \frac{1}{2}\Big[Q(\bar{\nu}_{i+1/2}) - \bar{\nu}_{i+1/2}\Big]\Delta v^n_{i+1/2} - \frac{1}{2}\Big[Q(\bar{\nu}_{i-1/2}) + \bar{\nu}_{i-1/2}\Big]\Delta v^n_{i-1/2} \\ &= v_i^n + D_{i+1/2}\Delta v^n_{i+1/2} - C_{i-1/2}\Delta v^n_{i-1/2} \end{aligned}$$

where

$$C_{i+1/2} = \frac{1}{2}\Big[Q(\bar{\nu}_{i+1/2}) + \bar{\nu}_{i+1/2}\Big]$$

$$D_{i+1/2} = \frac{1}{2}\Big[Q(\bar{\nu}_{i+1/2}) - \bar{\nu}_{i+1/2}\Big].$$

Since $C_{i+1/2} + D_{i+1/2} = Q(\bar{\nu}_{i+1/2})$ it follows from (2.12) and (2.13) that conditions (2.9) are satisfied and therefore the finite difference scheme is TVD.

Note that we can write any numerical flux in the form

$$\bar{f}(u,v) = \frac{1}{2}[f(u) + f(v)] - \frac{1}{2}d(u,v) \tag{2.14}$$

and for the sake of consistency we need

$$d(u,u) = 0.$$

We turn now to describing various forms of d(u,v) in (2.14). The most straightforward way to generate such functions is by

$$d(u,v) = |a|(u,v)(v-u) \tag{2.15a}$$

where one can choose

$$|a|(u,v) = \left|a\left(\frac{u+v}{2}\right)\right| \tag{2.15b}$$

or

$$|a|(u,v) = \frac{1}{2}\Big[|a(u)| + |a(v)|\Big]. \tag{2.15c}$$

Another form of $d(u,v)$ is

$$d(u,v) = sgn\left[a\left(\frac{u+v}{2}\right)\right](f(v) - f(u)) \tag{2.15d}$$

here $sgn(x)$ is the sign of x.

Theorem 2.4: Consider the case where $f(u) = \frac{1}{2}u^2$ and the function $d(u,v)$ is given by (2.15b). Then the 3-point finite difference scheme is TVD under the CFL like restriction

$$\lambda \max_i |a|_{i+1/2} \leq 1. \tag{2.16}$$

Proof: One can easily verify that

$$\bar{a}_{i+1/2} = \frac{1}{2}(u_i + u_{i+1})$$

The function $d_{i+1/2}$ is given by

$$d_{i+1/2} = \left|\frac{u_i + u_{i+1}}{2}\right|(u_{i+1} - u_i)$$

or

$$|a|_{i+1/2} = \left|\frac{u_i + u_{i+1}}{2}\right| = |\bar{a}_{i+1/2}|. \tag{2.17}$$

The numerical flux is given by

$$\bar{f}_{i+1/2} = \frac{1}{2}[f_i + f_{i+1} - |a|_{i+1/2}(v_{i+1} - v_i)]$$

or

$$\bar{f}_{i+1/2} = f_i - \frac{1}{2}\left[|a|_{i+1/2} - \bar{a}_{i+1/2}\right](v_{i+1} - v_i).$$

So we find

$$\begin{aligned} v_i^{n+1} &= v_i^n - \lambda(\bar{f}_{i+1/2} - \bar{f}_{i-1/2}) \\ &= v_i^n + \frac{\lambda}{2}\left[|a|_{i+1/2} - \bar{a}_{i+1/2}\right](v_{i+1} - v_i) \\ &- \frac{\lambda}{2}\left[|a|_{i-1/2} + \bar{a}_{i-1/2}\right](v_i - v_{i-1}) \\ &= v_i^n + D_{i+1/2}\Delta v_{i+1/2}^n - C_{i-1/2}\Delta v_{i-1/2}^n \end{aligned}$$

where

$$C_{i+1/2} = \frac{\lambda}{2}\left[|a|_{i+1/2} + \bar{a}_{i+1/2}\right]$$

$$D_{i+1/2} = \frac{\lambda}{2}\Big[|a|_{i+1/2} - \bar{a}_{i+1/2}\Big].$$

Since $C_{i+1/2} + D_{i+1/2} = \lambda|a|_{i+1/2}$ it follows from relations (2.16) and (2.17) that conditions (2.9) are satisfied and therefore the finite difference scheme is TVD.

We shall now derive a TVD scheme by adding only a limited amount of an antidiffusive flux to a first order scheme. For clarity of approach we first consider the linear scalar equation

$$u_t + au_x = 0, \qquad a = const. > 0. \tag{2.18}$$

The second order Lax-Wendroff scheme [9] may be written as

$$v_i^{n+1} = v_i^n - \nu\Delta v_{i-1/2}^n - \Delta_-\left\{\frac{1}{2}\nu(1-\nu)\Delta v_{i+1/2}^n\right\} \tag{2.19}$$

where here

$$\nu = a\frac{\Delta t}{\Delta x} < 1. \tag{2.20}$$

[We use the convention $\Delta_+ y_k = \Delta_- y_{k+1} = y_{k+1} - y_k$]. It is seen that (2.19) is in fact the result of a first order scheme

$$v_i^{n+1} = v_i^n - \nu\Delta v_{i-1/2}^n \tag{2.21}$$

with an additional term

$$-\Delta_-\left\{\frac{1}{2}(1-\nu)\nu\Delta v_{i+1/2}^n\right\} \tag{2.22}$$

added. That is, the numerical flux of the Lax-Wendroff scheme is that of the first order scheme (2.21) plus an antidiffusive flux.

Since it is well known that the Lax-Wendroff scheme is not TVD, we try to remedy this by adding only a limited amount of the antidiffusive flux to the first order scheme, i.e.

$$v_i^{n+1} = v_i^n - \nu \Delta v_{i-1/2}^n - \Delta_- \left\{ \phi_i \frac{1}{2}(1-\nu)\nu \Delta v_{i+1/2}^n \right\} \tag{2.23}$$

where ϕ_i is some form of limiter, taken to be nonnegative so as to maintain the sign of the antidiffusive flux. We take the limiter to be a function of consecutive gradients, i.e., $\phi_i = \phi(r_i)$ where

$$r_i = \frac{\Delta v_{i-1/2}}{\Delta v_{i+1/2}}. \tag{2.24}$$

Viewing the scheme (2.23), a possible choice of the coefficients $C_{i-1/2}$, $D_{i+1/2}$ in (2.8) is

$$\begin{aligned} C_{i-1/2} &= \nu + \frac{1}{2}(1-\nu)\nu \frac{\Delta_-\{\phi(r_i)\Delta v_{i+1/2}\}}{\Delta v_{i-1/2}} \\ D_{i+1/2} &= 0. \end{aligned} \tag{2.25}$$

Reorganization of the expression for $C_{i-1/2}$ gives

$$C_{i-1/2} = \nu + \frac{1}{2}\nu(1-\nu)\left\{ \phi(r_i)\frac{\Delta v_{i+1/2}}{\Delta v_{i-1/2}} - \phi(r_{i-1})\frac{\Delta v_{i-1/2}}{\Delta v_{i-1/2}} \right\}$$

or

$$C_{i-1/2} = \nu \left\{ 1 + \frac{1}{2}(1-\nu)\left[\frac{\phi(r_i)}{r_i} - \phi(r_{i-1}) \right] \right\}. \tag{2.26}$$

The TVD inequalities (2.9) reduce to

$$0 \le C_{i-1/2} \le 1. \tag{2.27}$$

Using equation (2.26), inequalities (2.27) are satisfied if

$$-\frac{2}{1-\nu} \le \frac{\phi(r_i)}{r_i} - \phi(r_{i-1}) \le \frac{2}{\nu}$$

or

$$\left|\frac{\phi(r_i)}{r_i} - \phi(r_{i-1})\right| \leq 2 \tag{2.28}$$

since $\nu < 1$.

In addition to requiring $\phi(r)$ to be nonnegative we also insist on

$$\phi(r) = 0, \qquad r \leq 0.$$

Hence we have the following theorem.

Theorem 2.5: For the scheme (2.23) to be TVD the limiter function $\phi(r)$ must lie in the shaded region of Fig.1.

In the case of the Lax-Wendroff scheme the limiter function $\phi(r) = 1$. The second order upwind scheme of Warming and Beam which is also non-TVD corresponds to the limiter function $\phi(r) = r$. It is noticed that the limiter functions $\phi(r)$ for both second order schemes pass through the point $\phi(1) = 1$, which is a general requirement for second order accuracy. So $\phi(r)$ is now confined to lie in the region shown in Fig.2.

In [5] Van Leer proposes a limiter

$$\phi_{VL} = \frac{|r| + r}{1 + |r|}. \tag{2.29}$$

Note that $\phi_{VL}(r) \geq 0$, with

$$\phi_{VL} = \begin{cases} 0, & \text{if } r \leq 0 \\ 2r/(1+r), & \text{if } r > 0 \end{cases}$$

and also satisfies the symmetry property

$$\frac{\phi_{VL}(r)}{r} = \phi_{VL}\left(\frac{1}{r}\right).$$

This limiter is shown in Fig.3 and it is seen that it lies within the second order TVD region.

Chakravarthy and Osher [6] propose a limiter

$$\phi_{CO}(r) = max(0, min(r, \psi)) \tag{2.30}$$

for $1 \leq \psi \leq 2$, and this limiter is sketched in Fig.4. Note, however, that it is not symmetric, i.e.,

$$\frac{\phi_{CO}(r)}{r} \neq \phi_{CO}\left(\frac{1}{r}\right)$$

unless $\psi = 1$.

C. Angelopoulos [8] proposes a discontinuous limiter which is shown in Fig.5.

We now propose an extension of the above ideas to the nonlinear equation (2.1). We take the underlying first order scheme to be an E-scheme and add both limited positive and negative fluxes. Consider the following first order scheme defined by the numerical flux

$$\bar{f}^{(1)}_{i+1/2} = \frac{1}{2}(f_i + f_{i+1}) - \frac{1}{2}|a|_{i+1/2}(v_{i+1} - v_i). \tag{2.31}$$

The second order TVD numerical flux is given by

$$\begin{aligned} \bar{f}^{(2)}_{i+1/2} = \bar{f}^{(1)}_{i+1/2} &+ \frac{1}{2}\phi^+_{i-1/2}a^+_{i-1/2}(v_i - v_{i-1}) \\ &- \frac{1}{2}\phi^-_{i+3/2}a^-_{i+3/2}(v_{i+2} - v_{i+1}) \end{aligned} \tag{2.32}$$

where

$$a^+ = \frac{1}{2}(a + |a|) \qquad and \qquad a^- = \frac{1}{2}(a - |a|).$$

The limiter is defined as

$$\phi^+_{i-1/2} = \phi(r^+_{i-1/2}) = \phi\left[\frac{a^+_{i+1/2}(v_{i+1} - v_i)}{a^+_{i-1/2}(v_i - v_{i-1})}\right]$$

$$\phi^-_{i+3/2} = \phi(r^-_{i+3/2}) = \phi\left[\frac{a^-_{i+1/2}(v_{i+1} - v_i)}{a^-_{i+3/2}(v_{i+2} - v_{i+1})}\right]. \tag{2.33}$$

Now, we briefly describe the above approach of extending the scalar scheme to systems of two linear equations

$$u_t + Au_x = 0, \qquad A = const. \tag{2.34}$$

The basic idea is to extend the scalar scheme to the system case by applying it "scalarly" to each of the characteristic variables.

Let

$$S = (R^1, R^2) \tag{2.35}$$

be a matrix, the columns of which are the right eigenvectors of the matrix A in (2.34). Then

$$A = S\Lambda S^{-1}, \qquad \Lambda_{ij} = a^i \delta_{ij} \tag{2.36}$$

where a^i are the eigenvalues of the matrix A. We define characteristic variables $W = (W^k)$ by

$$W = S^{-1}u. \tag{2.37}$$

It follows from (2.37) that (2.34) decouples into 2 scalar characteristic equations

$$\begin{aligned} (W^1)_t + a^1(W^1)_x &= 0 \\ (W^2)_t + a^2(W^2)_x &= 0. \end{aligned} \tag{2.38}$$

This offers a natural way of extending a scalar scheme to a constant coefficient system of equations (2.34) by applying it "scalarly" to each of the 2 scalar characteristic equations (2.38).

Equations (2.38) are solved as follows:

$$
\begin{aligned}
(W^1)_i^{n+1} &= (W^1)_i^n - \lambda(\bar{f}^{(2)}_{i+1/2} - \bar{f}^{(2)}_{i-1/2}) \\
(W^2)_i^{n+1} &= (W^2)_i^n - \lambda(\bar{\bar{f}}^{(2)}_{i+1/2} - \bar{\bar{f}}^{(2)}_{i-1/2})
\end{aligned}
\tag{2.39}
$$

where

$$
\begin{aligned}
\bar{f}^{(2)}_{i+1/2} = \bar{f}^{(1)}_{i+1/2} &+ \frac{1}{2}\phi^{+}_{i-1/2}(a^1)^{+}_{i-1/2}(W_i^1 - W_{i-1}^1) \\
&- \frac{1}{2}\phi^{-}_{i+3/2}(a^1)^{-}_{i+3/2}(W_{i+2}^1 - W_{i+1}^1)
\end{aligned}
$$

$$
\begin{aligned}
\bar{f}^{(1)}_{i+1/2} =& \frac{1}{2}[(a^1W^1)_i + (a^1W^1)_{i+1}] \\
&- \frac{1}{2}|a^1|_{i+1/2}(W_{i+1}^1 - W_i^1)
\end{aligned}
\tag{2.40}
$$

and similarly for $\bar{\bar{f}}^{(2)}_{i+1/2}$.

Note that eventhough the eigenvalue a^1 is constant we keep the index $i-1/2$ or $i+3/2$.

In the original variables, the scheme (2.39) takes the form

$$
\begin{aligned}
(u^1)_i^{n+1} = S^{11}&[(W^1)_i^n - \lambda(\bar{f}^{(2)}_{i+1/2} - \bar{f}^{(2)}_{i-1/2})] \\
&+ S^{12}[(W^2)_i^n - \lambda(\bar{\bar{f}}^{(2)}_{i+1/2} - \bar{\bar{f}}^{(2)}_{i-1/2})]
\end{aligned}
$$

$$
\begin{aligned}
(u^2)_i^{n+1} = S^{21}&[(W^1)_i^n - \lambda(\bar{f}^{(2)}_{i+1/2} - \bar{f}^{(2)}_{i-1/2})] \\
&+ S^{22}[(W^2)_i^n - \lambda(\bar{\bar{f}}^{(2)}_{i+1/2} - \bar{\bar{f}}^{(2)}_{i-1/2})].
\end{aligned}
\tag{2.41}
$$

One can easily verify that

$$
\begin{aligned}
S^{11}(W^1)_i^n + S^{12}(W^2)_i^n &= (u^1)_i^n \\
S^{21}(W^1)_i^n + S^{22}(W^2)_i^n &= (u^2)_i^n
\end{aligned}
\tag{2.42}
$$

Also we have

$$
\begin{aligned}
S^{11}\bar{f}^{(1)}_{i+1/2} + S^{12}\bar{\bar{f}}^{(1)}_{i+1/2} = &\frac{1}{2}S^{11}[a^1(W^1)^n_i + a^1(W^1)^n_{i+1}] \\
&-\frac{1}{2}S^{11}|a^1|[(W^1)^n_{i+1} - (W^1)^n_i] \\
&+\frac{1}{2}S^{12}[a^2(W^2)^n_i + a^2(W^2)^n_{i+1}] \\
&-\frac{1}{2}S^{12}|a^2|[(W^2)^n_{i+1} - (W^2)^n_i].
\end{aligned}
\tag{2.43}
$$

We know that

$$Au = S\Lambda S^{-1}u = S\Lambda W$$

or

$$
\begin{aligned}
f^1 &= S^{11}a^1W^1 + S^{12}a^2W^2 \\
f^2 &= S^{21}a^1W^1 + S^{22}a^2W^2
\end{aligned}
$$

where f^1 and f^2 are the components of the flux.

Similarly

$$|A|u = S|\Lambda|S^{-1}u = S|\Lambda|W$$

or

$$\sum_k |A|^{1k}u^k = S^{11}|a^1|W^1 + S^{12}|a^2|W^2.$$

Therefore equation (2.43) can be written as follows

$$S^{11}\bar{f}^{(1)}_{i+1/2} + S^{12}\bar{\bar{f}}^{(1)}_{i+1/2} = \frac{1}{2}(f^1_i + f^1_{i+1}) - \frac{1}{2}\sum_k |A|^{1k}(u^k_{i+1} - u^k_i).$$

We note that so far we have made use of the linearity of the system (2.34). We derived a first order scheme for the system (2.34). In this scheme the numerical flux is given by

$$\tilde{f}^{(1)}_{i+1/2} = \frac{1}{2}(f_i + f_{i+1}) - \frac{1}{2}|A|_{i+1/2}(u_{i+1} - u_i) \tag{2.44}$$

and this formula is used for the nonlinear case.

The rest of the terms in equation (2.41) are treated similarly. As an example consider the term

$$S^{11}\phi^{+}_{i-1/2}(a^1)^{+}_{i-1/2}(W^1_i - W^1_{i-1})$$

which, in the general case, is written as

$$(S^{11})_i\phi^{+}_{i-1/2}(a^1)^{+}_{i-1/2}\left[\sum_k (S^{-1})^{1k}_i(u^k_i - u^k_{i-1})\right].$$

3. NUMERICAL EXPERIMENTS

In this section we present some numerical experiments that demonstrate the performance of the proposed scheme. We consider now a Riemann problem for the Euler equations of a polytropic gas. Our set of data is $\rho_1 = 2.0$, $p_1 = 2.0$, $u_1 = 0.0$, $\rho_4 = 1.0$, $p_4 = 1.0$ and $u_4 = 0.0$. In all of the calculations $\Delta x = 0.01$ and $\frac{\Delta t}{\Delta x} = 0.1$. In Figs 6, 7 and 8 we show the results obtained by a high resolution scheme described in section 2 using the flux limiters (2.29), (2.30) and the discontinuous limiter respectively.

4. CONCLUDING REMARKS

We have derived high resolution schemes by adding a limited antidiffusive flux to a general entropy-satisfying first order scheme. Constraints on the limiters have been obtained so that the resulting scheme is TVD. Flux limiters used by Van Leer and the author have been studied and shown to be equivalent in the cases considered. The flux limiter used by Chakravarthy and Osher has also been investigated and shown to satisfy the TVD constraints and to exhibit results nearly as good as the previous limiters.

APPENDIX

In one spatial dimension, the Euler equations of gas dynamics can be written in the conservative form as

$$\frac{\partial U}{\partial t} + \frac{\partial F(U)}{\partial x} = 0$$

where

$$U = \begin{pmatrix} \rho \\ m \\ e \end{pmatrix},$$

$$F = \begin{pmatrix} m \\ \frac{m^2}{\rho} + P \\ (e + P)\frac{m}{\rho} \end{pmatrix}.$$

Here U is the vector of conservative variables, F is the flux vector, and $m = \rho u$. The primitive variables are the density ρ, the velocity u, and the pressure P. The total energy per unit volume e, is defined as

$$e = \rho\epsilon + \rho\frac{u^2}{2}$$

with ϵ as the internal energy per unit mass. The pressure P for a perfect gas is defined as

$$P = (\gamma - 1)\left(e - \frac{m^2}{2\rho}\right)$$

where γ is the ratio of specific heats.

Let A denote the Jacobian matrix $\partial F(U)/\partial U$ whose eigenvalues are

$$(a^1, a^2, a^3) = (u - c, u, u + c)$$

where c is the local speed of sound. The right eigenvectors of A form the matrix $R = (R^1, R^2, R^3)$ given by

$$R = \begin{pmatrix} 1 & 1 & 1 \\ u-c & u & u+c \\ H-uc & \frac{1}{2}u^2 & H+uc \end{pmatrix}$$

$$H = \frac{c^2}{\gamma - 1} + \frac{1}{2}u^2.$$

ACKNOWLEDGEMENTS

This research was supported by the Greek Air Force Office of Scientific Research.

REFERENCES

[1]. P. Lax, "Hyperbolic systems of conservation laws and the mathematical theory of shock waves", CBMS Regional Conference in Applied Mathematics 11, SIAM, 1972.

[2]. O.A. Oleinik, "Uniqueness and stability of the generalized solution of the Cauchy problem for a quasi-linear equation", Amer. Math. Soc. Transl., Ser. 2, 33 (1964), 285-290.

[3]. I. Gelfand, "Some problems in the theory of quasilinear equations", Amer. Math. Soc. Transl., ser. 2, (1983), 295-381.

[4]. S. Kruzkov, "First-order quasilinear equations with several space variables", Math. USSR sb, 10 (1970), 217-273.

[5]. B. Van Leer, "Towards the ultimate conservative difference scheme, II. Monotonicity and conservation combined in a second order scheme", J. Comp. Phys., 14 (1974), pp. 361-370.

[6]. S. Chakravarthy and S. Osher, "High resolution applications of the Osher upwind scheme for the Euler equations", AIAA paper presented at 6th CFD conference 1983.

[7]. A. Harten, "High resolution schemes for hyperbolic conservation laws", J. Comp. Phys., 49 (1983), 357-393.

[8]. C.D. Angelopoulos, "Shock modelling in transonic and supersonic flow", Ph.D. thesis, London University, 1991.

[9]. P. Lax and B. Wendroff, "Systems of conservation laws", Comm. Pure Appl. Math., 13 (1960), pp. 217-237.

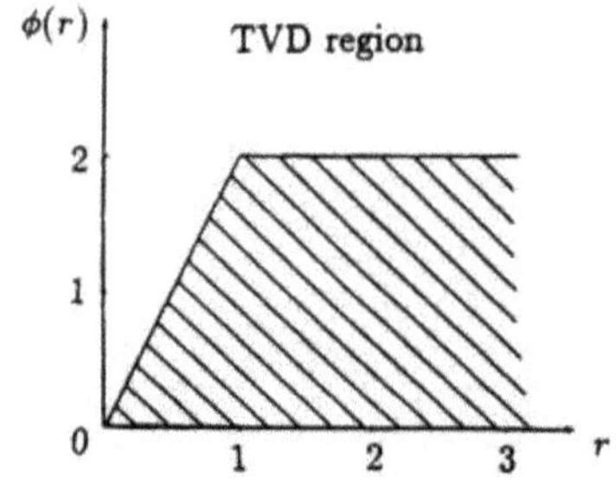

Fig.1: TVD region.

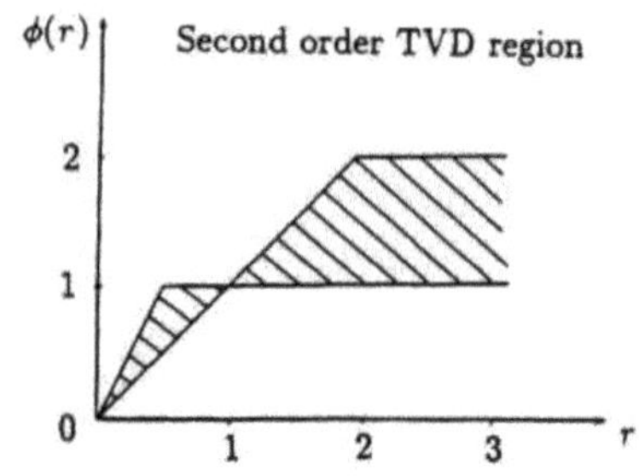

Fig.2: Second order TVD region.

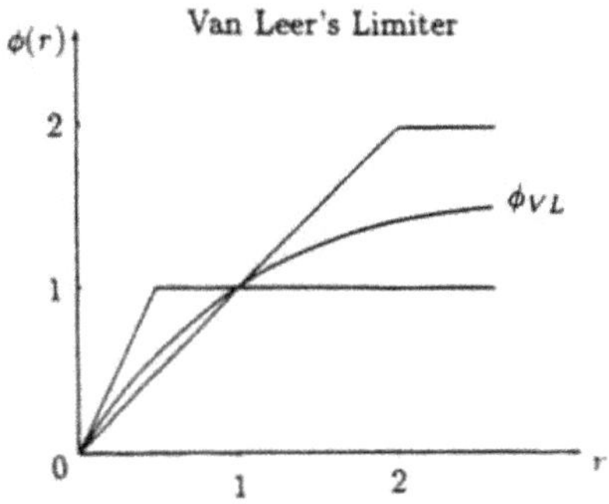

Fig.3: Van Leer's limiter.

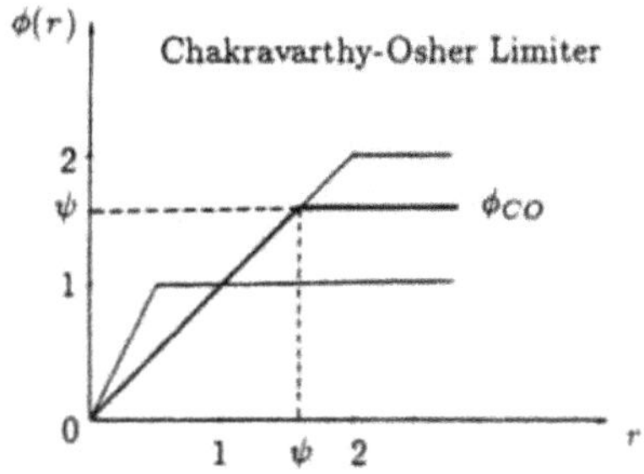

Fig.4: Chakravarthy-Osher limiter.

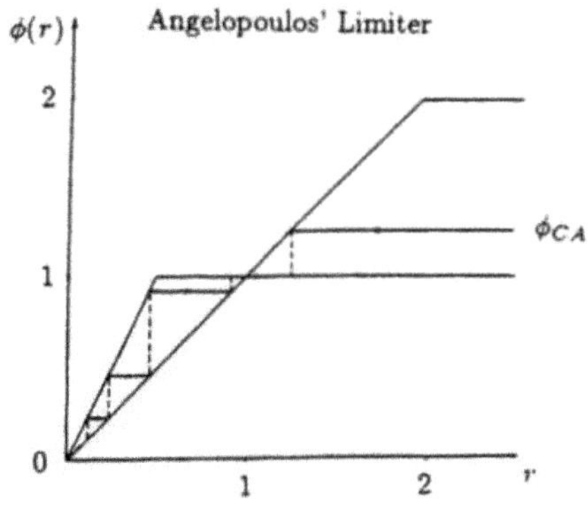

Fig.5: Angelopoulos' limiter.

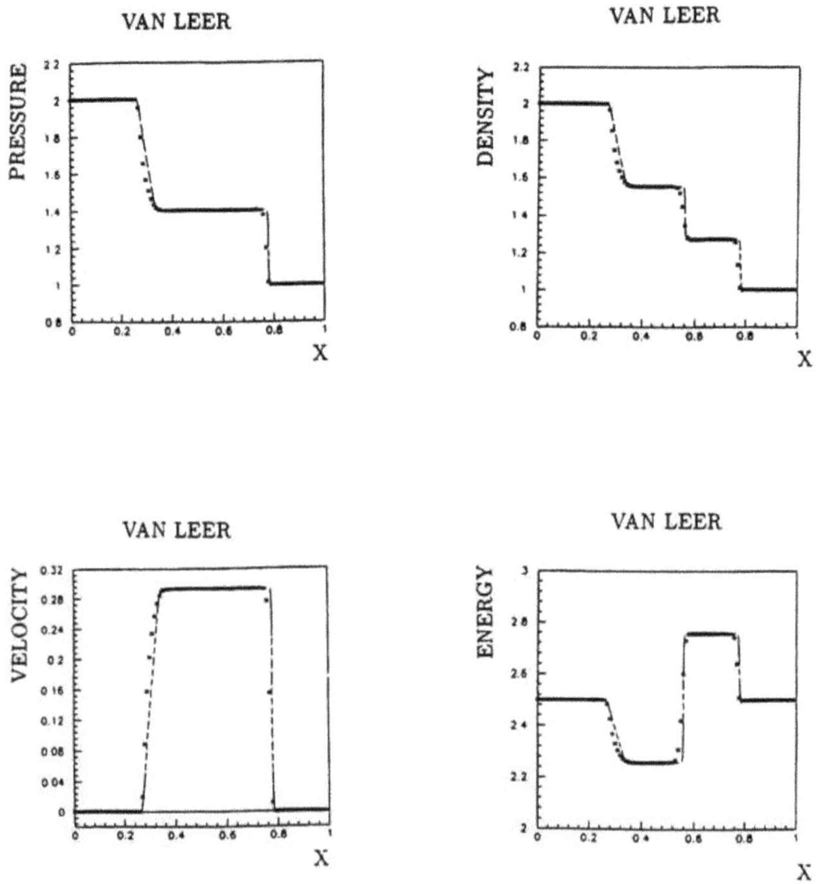

Fig.6: High resolution scheme described in §2 using Van Leer's limiter.

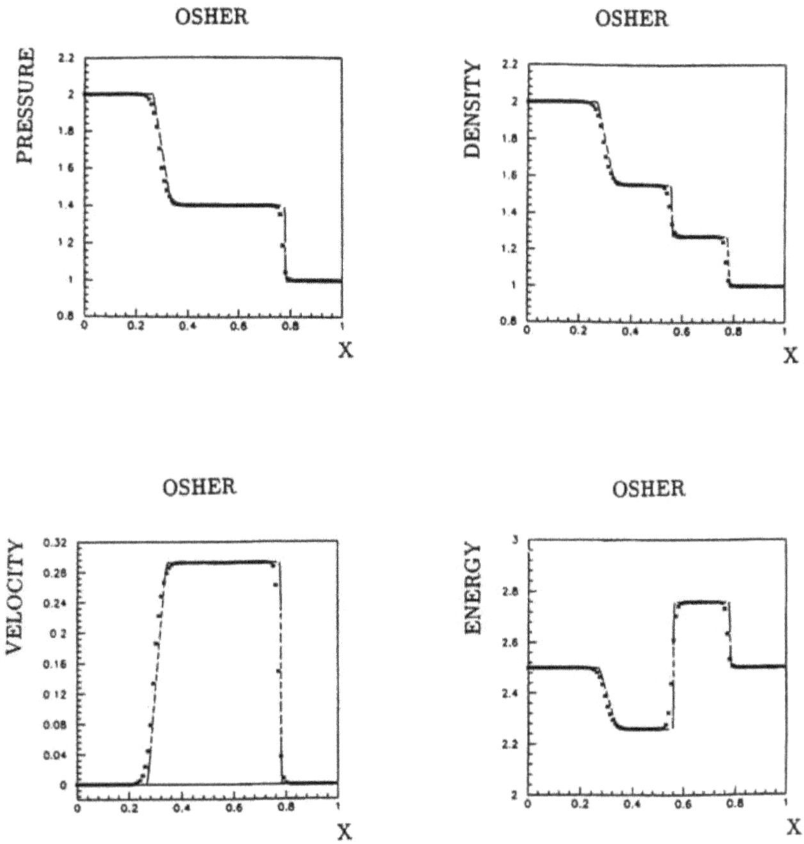

Fig.7: High resolution scheme described in §2 using Chakravarthy-Osher's limiter.

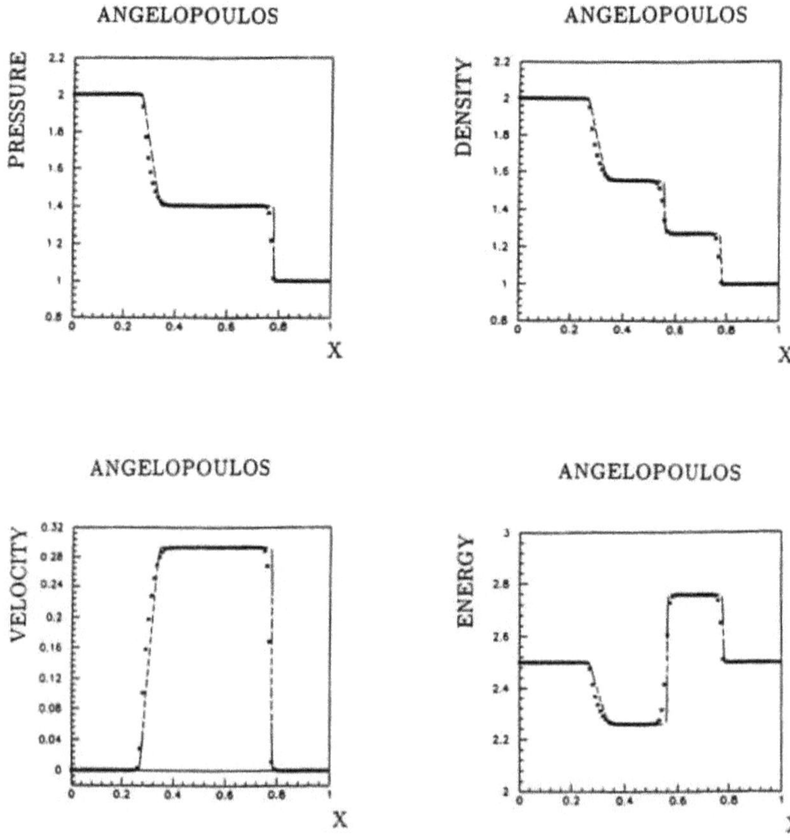

Fig.8: High resolution scheme described in §2 using Angelopoulos' limiter.

PARALLEL ALGORITHMS IN NUMERICAL LINEAR ALGEBRA

D.J. Evans
Parallel Algorithms Research Centre
Loughborough University of Technology
Loughborough, Leicestershire, U.K.

ABSTRACT

New parallel algorithms are proposed for the solution of large linear systems of tridiagonal equations which occur repeatedly in the numerical solution of partial differential equations. The extensions to block tridiagonal linear systems are also discussed.

KEYWORDS: Cyclic odd-even, stride reduction algorithms, tridiagonal linear systems, recursive striding algorithm.

C.R. CATEGORIES: G1.8

1. INTRODUCTION

The numerical solution of elliptic partial differential equations is an important problem in many fields of science and engineering. In certain cases, the structure of the coefficient matrix in the linear system allows the use of methods of reduced complexity compared to the standard direct or iterative solvers. This is the case in particular for linear, separable, self-adjoint elliptic equations defined on rectangular domains, with appropriate boundary conditions. Several techniques have been developed in the past few decades to solve these problems much faster than the traditional iterative methods [5]. For example, the solution of Poisson's equation on a rectangle discretized with an $n \times n$ grid entails an asymptotic operation count of $0(n^3 \log n)$ and $0(n^2 \log^2 n)$ for the (iterative) SOR and ADI methods respectively, but only $0(n^2 \log n)$ for the cyclic reduction and FFT-based methods.

This paper extends the well known cyclic reduction (CR) algorithm for the efficient solution of tridiagonal linear systems. Sets of such equations occur frequently in computations involving the numerical solution of differential equations.

In the proposed new methods the well known odd-even reduction strategy is extended to reduction strides of 3 and 4 resulting in efficient algorithms possessing considerable inherent parallelism. Numerical experiments show that the Quad Stride algorithm is the more efficient.

A new recursive stride (RS) procedure is presented for the LU decomposition of a positive definite tridiagonal matrix of order n. The new strategy involves $\log_2(n/2)$ steps and is superior to recursive doubling [6].

We consider in the rectangle,

$$\Omega = \{ (x,y) : x_1 \le x \le x_2, y_1 \le y \le y_2 \}.$$

the linear elliptic boundary value problems defined by,

$$\Delta u = \frac{\partial^2 u}{\partial x^2} + \frac{\partial^2 u}{\partial y^2} = f(x,y) \text{ in } \Omega ,$$

with the following boundary conditions,

- *Dirichlet conditions* $u(x,y) = g(x,y)$ on $\partial\Omega$,
- *Neumann conditions* $\frac{\partial u}{\partial n} = g(x,y)$ on $\partial\Omega$ $(f > 0)$,
- *Mixed conditions* $\alpha u + \beta \frac{\partial u}{\partial n} = g(x,y)$ on $\partial\Omega$,
- *Periodic conditions* $u(x,y) = g_1(x,y)$ on $x=x_1$ and $x=x_2$, $u(x,y_1) = u(x,y_2)$.

Let M and N be positive integers. We place a rectangular mesh $\Omega_{h,k} = \{(mh,nk), m=0,1,\dots,M+1; n=0,1,\dots,N+1\}$ over Ω where,

$$h = (x_2 - x_1) / (M+1) , \quad k=(y_2 - y_1) / (N+1) ,$$

are the grid spacing in the x and y directions, respectively.

For all functions u(x,y) defined at the mesh points of $\Omega_{h,k}$ let,

$$u_{m,n} = u(mh,nk) .$$

Using the standard second-order five-point finite difference approximation, with natural ordering of grid points, we obtain the following M × N linear system,

$$\frac{u_{m+1,n} - 2u_{m,n} + u_{m-1,n}}{h^2} + \frac{u_{m,n+1} - 2u_{m,n} + u_{m,n-1}}{k^2} = f(mh,nk) ,$$

where the boundary data are defined as,

$$u_{0,n} = g_{0,n} , u_{M+1,n} = g_{M+1,n} , \qquad n=0,1,\dots,N+1 ,$$
$$u_{m,0} = g_{m,0}, u_{m,N+1} = g_{m,N+1}, \qquad m=0,1,\dots,M+1 .$$

Let,

$$\gamma = k/h ,$$

then it can be readily shown that the resulting sets of linear equations can be expressed in block matrix form as,

$$C u = d ,$$

where C is the following (N×N) block matrix,

$$C = \begin{bmatrix} B & A & & & O \\ A & B & A & & \\ & \ddots & \ddots & \ddots & \\ & O & & \ddots & A \\ & & & A & B \end{bmatrix} ,$$

with $A = -I_{M\times M}$ and B is the M×M tridiagonal matrix,

$$B = \begin{bmatrix} b & a & & & & \\ a & b & a & & & O \\ & \ddots & \ddots & \ddots & & \\ & & \ddots & \ddots & \ddots & \\ & O & & \ddots & \ddots & a \\ & & & & a & b \end{bmatrix},$$

with $b = 1+\gamma^2$ and $a = -\gamma$.

For periodic boundary conditions the matrix C is a (N×N) block circulant matrix and B a (M×M) point circulant matrix.

2. THE CYCLIC ODD-EVEN REDUCTION METHOD

2.1 The Symmetric Constant-Diagonal Case

In this section we examine the tridiagonal equation solvers, operating on a set of n linear equations of the form,

$$A \, . \, u = d \ , \tag{1}$$

where A is a tridiagonal matrix of the form,

$$A = \begin{bmatrix} b & a & & & & \\ a & b & a & & & \\ & a & b & a & & O \\ & & \ddots & \ddots & \ddots & \\ & O & & \ddots & \ddots & a \\ & & & & a & b \end{bmatrix},$$

which, in shorthand notation, we denote as the (...,a,b,a,...) case.

We now explore a technique known as *cyclic odd-even reduction,* [2] which possesses advantages for transforming a sequential calculation into a highly parallel one, and appears to be the best algorithm for the *symmetric constant - tridiagonal* case.

In the following discussion, we assume that n can take any of the following values, i.e. 2^m-1, 2^m, 2^m+1, where m is any positive integer.

The general procedure for the *cyclic odd-even reduction* algorithm, is the following; let us consider three adjacent rows of A, i.e.,

$$\begin{array}{lccccc} i-1 & a & b & a & & \\ i & & a & b & a & \\ i+1 & & & a & b & a. \end{array} \tag{2}$$

A multiple of the middle row i is added to the summation of the i-1, i+1 rows, thus obtaining the form (..., a',O,b',O,a',...). This operation creates a tridiagonal system consisting of 2^{m-1}-1 or 2^{m-1} or 2^{m-1}+1 only, *even* or *odd* rows of the original matrix A.

For example, if we consider that, originally, the number of equations is: $n = 2^m$-1, and the new system consists of the *even* rows of the original matrix A, 2^{m-1}-1 the number, then although the *odd* rows have been eliminated, the *odd* unknowns can be computed from the *even* unknowns, by a back substitution process. Repeating the process to the new system of $2^{m-1}-1$ equations, involving just the *even* unknowns, we can eliminate every other row, thus obtaining a set of $2^{m-2}-1$ equations involving unknowns with subscripts multipes of 4. We can repeat this process, until we obtain a single equation for $u_{2^{m-1}}$, which can be easily solved.

Then, *back-substituting*, we can compute the eliminated unknowns in the reverse order in which they were eliminated.

The *cyclic odd-even reduction* algorithm reaches a significant speed increase when the *a* coefficients are initially equal to *unity*, i.e., the shorthand notation now has the form (...,1,b,1,...), which remain equal to unity throughout the computation, thus reducing the number of additions and multiplications per iteration.

The *symmetric constant-tridiagonal* system, given in (1), can be *normalized* in the (...,1,b,1,...) form, dividing by a, to normalize b and d, to produce the *symmetric constant-diagonal form*.

This can be done simultaneously, in most of the parallel computers, since the diagonal elements are a constant number.

More specifically, to describe the algorithm, let us consider the case that $n = 2^m-1$, where $m = 3$, i.e., the original matrix A consists of 7 equations. In this case the normalised matrix equation, given in (1) has the form,

$$\begin{array}{c} 1\\2\\3\\4\\5\\6\\7 \end{array}
\begin{bmatrix} b & 1 & & & & & \\ 1 & b & 1 & & & O & \\ & 1 & b & 1 & & & \\ & & 1 & b & 1 & & \\ & & & 1 & b & 1 & \\ & O & & & 1 & b & 1 \\ & & & & & 1 & b \end{bmatrix} \cdot \begin{bmatrix} u_1\\u_2\\u_3\\u_4\\u_5\\u_6\\u_7 \end{bmatrix} = \begin{bmatrix} d_1\\d_2\\d_3\\d_4\\d_5\\d_6\\d_7 \end{bmatrix} \tag{3}$$

When, the procedure described previously, is applied, (e.g. for the *even* rows), we multiply equations 2,4,6 by -b, adding the two adjacent rows to each of them. Then, the system (3) becomes,

$$\begin{array}{c} 1\\2\\3\\4\\5\\6\\7 \end{array}
\begin{bmatrix} b & 1 & & & & & \\ 0 & b^{[2]} & 0 & 1 & & O & \\ & 1 & b & 1 & & & \\ & 1 & 0 & b^{[2]} & 0 & 1 & \\ & & & 1 & b & 1 & \\ & O & & 1 & 0 & b^{[2]} & 0 \\ & & & & & 1 & b \end{bmatrix} \cdot \begin{bmatrix} u_1\\u_2\\u_3\\u_4\\u_5\\u_6\\u_7 \end{bmatrix} = \begin{bmatrix} d_1\\d_2^{[2]}\\d_3\\d_4^{[2]}\\d_5\\d_6^{[2]}\\d_7 \end{bmatrix} , \tag{4}$$

where,

$$b^{[2]} = 2 - b^2, \tag{5}$$

$$d_j^{[2]} = d_{j-1} - b.d_j + d_{j+1}\text{, for } j = 2,4,6\,. \tag{6}$$

Since the *even* rows 2,4,6 in the system (4) are independent of the *odd* rows, they may be separated as follows:

$$\begin{matrix} 1(2) \\ 2(4) \\ 3(6) \end{matrix} \begin{bmatrix} b^{[2]} & 1 & 0 \\ 1 & b^{[2]} & 1 \\ 0 & 1 & b^{[2]} \end{bmatrix} \cdot \begin{bmatrix} u_2 \\ u_4 \\ u_6 \end{bmatrix} = \begin{bmatrix} d_2^{[2]} \\ d_4^{[2]} \\ d_6^{[2]} \end{bmatrix} \tag{7}$$

Applying the above process, once more, to the system (7), i.e. multiplying the second row of the (7) system by $-b^{[2]}$, and adding the first and third rows, the system (7) becomes,

$$\begin{matrix} 1(2) \\ 2(4) \\ 3(6) \end{matrix} \begin{bmatrix} b^{[2]} & 1 & 0 \\ 0 & b^{[3]} & 0 \\ 0 & 1 & b^{[2]} \end{bmatrix} \cdot \begin{bmatrix} u_2 \\ u_4 \\ u_6 \end{bmatrix} = \begin{bmatrix} d_2^{[2]} \\ d_4^{[3]} \\ d_6^{[2]} \end{bmatrix} \tag{8}$$

where,

$$b^{[3]} = 2 - (b^{[2]})^2, \tag{9}$$

$$d_4^{[3]} = d_2^{[2]} - b^{[2]}.d_4^{[2]} + d_6^{[2]} \tag{10}$$

Separating again the second row of the (8) system we obtain,

$$b^{[3]}.\, u_4 = d_4^{[3]}\,, \tag{11}$$

which can be easily solved, thus finding u_4.

By a process of *back-substitution*, and in terms of u_4, the first and third rows of the system (8), may be written as,

$$b^{[2]}\,.\, u_2 = d_2^{[2]} - u_4\,, \tag{12}$$

$$b^{[2]}\,.\, u_6 = d_6^{[2]} - u_4. \tag{13}$$

Therefore, we can easily calculate u_2 and u_6.

Continuing the back-substitution, in the same way, to the system (4), we can calculate u_1, u_3, u_5, u_7 in terms of u_2, u_4, u_6, thus finding the solution of the matrix equation (3).

The above described process, can be applied to any of the values for n, (i.e. 2^m-1, 2^m, 2^m+1, where m is any integer), choosing each time, the *even* or *odd* rows of the system to work with.

The solution stages for the Cyclic Odd-Even algorithm are illustrated in Figure 1.

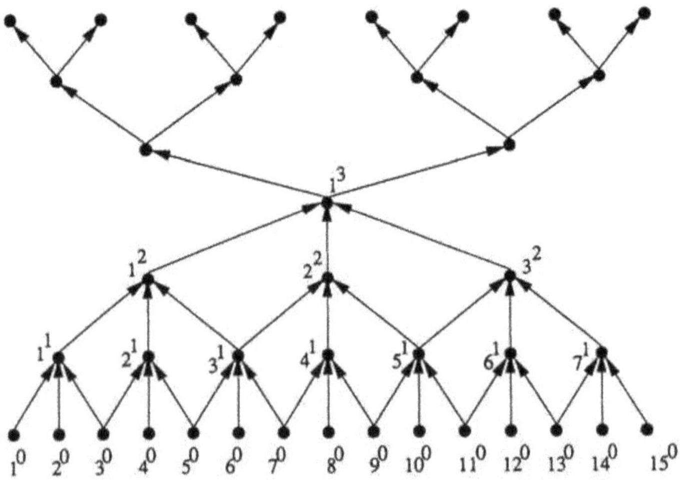

FIGURE 1: The Communication topology of odd-even cyclic reduction

2.2 Stride of 3 Cyclic Reduction Algorithm

In the previous section it can be seen that the Cyclic Reduction (C.R.) algorithm consists of successive reductions of the system (1) to similar reduced systems in a 'stride of 2'.Similarly we can consider forming a C.R. algorithm with reduced systems in a 'stride of 3' in the following way.

We now consider the reduced C.R. algorithm given by equation (1). Thus we have the equations,

$$u_1 + b^{[2]}u_3 + u_5 = d_3^{[2]} \tag{14}$$

$$u_3 + b^{[2]}u_5 + u_7 = d_5^{[2]} \tag{15}$$

In addition we have the equation,

$$u_3 + bu_4 + u_5 = d_4 , \tag{16}$$

from equation (3).

By adding equations (14) and (15) we obtain,

$$u_1 + (1+b^{[2]})u_3 + (1+b^{[2]})u_5 + u_7 = d_3^{[2]} + d_5^{[2]} \tag{17}$$

Then by multiplying (16) by $(1+b^{[2]})$ and subtracting from (17) we obtain the final equation,

$$u_1 - b(1+b^{[2]})u_4 + u_7 = d_3^{[2]} + d_5^{[2]} - (1+b^{[2]})d_4 . \tag{18}$$

If we now rewrite equation (18) in the form,

$$u_1 + \bar{b}^{[2]}u_4 + u_7 = d_4^{[2]} \quad , \tag{19}$$

which is the representative equation for a Stride of 3 in the C.R. algorithm, where,

$$\bar{b}^{[2]} = -b(1+b^{[2]}) = b(b^2-3) , \tag{20a}$$

and, $$\bar{d}_4^{[2]} = d_3^{[2]} + d_5^{[2]} - (3-b^2)d_4 = d_2+d_4-bd_3+d_4+d_6-bd_5 - (3-b^2)d_4$$

$$= d_2+d_6-b(d_3+d_5) - (1-b^2)d_4 . \tag{20b}$$

So the solution procedure for $n=3^p-1$, for $p=2$, i.e. $n=8$ is as follows:

the system $Au = d$,

i.e.,

$$\begin{bmatrix} b & 1 & & & & & & \\ 1 & b & 1 & & & & & \\ & 1 & b & 1 & & O & & \\ & & 1 & b & 1 & & & \\ & & & 1 & b & 1 & & \\ & O & & & 1 & b & 1 & \\ & & & & & 1 & b & 1 \\ & & & & & & 1 & b \end{bmatrix} \cdot \begin{bmatrix} u_1 \\ u_2 \\ u_3 \\ u_4 \\ u_5 \\ u_6 \\ u_7 \\ u_8 \end{bmatrix} = \begin{bmatrix} d_1 \\ d_2 \\ d_3 \\ d_4 \\ d_5 \\ d_6 \\ d_7 \\ d_8 \end{bmatrix} , \tag{21}$$

is reduced to $A^{[2]}u^{[2]} = d^{[2]}$,

$$\begin{bmatrix} \bar{b}^{[2]} & 1 \\ & \\ 1 & \bar{b}^{[2]} \end{bmatrix} \begin{bmatrix} u_2 \\ \\ u_6 \end{bmatrix} = \begin{bmatrix} \bar{d}_3^{[2]} \\ \\ \bar{d}_6^{[2]} \end{bmatrix} , \tag{22}$$

which can be solved for u_3 and u_6, i.e.,

$$u_6 = [\bar{d}_3^{[2]} - \bar{b}^{[2]}\bar{d}_6^{[2]}]/(1-b^{[2]^2}), \quad u_3 = [\bar{d}_6^{[2]} - \bar{b}^{[2]}\bar{d}_3^{[2]}]/(1-b^{[2]^2}) \tag{23}$$

Then, the remaining elements of the solution vector u can be obtained by solving the 3 subsystems,

$$\begin{bmatrix} b & 1 \\ 1 & b \end{bmatrix} \begin{bmatrix} u_1 \\ u_2 \end{bmatrix} = \begin{bmatrix} d_1 \\ d_2 - u_3 \end{bmatrix} , \tag{24a}$$

$$\begin{bmatrix} b & 1 \\ 1 & b \end{bmatrix} \begin{bmatrix} u_4 \\ u_5 \end{bmatrix} = \begin{bmatrix} d_4 - u_3 \\ d_5 - u_6 \end{bmatrix} , \tag{24b}$$

and,

$$\begin{bmatrix} b & 1 \\ 1 & b \end{bmatrix} \begin{bmatrix} u_7 \\ u_8 \end{bmatrix} = \begin{bmatrix} d_7 - u_6 \\ d_9 \end{bmatrix} , \tag{24c}$$

by simple elimination.

A comparison of the 'Stride of 3' and the Cyclic Odd Even algorithms for the solution of n equations is given by the ratio,

$$\log 3/\log 2 = 1.585, \tag{25}$$

which should be some indication of the efficiency to be obtained from the new algorithm.

The solution stages for the Stride of 3 algorithm are given in Figure 2.

Finally, it can be shown that the computational complexity of the 'Stride of 3' reduction algorithm is 14 operations/reduction stage for $\log_3 n$ reductions in comparison to 9 operations/reduction stage for $\log_2 n$ reductions for the cyclic odd-even algorithm which makes the 'Stride of 3' reduction algorithm slightly more efficient. This has been achieved by increasing the granularity of the problem.

2.3 Block Tridiagonal Linear Systems

In this section we outline the 'Stride of 3' reduction algorithm (SR3) for the block tridiagonal linear systems [2], [3], [4].

Given the block tridiagonal matrix derived from the finite difference discretisation of the Poisson Equation on a square grid of points in the unit square we have for $n=2\times3^2-1=17$, the following system of equations to solve,

$$Au = d , \tag{26}$$

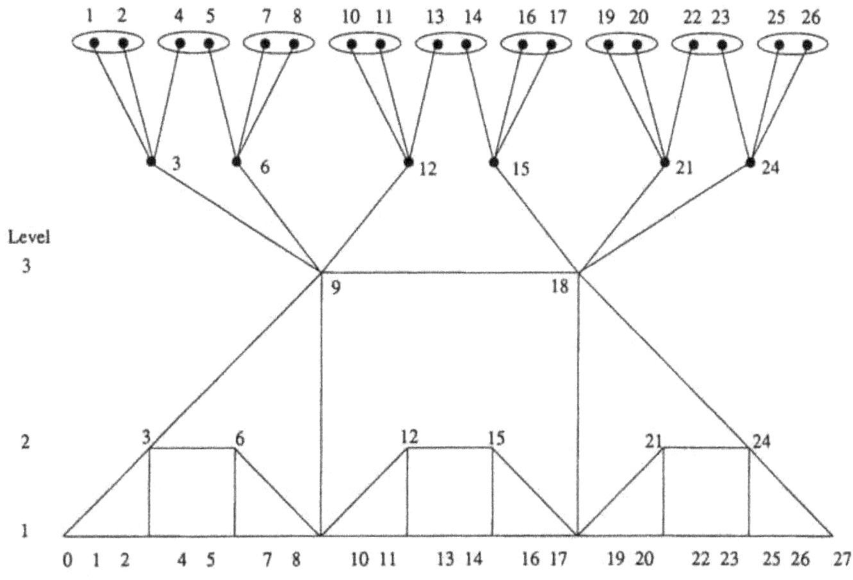

FIGURE 2: Communication topology Stride of 3 for $n=3^3-1$ (26)

or,

$$\begin{bmatrix} B & I & & & & \\ I & B & I & & O & \\ & \ddots & \ddots & \ddots & & \\ & & & & & \\ & O & & I & B & I \\ & & & & I & B \end{bmatrix} \begin{bmatrix} u_1 \\ u_2 \\ \vdots \\ u_{16} \\ u_{17} \end{bmatrix} = \begin{bmatrix} d_1 \\ d_2 \\ \vdots \\ d_{16} \\ d_{17} \end{bmatrix},$$

where,

$$B = \begin{bmatrix} -4 & +1 & & & & \\ +1 & -4 & +1 & & O & \\ & \ddots & \ddots & \ddots & & \\ & O & & +1 & -4 & +1 \\ & & & & +1 & -4 \end{bmatrix}, \tag{27}$$

and I is the (17×17) unit matrix.

We now consider the Stride of 3 Reduction (SR3) Algorithm which is given by the representative equation,

$$u_{i-3} + B^{[2]}u_i + u_{i+3} = d_i^{[2]}, \text{ for } i=3,6,9,12,15 \tag{28}$$

where,

$$B^{[2]} = B(B^2-3I), \tag{29}$$

and

$$d_i^{[2]} = d_{i-2} + d_{i+2} - B(d_{i-1}+d_{i+1}) - (I-B^2)d_i, \quad \text{for } i=3,6,9,12,15, \tag{30}$$

as given earlier.

These equations can be derived quite easily from 2 adjacent equations derived from the cyclic odd-even reduction algorithm, i.e.,

$$u_{i-2} + \hat{B}^{[2]}u_i + u_{i+2} = \hat{d}_i^{[2]}, \tag{31}$$

and,

$$u_i + \hat{B}^{[2]}u_{i+2} + u_{i+4} = \hat{d}_{i+2}^{[2]}, \tag{32}$$

with, $\hat{B} = 2I-B^2$ and $\hat{d}_i = d_{i-1} - Bd_i + d_{i+1}$,
combined with,

$$u_i + Bu_{i+1} + u_{i+2} = d_{i+1}, \tag{33}$$

derived from (26).

By eliminating u_i and u_{i+2} from equations (31), (32) and (33) we obtain the result given in (28) and (30).

Now by carrying on the application of (28) to equation (26) we obtain the following reduced system, where $B^{[2]}$ and $d^{[2]}$ are given by (29) and (30).

Now the solution to (26), can be obtained by applying the SR3 algorithm to obtain the system,

$$\begin{bmatrix} B^{[2]} & I & & & \\ I & B^{[2]} & I & & O \\ & I & B^{[2]} & I & \\ O & & I & B^{[2]} & I \\ & & & I & B^{[2]} \end{bmatrix} \begin{bmatrix} u_3 \\ u_6 \\ u_9 \\ u_{12} \\ u_{15} \end{bmatrix} = \begin{bmatrix} d_3^{[2]} \\ d_6^{[3]} \\ d_9^{[2]} \\ d_{12}^{[2]} \\ d_{15}^{[2]} \end{bmatrix} , \tag{34}$$

where $B^{[2]}$ and $d^{[2]}$ are given by (29) and (30).

Now the solution to (34), can be obtained by applying the SR3 algorithm again to obtain the system,

$$B^{[3]}u_9 = d_9^{[3]} , \tag{35}$$

where, $B^{[3]} = B^{[2]}(B^{[2]^2} - 3I)$,

and $d_i^{[3]} = d_{i-2}^{[2]} + d_{i+2}^{[2]} - B^{[2]}(d_{i-1}^{[2]} + d_{i+1}^{[2]} - (I - B^{[2]^2})d_i^{[2]}$, for i=9.

However the system $B^{[3]}u_9 = d_9^{[3]}$ etc. is not tridiagonal and appears to be not easily solvable but the solutions to these equations may be obtained by applying the solution to the tridiagonal problem repeatedly as follows.

In general, the matrices, $B^{[n]}$ satisfy the recursion relationships,

$$\begin{aligned} B^{[1]} &= B \\ B^{[2]} &= B^{[2]} [(B^{[1]2} - 3I] \\ &\cdot \\ &\qquad\qquad\qquad\qquad\qquad\qquad\qquad\qquad (36) \\ B^{[n]} &= B^{[n-1]} [(B^{[n-1]})^2 - 3I]. \end{aligned}$$

Further it can be seen that $B^{[n]}$ can be factorised as,

$$B^{[n-1]} (B^{[n-1]} - \sqrt{3}I) (B^{[n-1]} + \sqrt{3}I),$$

or

$$B^{[n]} = \prod_{k=1}^{3^{n-1}} G_3^{[n]} , n>1 , \tag{37}$$

where $G_k^{[n]}$ is a tridiagonal matrix given by,

$$G_k^{[n]} = B^{[n]} - l_k^{[n]} I , \tag{38}$$

and, $$l_k^{[n]} = 2\cos\left[\frac{(2k-1)p}{2\times 3^{n-1}}\right]. \tag{39}$$

Thus the equation (35) can be solved by applying the tridiagonal solver $G_k^{[n]}$ repeatedly, i.e. for u_9,

$$u_9 = \left[\prod_{k-1}^{3^{n-1}} G_k^{[3]}\right]^{-1} d_9^{[3]} \tag{40}$$

After the determination of u_9 the solution of the remaining elements can be determined by solving the subsystems,

$$\begin{bmatrix} B^{[2]} & I \\ I & B^{[2]} \end{bmatrix}\begin{bmatrix} u_3 \\ u_6 \end{bmatrix} = \begin{bmatrix} d_3 \\ d_6 \end{bmatrix}, \tag{41a}$$

$$\begin{bmatrix} B^{[2]} & I \\ I & B^{[2]} \end{bmatrix}\begin{bmatrix} u_{12} \\ u_{15} \end{bmatrix} = \begin{bmatrix} d_{12} \\ d_{15} \end{bmatrix}, \tag{41b}$$

in a similar manner as before, i.e.,

$$(I-B^{[2]^2})u_6 = d_3 - B^{[2]}d_6, \quad (I-B^{[2]^2})u_3 = d_6 - B^{[2]}d_3 , \tag{42}$$

with similar results for d_{12} and d_{15}.

Again the systems $(I-B^{[2]^2})$ are not tridiagonal but can be made to appear so by the application of the recursive procedure (37)-(39).

Finally the remaining elements can be obtained by solving the subsystems,

$$\begin{gathered}\begin{bmatrix} B & I \\ I & B \end{bmatrix}\begin{bmatrix} u_1 \\ u_2 \end{bmatrix} = \begin{bmatrix} d_1 \\ d_2 \end{bmatrix}, \begin{bmatrix} B & I \\ I & B \end{bmatrix}\begin{bmatrix} u_4 \\ u_5 \end{bmatrix} = \begin{bmatrix} d_4 \\ d_5 \end{bmatrix}, \begin{bmatrix} B & I \\ I & B \end{bmatrix}\begin{bmatrix} u_7 \\ u_8 \end{bmatrix} = \begin{bmatrix} d_7 \\ d_8 \end{bmatrix}, \\ \begin{bmatrix} B & I \\ I & B \end{bmatrix}\begin{bmatrix} u_{10} \\ u_{11} \end{bmatrix} = \begin{bmatrix} d_{10} \\ d_{11} \end{bmatrix}, \begin{bmatrix} B & I \\ I & B \end{bmatrix}\begin{bmatrix} u_{13} \\ u_{14} \end{bmatrix} = \begin{bmatrix} d_{13} \\ d_{14} \end{bmatrix}, \begin{bmatrix} B & I \\ I & B \end{bmatrix}\begin{bmatrix} u_{16} \\ u_{17} \end{bmatrix} = \begin{bmatrix} d_{16} \\ d_{17} \end{bmatrix},\end{gathered} \tag{43}$$

in a similar manner as previously shown.

3. THE GENERALISED CYCLIC AND STRIDE REDUCTION ALGORITHMS

In this section we consider similar reduction methods on general systems of n linear equations of the form,

$$A \cdot u = d , \tag{44}$$

where A is a non-periodic tridiagonal matrix, i.e.,

$$A = \begin{bmatrix} b_1 & c_1 & & & & \\ a_2 & b_2 & c_2 & & O & \\ & a_3 & b_3 & c_3 & & \\ & & \ddots & \ddots & \ddots & \\ & O & & & & c_{n-1} \\ & & & & a_n & b_n \end{bmatrix} \tag{45}$$

with $b_i > a_i+c_i$, $i=1,2,\ldots,n$.

3.1 The Generalised Cyclic Odd-Even Reduction Method

To solve a general tridiagonal matrix such as system (45) it is necessary to calculate the a', b', c' and y' for the new reduced system. If we consider three general adjacent rows,

$$\begin{array}{llll} i\text{-}1 & a_{i-1} & b_{i-1} & c_{i-1} \\ i & & a_i & b_i \quad c_i \;, \end{array} \tag{46}$$

unlike the previous examples it is not possible to simply add rows i-1 and i+1 to a multiple row i, as we now require two degrees of freedom to produce the desired eliminations. We shall now discuss one possible approach,

$$-\frac{a_i}{b_{i-1}}\,\text{row } i\text{-}1 + \text{row } i - \frac{c_i}{b_{i+1}}\,\text{row } i\text{+}1\,. \tag{47}$$

This achieves the required eliminations and gives our reduced system the following coefficients,

$$a'_i = -\frac{a_i a_{i-1}}{b_{i-1}}, \tag{48}$$

$$b'_i = b_i - a_i\frac{c_{i-1}}{b_{i-1}} - c_i\frac{a_{i+1}}{b_{i+1}}, \tag{49}$$

$$c'_i = -\frac{c_i c_{i+1}}{b_{i+1}}, \tag{50}$$

and

$$y'_i = y_i - \frac{a_i}{b_{i-1}}y_{i-1} - \frac{c_i}{b_{i+1}}y_{i+1}\,. \tag{51}$$

These eliminations are in such a way as to leave the ith row in the following form,

$$a'_i \quad 0 \quad b'_i \quad 0 \quad c'_i\,,$$

thus we obtain a tridiagonal system of equations consisting of either, only the odd, or the even rows of the original matrix. For example, if we consider that, originally, the number of equations is $n = 2^m-1$ (where m is a positive integer), and the new system consists of the even rows of the original matrix, $2^{m-1}-1$ in number, then although the odd rows have been eliminated, they can be computed from the even unknowns, by a backsubstitution process. Repeating the process to the new system of $2^{m-1}-1$ equations, involving just the even unknowns, we can eliminate every other row, thus obtaining a set of $2^{m-2}-1$ equations involving unknowns with subscripts that are multiples of 4. We can repeat this process, until we obtain a single equation for $u_{2^{m-1}}$, which can be easily solved.

Then, back-substituting, we can compute the eliminated unknowns in the reverse order in which they were eliminated.

3.2 The Generalised Stride of Three Algorithm

Although the reduction presented earlier perform significantly better on less general matrices the following derivation is included for the sake of completeness. The matrix system we shall consider is as follows,

$$\begin{bmatrix} b_1 & c_1 & & & & O \\ a_2 & b_2 & c_2 & & & \\ & a_3 & b_3 & c_3 & & \\ & & \ddots & \ddots & \ddots & \\ & O & & & & c_{n-1} \\ & & & & a_n & b_n \end{bmatrix} \begin{bmatrix} u_1 \\ u_2 \\ u_3 \\ \vdots \\ \vdots \\ u_n \end{bmatrix} = \begin{bmatrix} d_1 \\ d_2 \\ d_3 \\ \vdots \\ \vdots \\ d_n \end{bmatrix} \tag{52}$$

The first stage of a stride of three reduction, is the formation of Odd-Even reduced equations where the updates for the Odd-Even reduction method on a general tridiagonal matrix are presented in Section 3.1. Using the same notation as before, for the Odd-Even reduction updates, we shall consider the following equations,

$$\begin{array}{lllllll} 1 & a'_{i-1} & b'_{i-1} & \cdot & c'_{i-1} & & \\ 2 & & a_i & b_i & c_i & & \\ 3 & & a'_{i+1} & \cdot & b'_{i+1} & c'_{i+1} & \end{array} \tag{53}$$

We require two degrees of freedom to perform the necessary eliminations, if we multiply row 1 by R and row 3 by S then R,S have the following form,

$$R = \frac{a_i b'_{i+1} - c_i a'_{i+1}}{c'_{i-1} a'_{i+1} - b'_{i-1} b'_{i+1}}, \tag{54}$$

$$S = \frac{c_i b'_{i-1} - a_i c'_{i-1}}{c'_{i-1} a'_{i+1} - b'_{i-1} b'_{i+1}}, \tag{55}$$

Thus, the stride of three updates for a general matrix have the form,

$$\bar{a}_i = a'_{i-1} R, \tag{56}$$

$$\bar{b}_i = b_i, \tag{57}$$

$$\bar{c}_i = c'_{i+1} S, \tag{58}$$

$$\bar{d}_i = d_i + R\, d'_{i-1} + S\, d'_{i+1}. \tag{59}$$

From the above updates it can be seen that $\bar{b}_i$ has the same value as b_i (i.e. the diagonal elements do not alter their value during the reduction) hence if we assume that each equation is initially scaled by dividing throughout by the diagonal element (i.e. set the diagonal elements to unity) this simplifies the algebra considerably.

So, setting b_i to 1 for all i and substituting in the Odd-Even updates the stride of three reduction updates for a general tridiagonal matrix are as follows,

$$\bar{a}_i = \frac{-a_{i-2}a_{i-1}a_i(1-a_{i+2}c_{i+1})}{a_ia_{i+1}c_{i-1}c_i - (1-a_{i+1}c_i-a_{i+2}c_{i+1})(1-a_{i-1}c_{i-2}-a_ic_{i-1})}, \tag{60}$$

$$\bar{b}_i = 1, \tag{61}$$

$$\bar{c}_i = \frac{c_ic_{i+1}c_{i+2}(1-a_{i-1}c_{i-2})}{a_ia_{i+1}c_{i-1}c_i - (1-a_{i+1}c_i-a_{i+2}c_{i+1})(1-a_{i-1}c_{i-2}-a_ic_{i-1})}, \tag{62}$$

and,

$$\bar{d}_i = d_i + \frac{a_i(1-a_{i+2}c_{i+1})(d_{i-1}-a_{i-1}d_{i-2}-c_{i-1}d_i) + c_i(1-a_{i-1}c_{i-2})(d_{i+1}-a_{i+1}d_i-c_{i+1}d_{i+2})}{a_ia_{i+1}c_{i-1}c_i - (1-a_{i+1}c_i-a_{i+2}c_{i+1})(1-a_{i-1}c_{i-2}-a_ic_{i-1})} \tag{63}$$

These coefficients are considerably more complex than for the constant tridiagonal case and will impair the efficiency of the method.

4. STABILITY OF THE STRIDE REDUCTION ALGORITHM

It is well known that the Odd-Even reduction method is unstable and that the instability occurs in the update for the right-hand side of the system. Bunemann [1] presented an improved version of the algorithm in which he achieved greater stability by splitting the update into the following form,

$$d_i = b\alpha_i + \beta_i,$$

where α_i and β_i have a standard form for all i.

We now present a similar decomposition of the right-hand side for the stride of three reduction algorithm.

We need to rewrite $\bar{d}^{[2]}$ so that it is of the form,

$$\bar{d}_i^{[2]} = \bar{b}^{[2]} p_i^{[2]} + q_i^{[2]} \tag{64}$$

From (20a) we have, $\bar{b}^{[2]} = b(b^2-3)$ therefore,

$$\bar{d}_i^{[2]} = b(b^2-3)\, p_i^{[2]} + q_i^{[2]} \tag{65}$$

From (20b) we have,

$$\bar{d}_i^{[2]} = d_{i-2} + d_{i+2} - b(d_{i-1} + d_{i+1}) - (1 - b^2)d_i . \tag{66}$$

If we let $q_i^{[1]} = y_i \ \forall$ i, (66) becomes,

$$\bar{d}_i^{[2]} = q_{i-2} + q_{i+2} - b(q_{i-1} + q_{i+1}) - (1 - b^2)q_i \,. \tag{67}$$

Combining (65) and (67) we obtain,

$$b(b^2 - 3)\, p_i^{[2]} + q_i^{[2]} = q_{i-2} + q_{i+2} - b(q_{i-1} + q_{i+1}) - (1 - b^2)q_i \,. \tag{68}$$

If we let $b^3 p_i^{[2]} = b^2 q_i$ (i.e. $p_i^{[2]} = b^{-1}\, q_i^{[1]}$) then (68) becomes,

$$-3bp_i^{[2]} + q_i^{[2]} = q_{i-2} + q_{i+2} - b(q_{i-1} + q_{i+1}) - q_i$$

$$\therefore\; q_i^{[2]} = q_{i-2} + q_{i+2} - b(q_{i-1} + q_{i+1}) - q_i + 3bp_i^{[2]}$$

So if $\bar{d}_i^{[2]} = \bar{b}^{[2]}\, p_i^{[2]} + q_i^{[2]}$ then,

$$p_i^{[2]} = b^{-1}\, q_i^{[1]} \,,$$

$$q_i^{[2]} = q_{i-2} + q_{i+2} - b(q_{i-1} + q_{i+1}) - q_i + 3b\, p_i^{[2]}$$

Similarly we require that,

$$\bar{d}_i^{[3]} = \bar{b}^{[3]}\, p_i^{[3]} + q_i^{[3]}$$

From (20a) we have, $\bar{b}^{[3]} = b^{[2]}((b^{[2]})^2 - 3)$ therefore,

$$\bar{d}_i^{[3]} = b^{[2]}((b^{[2]})^2 - 3)\, p_i^{[3]} + q_i^{[3]} \tag{69}$$

From (20b) we have,

$$\bar{d}_i^{[3]} = \bar{d}_{i-2s}^{[2]} + \bar{d}_{i+2s}^{[2]} - b^{[2]}\,(\bar{d}_{i-s}^{[2]} + \bar{d}_{i+2s}^{[2]}) - (1-(b^{[2]})^2)\, \bar{d}_i^{[2]} \,, \tag{70}$$

where s represents the separation between the equations in the reduced system, i.e. in this case s=3.

Substituting (64) into (70) we have,

$$\bar{d}_i^{[3]} = b^{[2]}p_{i-2s}^{[2]} + q_{i-2s}^{[2]} + b^{[2]}p_{i+2s}^{[2]} + q_{i+2s}^{[2]} - b^{[2]}\,(b^{[2]}p_{i-s}^{[2]} + q_{i-s}^{[2]} +$$

$$b^{[2]}p_{i+s}^{[2]} + q_{i+s}^{[2]}) - (1-(b^{[2]})^2)\,(b^{[2]}p_i^{[2]} + q_i^{[2]}$$

$$= b^{[2]}(p_{i-2s}^{[2]} + p_{i+2s}^{[2]} - b^{[2]}(p_{i-s}^{[2]} + p_{i+s}^{[2]}) - p_i^{[2]}) + (b^{[2]})^3\, p_i^{[2]} + (b^{[2]})^2\, q_i^{[2]}$$

$$+ q_{i-2s}^{[2]} + q_{i+2s}^{[2]} - b^{[2]}(q_{i-s}^{[2]} + q_{i+s}^{[2]}) - q_i^{[2]} \tag{71}$$

Combining (69) and (71),

$$b^{[2]}((b^{[2]})^2 - 3)\, p_i^{[3]} + q_i^{[3]} = b^{[2]}(p_{i-2s}^{[2]} + p_{i+2s}^{[2]} - b^{[2]}(p_{i-s}^{[2]} + p_{i+s}^{[2]}) - p_i^{[2]}) + (b^{[2]})^3 p_i^{[2]}$$

$$+ (b^{[2]})^2 q_i^{[2]} + q_{i-2s}^{[2]} + q_{i+2s}^{[2]} - b^{[2]}(q_{i-s}^{[2]} + q_{i+s}^{[2]}) - q_i^{[2]} \qquad (72)$$

Let,

$$q_i^{[3]} = q_{i-2s}^{[2]} + q_{i+2s}^{[2]} - b^{[2]}(q_{i-s}^{[2]} + q_{i+s}^{[2]}) - q_i^{[2]} + 3b^{[2]} p_i^{[3]}$$

Thus, (72) becomes,

$$(b^{[2]})^3 p_i^{[3]} = b^{[2]}(p_{i-2s}^{[2]} + p_{i+2s}^{[2]} - b^{[2]}(p_{i-s}^{[2]} + p_{i+s}^{[2]}) - p_i^{[2]}) + (b^{[2]})^3 p_i^{[2]} + (b^{[2]})^2 q_i^{[2]},$$

$$\therefore\ p_i^{[3]} = (b^{[2]})^3 p_i^{[2]} + (b^{[2]})^{-1} (b^{[2]})^3 q_i^{[2]} + (b^{[2]})^{-2}(p_{i-2s}^{[2]} + p_{i+2s}^{[2]} - b^{[2]}(p_{i-s}^{[2]} + p_{i+s}^{[2]}) - p_i^{[2]}).$$

(73)

So, if $\bar{d}_i^{[3]} = \bar{b}^{[3]} p_i^{[3]} + q_i^{[3]}$, then,

$$q_i^{[3]} = q_{i-2s}^{[2]} + q_{i+2s}^{[2]} - b^{[2]}(q_{i-s}^{[2]} + q_{i+s}^{[2]}) - q_i^{[2]} + 3b^{[2]} p_i^{[3]}$$

$$p_i^{[3]} = p_i^{[2]} + (b^{[2]})^{-1} (b^{[2]})^3 q_i^{[2]} + (b^{[2]})^{-2}(p_{i-2s}^{[2]} + p_{i+2s}^{[2]} - b^{[2]}(p_{i-s}^{[2]} + p_{i+s}^{[2]}) - p_i^{[2]}).$$

Now we will give the generalised expressions,

$$\bar{d}_i^{[j]} = \bar{b}^{[j]} p_i^{[j]} + q_i^{[j]}, \qquad (74)$$

where,

$$q_i^{[j+1]} = q_{i-2s}^{[j]} + q_{i+2s}^{[j]} - b^{[j]}(q_{i-s}^{[j]} + q_{i+s}^{[j]}) - q_i^{[j]} + 3b^{[j]} p_i^{[j+1]},$$

$$p_i^{[j+1]} = p_i^{[j]} + (b^{[j]})^{-1} q_i^{[j]} + (b^{[j]})^{-2} (p_{i-2s}^{[j]} + p_{i+2s}^{[j]} - b^{[j]}(p_{i-s}^{[j]} + p_{i+s}^{[j]}) - p_i^{[j]}),$$

for j=1,2,... and where,

$$q_i^{[1]} = d_i\ \forall\ i,\ \ p_i^{[1]} = 0\ \forall\ i,\ \ s = 3^{j-1}, \text{ for } j=1,2,\ldots.$$

This proves the stability of the Stride of 3 algorithm.

5. STRIDE OF 4 CYCLIC REDUCTION ALGORITHM

In the previous section it can be seen that the Cyclic Reduction (CR) Odd-Even algorithm consists of successive reductions of the original system (1) to similar reduced systems in a 'stride of 2' fashion.

We can consider forming a C.R. algorithm with reduced systems in a 'stride of 4' algorithm in the following way. This will avoid some of the overheads involved in the reduction stages and improve the efficiency of the algorithm by increasing the granularity of the solution process.

We now consider the reduced C.R. algorithm given by equations (5) and (6) to the equation (1).

After the first reduction of the C.R. algorithm we have the equations,

$$u_1 + b^{[2]}u_3 + u_5 \qquad\qquad = d_3^{[2]} \tag{75}$$

$$u_3 + b^{[2]}u_5 + u_7 \qquad = d_5^{[2]} \tag{76}$$

$$u_5 + b^{[2]}u_7 + u_9 = d_7^{[2]} \tag{77}$$

where,

$$b^{[2]} = 2 - b^{[1]^2},\ b^{[1]} = b,$$

$$d_j^{[2]} = d_{j-1} - b^{[1]}d_j + d_{j+1}\,. \tag{78}$$

A further reduction stage is now carried out by multiplying equation (76) by $-b^{[2]}$ and adding to equations (75) and (77) to obtain,

$$u_1 + b^{[3]}u_5 + u_9 = d_5^{[3]} \tag{79}$$

and similar equations with a 'stride of 4', where,

$$b^{[3]} = 2 - b^{[2]^2},$$

$$d_j^{[3]} = d_{j-2}^{[2]} - b^{[2]}d_j^{[2]} + d_{j+2}^{[2]}\,. \tag{80}$$

Alternatively we can easily avoid the initial reduction stages and proceed to equations (79) by reformulating $b^{[3]}$ and $d_j^{[3]}$ as,

$$b^{[3]} = 2 - (2-b^{[1]^2})^2 = 4b^2 - b^4 - 2\,, \tag{81}$$

and,

$$d_j^{[3]} = d_{j-1} + d_{j+3} - b(d_{j-2} + d_{j+2}) - (1-b^2)\,(d_{j-1} + d_{j+1}) + b(2-b^2)d_j\,. \tag{82}$$

The solution process for $n = 4^p-1$ for $p=2$, i.e. $n=15$ is as follows. The system,

$$Au = d\,,$$

$$\begin{bmatrix}
b & 1 \\
1 & b & 1 \\
 & 1 & b & 1 \\
 & & 1 & b & 1 \\
 & & & 1 & b & 1 & & & & & O \\
 & & & & 1 & b & 1 \\
 & & & & & 1 & b & 1 \\
 & & & & & & 1 & b & 1 \\
 & & & & & & & 1 & b & 1 \\
 & & & & & & & & 1 & b & 1 \\
 & & & & & & & & & 1 & b & 1 \\
 & & & & O & & & & & & 1 & b & 1 \\
 & & & & & & & & & & & 1 & b & 1 \\
 & & & & & & & & & & & & 1 & b & 1 \\
 & & & & & & & & & & & & & 1 & b
\end{bmatrix}, \tag{83}$$

is reduced to,

$$\begin{bmatrix} b^{[3]} & 1 & \\ 1 & b^{[3]} & 1 \\ & 1 & b^{[3]} \end{bmatrix} \begin{bmatrix} u_4 \\ u_8 \\ u_{12} \end{bmatrix} = \begin{bmatrix} d_4^{[3]} \\ d_8^{[3]} \\ d_{12}^{[3]} \end{bmatrix}, \tag{84}$$

by the reduction given by equations (81) and (82).

The solution of equation (84) can be obtained explicitly in the form,

$$\begin{bmatrix} u_4 \\ u_8 \\ u_{12} \end{bmatrix} = \frac{1}{b^{[3]}(b^{[3]^2}-2)} \begin{bmatrix} b^{[3]^2}-1 & -b^{[3]} & 1 \\ -b^{[3]} & b^{[3]^2} & -b^{[3]} \\ 1 & -b^{[3]} & b^{[3]^2}-1 \end{bmatrix} \begin{bmatrix} d_4^{[3]} \\ d_8^{[3]} \\ d_{12}^{[3]} \end{bmatrix}, \tag{85}$$

to give the values u_4, u_8 and u_{12}.

Then the remaining elements of the solution vector u can be obtained by solving the 4 subsystems,

$$\begin{bmatrix} b & 1 & \\ 1 & b & 1 \\ & 1 & b \end{bmatrix} \begin{bmatrix} u_1 \\ u_2 \\ u_3 \end{bmatrix} = \begin{bmatrix} d_1 \\ d_2 \\ d_3-u_4 \end{bmatrix}, \tag{86a}$$

$$\begin{bmatrix} b & 1 & \\ 1 & b & 1 \\ & 1 & b \end{bmatrix} \begin{bmatrix} u_5 \\ u_6 \\ u_7 \end{bmatrix} = \begin{bmatrix} d_5-u_4 \\ d_6 \\ d_7-u_8 \end{bmatrix}, \tag{86b}$$

$$\begin{bmatrix} b & 1 & \\ 1 & b & 1 \\ & 1 & b \end{bmatrix} \begin{bmatrix} u_9 \\ u_{10} \\ u_{11} \end{bmatrix} = \begin{bmatrix} d_9-u_8 \\ d_{10} \\ d_{11}-u_{12} \end{bmatrix}, \tag{86c}$$

and

$$\begin{bmatrix} b & 1 & \\ 1 & b & 1 \\ & 1 & b \end{bmatrix} \begin{bmatrix} u_{13} \\ u_{14} \\ u_{15} \end{bmatrix} = \begin{bmatrix} d_{13}-u_{12} \\ d_{14} \\ d_{15} \end{bmatrix}, \tag{86d}$$

which can be solved in parallel for the different righthand sides using a similar form of the inverse matrix given by equation (85), i.e.,

$$\begin{bmatrix} u_1 \\ u_2 \\ u_3 \end{bmatrix} = \frac{1}{b(b^2-2)} \begin{bmatrix} b^2-1 & -b & 1 \\ -b & b^2 & -b \\ 1 & -b & b^2-1 \end{bmatrix} \begin{bmatrix} d_1 \\ d_2 \\ d_3 \end{bmatrix} \tag{87}$$

A rough comparison of the 'stride of 4' and the cyclic Odd-Even algorithm for the solution of n equations is given by the ratio log 4/log 2 = 2 with the amount of work at each solution stage increasing from 9 to 15 operations per point, with some additional precalculations.

The solution stages for the Quad Stride algorithm are given in Figure 3.

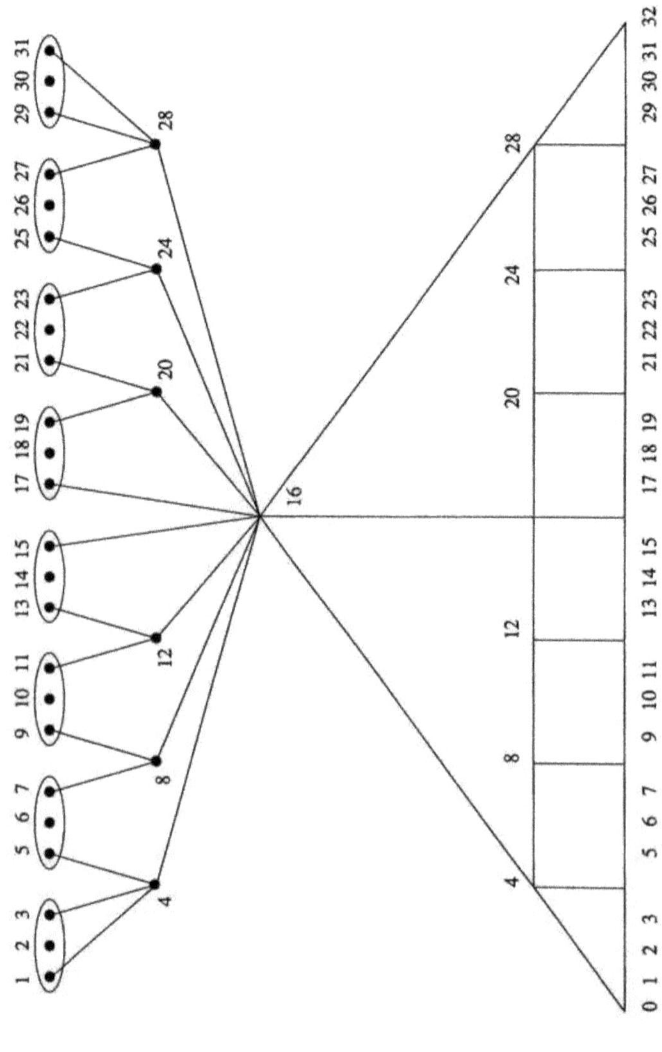

FIGURE 3: Communication topology Stride of 4 for $n=2^5-1$ (31)

6. NUMERICAL RESULTS

The three methods considered were applied to the symmetric constant-tridiagonal matrices of size 2^m-1, 3^m-1 and 4^m-1 for the parallel Odd-Even, parallel stride of three and parallel stride of four reduction methods respectively. All methods produced accurate solutions to the problems that they were tested on. The computer used was the Balance 8000 multiprocessor at P.A.R.C.

We show in Figure 4 a comparison of the times obtained for the parallel Odd-Even and the parallel stride of four reduction methods for n=4095 and for the parallel stride of three for n=6560.

FIGURE 4: Graph of the computation time

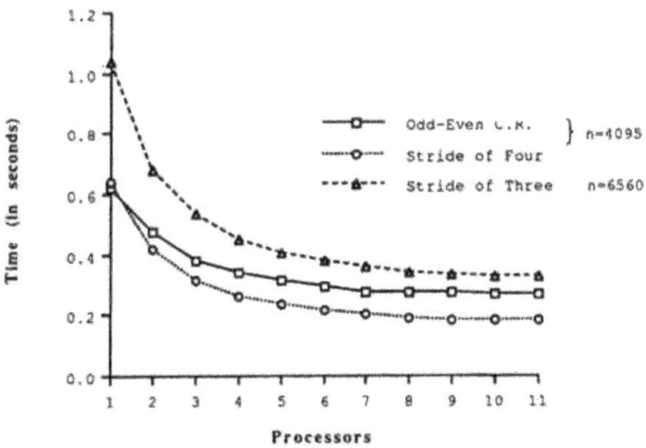

In Figure 5 we show the corresponding speed-ups obtained for the parallel algorithms for the three methods. Figures 6,7 and 8 show the speed-up values obtained for the parallel Odd-Even, parallel stride of three and parallel stride of four reduction methods respectively as the number of processors varies.

FIGURE 5: Graph of speed-up values

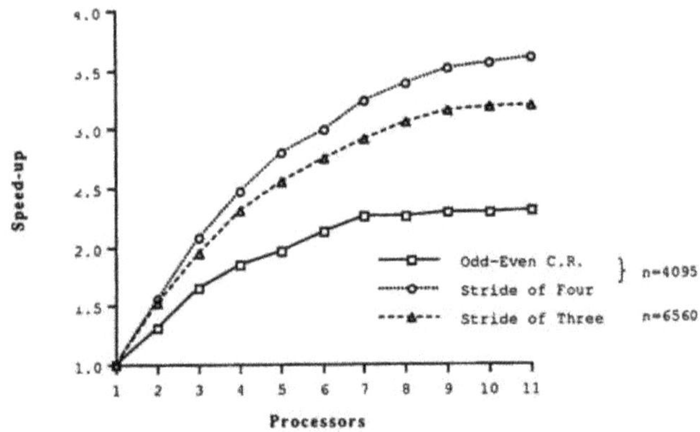

FIGURE 6: Speed-up for the C.R. method on nxn system of equations

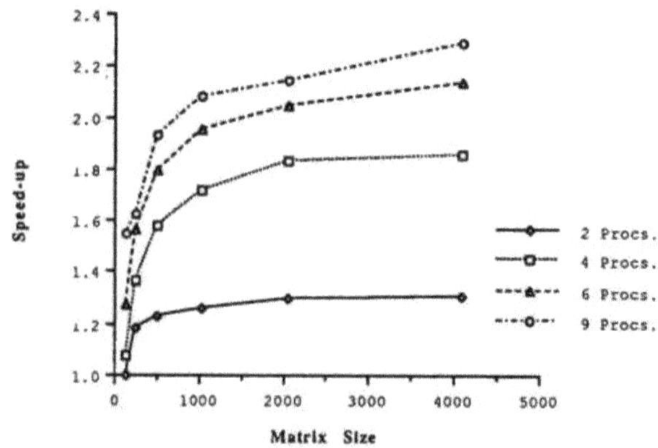

From these results it can be noted that the parallel strides of three and four in comparison with the parallel C.R. method gives a better speed-up as the number of processors increases. The poor speed-ups for small values of n is caused by the fact that in the first stages of the elimination process and also in the initial stages of obtaining the solution most of the processors remain idle. This means not sufficient computational work is available for all the processors. In conclusion, the three methods have been shown to be viable parallel algorithms with the parallel stride of four as the most efficient one.

FIGURE 7: Speed-up for the stride of 3 on nxn system of equations

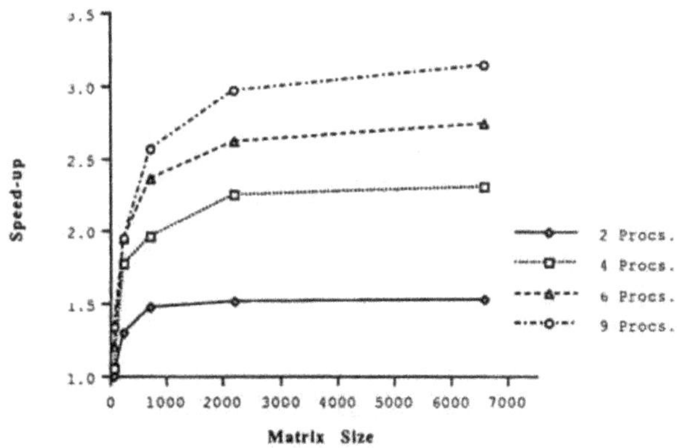

FIGURE 8: Speed-up for the stride of 4 on nxn system of equations

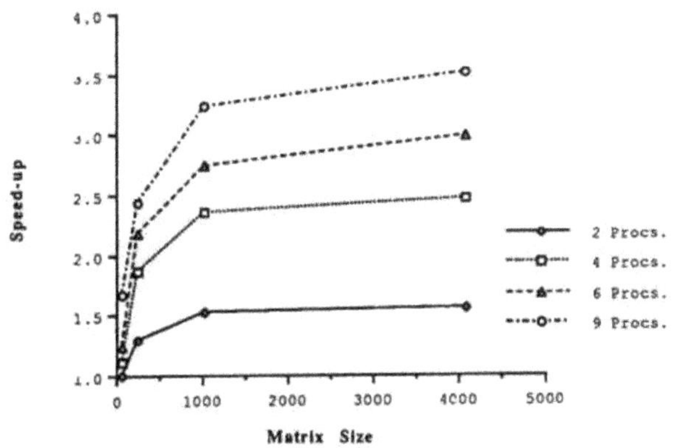

7. THE RECURSIVE DOUBLING METHOD

The parallel triangular decomposition of a positive definite tridiagonal matrix into L and U triangular factors involves nonlinear recurrence relations which can be solved after transformation by the well known recursive doubling technique. Here it is shown that a new and faster Recursive Stride procedure can be developed to derive the LU factors.

We consider the well known triangular decomposition of a tridiagonal matrix A as follows,

Let,

$$A \equiv \begin{bmatrix} d_1 & f_1 & & & \\ e_2 & d_2 & f_2 & & O \\ & \ddots & \ddots & \ddots & \\ O & & \ddots & \ddots & f_{n-1} \\ & & & e_n & d_n \end{bmatrix} \equiv LU, \tag{88}$$

where,

$$L = \begin{bmatrix} 1 & & & \\ m_2 & 1 & & O \\ & \ddots & \ddots & \\ O & & m_n & 1 \end{bmatrix}, \quad U = \begin{bmatrix} u_1 & f_1 & & \\ & u_2 & f_2 & O \\ & & \ddots & \ddots \\ O & & & f_{n-1} \\ & & & u_n \end{bmatrix} \tag{89}$$

By equating the terms in (88) and (89) we can evaluate the u_i, $i=1,2,\ldots,n$ from the scheme,

$$u_1 = d_1, \quad u_i = d_i - e_i f_{i-1} / u_{i-1}, \quad 2 \le i \le n, \tag{90}$$

and then evaluate the m_i, $i=2,3,\ldots,n$, from

$$m_i = e_i / u_{i-1}, \quad i=2,3,\ldots,n. \tag{91}$$

Now from eqn.(91) it can be observed that the quantities m_i are independent of n and so can be evaluated in parallel when the u_i quantities are known. To determine u_i, $i=2,3,\ldots,n$, the serial procedure eqn. (89) can be transformed into a parallel iterative strategy of the form,

$$\begin{aligned} u_i^{(0)} &= d_i, \\ u_i^{(j)} &= d_i - e_i f_{i-1} / u_{i-1}^{(j-1)}, \quad i=2,3,\ldots,n, \end{aligned} \tag{92}$$

where $u_i^{(j)}$ is the jth iterate of u_i. For computers with vector operations which are faster than their scalar counterparts, this process can be acceptable.

However, a recursive doubling (RD) strategy can also be applied to eqn.(89) as follows.

We define the quantities, q_i, $i=0,1,2,\ldots,n$, defined as,

$$q_0 = 1, \quad q_1 = d_1, \quad q_i = u_i q_{i-1}, \quad 2 \le i \le n. \tag{93}$$

Then we have,

$$u_i = q_i / q_{i-1}, \quad i=1,2,\ldots,n. \tag{94}$$

Hence from eqn. (89) we have,

$$q_i = d_i q_{i-1} - e_i f_{i-1} q_{i-2}, \quad 2 \le i \le n, \tag{95}$$

which is a second order recurrence relationship.

This can be converted to one of first order by defining,

$$Q_j = \begin{bmatrix} q_j \\ q_{j-1} \end{bmatrix}, \quad D_i = \begin{bmatrix} a_i & b_i \\ 1 & 0 \end{bmatrix}, \text{ for } \begin{matrix} 1 \le j \le n \\ 2 \le i \le n \end{matrix}, \tag{96}$$

where $a_i = d_i$ and $b_i = -e_i f_{i-1}$, from which we obtain, for eqn.(89) the alternative product relation,

$$Q_i = D_i D_{i-1} \dots D_2 Q_1, \; 2 \le i \le n. \tag{97}$$

Now Q_N can be calculated by the fan-in technique in $0(\log_2 n)$ steps and simultaneously we obtain the products Q_i, $i=2,3,\dots,n$.

The u_i quantities are then determined from eqn.(93) in $0(\log_2 n)$ steps.

8. A RECURSIVE STRIDE STRATEGY

A faster solution strategy can be obtained if we consider a larger stride in the recursive sequence given by eqn.(95).

First we rewrite eqn.(95) at the points i, i-1 and i-2 respectively, i.e.,

$$q_i = d_i q_{i-1} - e_i f_{i-1} q_{i-2}, \tag{98}$$

$$q_{i-1} = d_{i-1} q_{i-2} - e_{i-1} f_{i-2} q_{i-3}, \tag{99}$$

$$q_{i-2} = d_{i-2} q_{i-3} - e_{i-2} f_{i-3} q_{i-4}. \tag{100}$$

Now the quantity q_{i-1} can be eliminated by multiplying eqn.(97) by d_i and substituting in (97) to obtain,

$$q_i = (d_i d_{i-2} - e_i f_{i-1}) q_{i-2} - d_i e_{i-i} f_{i-2} q_{i-3}. \tag{101}$$

Finally, q_{i-3} can be eliminated by using eqn.(100) to obtain once again a second order recurrence relation,

$$q_i = (d_i d_{i-2} - e_i f_{i-1} - d_i e_{i-1} f_{i-2} / d_{i-2}) q_{i-2} - (d_i e_{i-1} f_{i-2} e_{i-2} f_{i-3} / d_{i-2}) q_{i-4}, \tag{102}$$

but now with a larger stride (of 2). We term this process the Recursive Stride (RS2) algorithm. It is easy to see that the process can be developed further with a larger stride, i.e. RS(k).

Once again this can be converted to one of first order by defining the quantities,

$$\hat{Q}_j = \begin{bmatrix} q_j \\ q_{j-2} \end{bmatrix}, \quad \hat{D}_i = \begin{bmatrix} \hat{a}_i & \hat{b}_i \\ 1 & 0 \end{bmatrix}, \text{ for } \begin{matrix} j=1(2)n \\ i=4(2)n \end{matrix},$$

where $\hat{a}_i = (d_i d_{i-2} - e_i f_{i-1} - d_i e_{i-1} f_{i-2} / d_{i-2})$,

$\hat{b}_i = (-d_i e_{i-1} f_{i-2} e_{i-2} f_{i-3} / d_{i-2})$.

Finally, the product relationship, to determine $\hat{Q}_i$ is given by,

$$\hat{Q}_i = \hat{D}_i \hat{D}_{i-2} \cdots \hat{D}_4 \hat{Q}_2 \quad , i=4(2)n ,$$

Thus $\hat{Q}_N$ can be obtained which can be calculated by the fan-in technique in $o(\log_2 n/2)$ steps.

Thus the Recursive Stride (RS2) algorithm is twice as fast as the Recursive Doubling (RD) method.

REFERENCES

[1] O. Buneman, A Compact Noniterative Poisson Solver, SUIPR Rep. 294, Inst. of Plasma Research Stanford Univ., California, (1969).

[2] B.L. Buzbee, G.H. Golub, and C.W. Nielson, On Direct Methods for Solving Poisson's Equations, SINUM 7, (1970), 627.

[3] P. Concus and G.H. Golub, Use of Fast Direct Methods for the Efficient Numerical Solution of Non-separable Elliptic Equations, SINUM, 10, (1973), 1103.

[4] F.W. Dorr, The Direct Solution of the Discrete Poisson Equation in a Rectangle, SIAM Rev. 12, (1970), 248.

[5] R. Hockney, A Fast Direct Solution of Poisson's Equation using Fourier Analysis, J.A.C.M., 12, (1965), 95.

[6] Kogge, P.M., [1974], Parallel Solution of Recurrence Problems, IBM Jour.Res.Dev., pp.138-148.

THE NUMERICAL SOLUTION OF LARGE FINITE ELEMENT SYSTEMS BY EXPLICIT PRECONDITIONING SEMI-DIRECT METHODS

ELIAS A. LIPITAKIS and GEORGE A. GRAVVANIS

Department of Informatics
Athens University of Economics
Athens, Greece

Abstract

A class of Approximate Inverse Matrix (AIM) algorithmic methods, based on the concept of "root-free" sparse factorization procedures, is introduced for computing explicitly approximate inverses of large sparse symmetric matrices of irregular structure, without inverting the decomposition factors. Explicit first and second-order semi-direct methods, in conjuction with modified forms of the AIM techniques, are presented for solving numerically large finite-element systems on multiprocessor systems. Application of the new methods on a 2D elliptic boundary-value problem is discussed and numerical results are given.

1. INTRODUCTION

In recent years research efforts are focused on the development, testing and analysis of new numerical algorithms that are needed to effectively exploit multiprocessor systems. There are several principles involved in the parallel algorithm design, i.e. vectorization, divide and conquer-partitioning, recursive doubling, conversion of implicit forms to explicit forms as well as the designing of complete new parallel algorithms. Among these principal parallel methods that can be classified as fundamental techniques of the "Parallel Numerical Analysis" the conversion of implicit forms to explicit forms offers substantial advantages on the grounds of exploitation of our previous long sequential computational experience.

The implicit preconditioning methods have been effectively used for the efficient solution of large sparse linear systems of algebraic equations. The implicitness of these methods has been introduced in sequential computations mainly for stability reasons (e.g. numerical solution of PDE's and ODE's). By taking out this implicitness and using corresponding explicit forms, all the available processors of a system can be effectively used for solving a certain class of problems at the same time.

It is known that the preconditioning methods incorporate a high-degree of implicitness in their solution processes due to the sequential nature of the forward-back substitution schemes and therefore this property tends to make them inappropriate for solution on the new multiprocessor systems. The derivation of suitable parallel methods was the main objective for which several forms of the approximate inverse (pseudo-inverse) of a given matrix, based on appropriate splittings and used in the place of the approximate factorization, have been recently proposed [2,5,9,10,19].

Our main motive for the derivation of the new Approximate Inverse Matrix techniques lies on the fact that they can be efficiently used in conjuction with explicit iterative schemes leading to effective composite semi-direct solution methods, which possess a high degree of explicitness and therefore are particularly suitable for solving large linear systems on parallel and vector processors.

In Section 2 we introduce the Approximate Inverse Matrix (AIM) techniques. These new techniques are originated from known [7,24] algorithmic procedures of inverting a real (nxn) symmetric matrix A of irregular structure, which can be factorized as $A=LDL^T$ (where L and D are properly defined matrices) by determining the elements of A^{-1} without inverting the factors L and D. In Section 3 we present a simple (optimized) storage scheme which, in conjuction with banded forms of the AIM algorithm, leads to a class of (optimized) "pseudoinverses" particularly effective for the efficient solution of large order linear systems resulting from the finite-element discretization of elliptic boundary-value problems. In Section 4 explicit first and second order (stationary/non-stationary) iterative methods, based on the derived AIM techniques, are presented. Finally, in Section 5 the applicability of the new parallel methods on a 2D elliptic boundary-value problem is discussed and numerical results are given.

2. APPROXIMATE INVERSE MATRIX METHODS

In this section we present the Approximate Inverse Matrix methods, based on an algorithmic procedure of inverting a real (nxn) matrix A (which can be factorized as $A=LDL^T$, where L, D are properly defined sparse triangular matrices) by computing the elements of the inverse A^{-1} without inverting the decomposition factors L and D [7,24]. Similar techniques, as the Product form of the Inverse and related minimization techniques of the fill-in (non-zero) terms, which occur during the

Gauss-Jordan elimination procedure, are of significant importance in large scale Linear Programming applications [1,21].

Let us consider the linear system resulting from the Finite-Element (FE) discretization of an elliptic boundary-value problem, i.e.

$$A\,u = s \,, \tag{2.1}$$

where A is a non-singular large sparse symmetric, positive definite, diagonally-dominant (nxn) matrix of irregular structure (where all the off-centre band terms are grouped in a regular band of width l), viz.

$A \equiv$

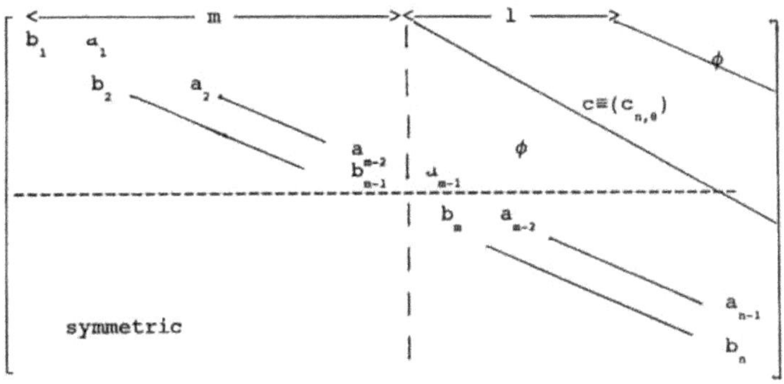

(2.1')

while u and s are n-column vectors which consist respectively of the unknown FE solution vector and the source-term of the given PDE plus the known boundary-values.

Approximate factorization methods can be obtained by considering the simple replacement of the coefficient matrix A by a matrix (A+Π) such that

$$A + \Pi = L_r D_r L_r^T \,, \tag{2.2}$$

where L_r and D_r are sparse strictly lower and diagonal matrices of the form:

$L_r \equiv$

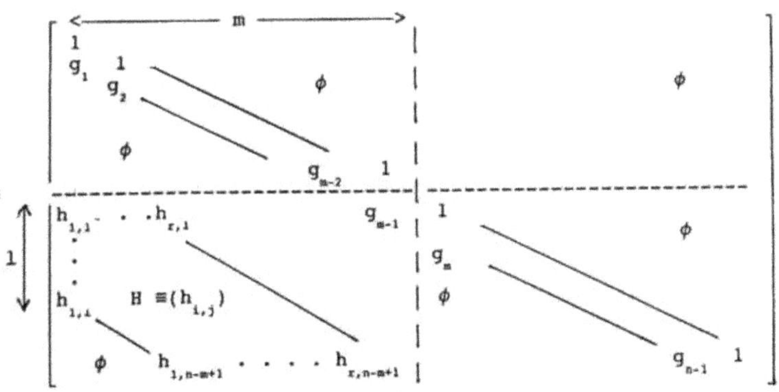

(2.3)

and $$D_r = \text{diag}(d_1, d_2, \ldots, d_n) \; . \tag{2.3a}$$

The choice of Π, assuming that the matrix $(A+\Pi)$ can be factored in sparse triangular matrices leading to "easily" solvable systems $(L_r D_r L_r^T)u=s$, proves to be of great importance, cf.[14], for the rate of convergence of accelerated semi-direct methods used for the solution of the linear system (2.1). The elements of the decomposition factors L_r and D_r can be defined by the FERFC algorithm [16]. The storage requirements of the FERFC algorithm is $\cong (r+2l+2)n$ words, while the computation work required by this algorithm is $\cong [(r+l-1)(r+l)/2+4]n$ multiplicative operations, cf.[16].

However, it is known that the implicit approximate factorization solution methods, due to the sequential nature of the forward and back-substitution schemes involved in the solution process, are not particularly suitable for parallel computations. In order to derive suitable parallel numerical methods for use on vector or parallel processing systems, we revert our attention back to the existing explicit point methods and investigate their application for solving elliptic boundary-value problems on parallel processors [3].

Let us assume that M_r, a non-singular real (nxn) matrix, is the inverse of $(A+\Pi)$, i.e.

$$M_r = (L_r D_r L_r^T)^{-1} \; , \qquad r\in[1,m-1] \; , \tag{2.3}$$

where r is the fill-in parameter, i.e. the number of outermost off-diagonal entries retained in semi-bandwidth m. Note that if r=m-1 entries have been retained in semibandwidth m, then $L_r \equiv L$ and $D_r \equiv D$, where L is a lower triangular matrix retaining (m-1) outermost off-diagonal entries and D is a diagonal matrix. In this case the coefficient matrix A is factorized exactly, i.e. $A=LDL^T$ and then the exact inverse can be obtained with the disadvantage the computational work involved being prohibitevely high for the proposed strategy. The elements of the exact inverse then can be defined by an explicit algorithmic procedure (direct fast solver FERFC), cf.[15].

The AIM algorithm is an effective method for determining the elements of the inverse $M_r \equiv (\mu_{ij}^r)$, $i,j \in [1,n]$, (up to a certain order) without computing the elements of L_r^{-1} and D_r^{-1}. However, there are certain unfavourable features especially in the case of (very) large systems, i.e. the prohibitevely high storage requirements and computational work as well as the loss of sparseness of the original matrix A, which need to be eliminated. Thus, it is a reasonable approach to redesign the AIM algorithm such that only δl, $(\delta l\in[1,n])$ diagonals are retained in the lower triangular part of M_r, the remaining elements being just not computed at all, i.e.

$$M_r^{\delta l} = (L_r D_r L_r^T)^{-1} , \quad r\in[1,m-1], \quad \delta l\in[1,n] . \tag{2.4}$$

The inverse matrix in this case can be stored compactly in δl-vector spaces. Hence, an approximate banded form of the inverse M_r is achieved in which both the sparseness is relatively retained and storage requirements are substantially reduced.

The banded AIM algorithm (henceforth called the BAIFEM algorithm) can be expressed in the following compact form:

Let us assume that

$$rl = r+1; \quad ml = m+1; \quad mr = m-r ; \tag{2.5}$$

and

$$nmr = n-mr; \quad nml = n-ml . \tag{2.6}$$

Then, the elements of the n-row and column of the inverse M_r can be computed as follows:

$$\mu_{n,n} = 1/d_n \tag{2.7}$$

$$\mu_{n,j} = -g_j\mu_{n,j+1} - \sum_{\lambda=0}^{nmr-j} h_{rl-1-\lambda,j+1-r+\lambda}\mu_{n,j+mr+\lambda} ,$$

$$\underline{\text{for}} \quad j=n-1,\dots,rl \quad \text{and} \quad j>n-\delta l , \tag{2.8}$$

$$\mu_{n,j} = -g_j\mu_{n,j+1} - \sum_{\substack{\lambda=j+1-r\\ \lambda>0}}^{l} h_{j,\lambda}\mu_{n,m+\lambda-1} - \sum_{\kappa=0}^{n-ml} h_{j-1-\kappa,l+1+\kappa}\mu_{n,ml+\kappa} ,$$

$$\underline{\text{for}} \quad j=rl-1,\dots,nml+2 \quad \text{and} \quad j>n-\delta l , \tag{2.9}$$

$$\mu_{n,j} = -g_j\mu_{n,j+1} - \sum_{\substack{\lambda=j+1-r\\ \lambda>0}}^{l} h_{j,\lambda}\mu_{n,m+\lambda-1} - \sum_{\kappa=1}^{j-1} h_{j-\kappa,l+\kappa}\mu_{n,ml+\kappa-1} ,$$

$$\underline{\text{for}} \quad j=nml+1,\dots,n-\delta l , \tag{2.10}$$

where $\delta l \in [m,n)$.

$$\mu_{j,n} = \mu_{n,j} , \quad \underline{\text{for}} \quad j=n-1,\dots,n-\delta l . \tag{2.11}$$

The elements of the n-1,...,1 rows and columns of M_r can be obtained as follows:

<u>For</u> $i = n-1, n-2,\dots,1$

    <u>if</u> $j>nmr$ <u>then</u>

        <u>if</u> $i=j$ <u>then</u>

$$\mu_{i,j} = 1/d_j \quad g_j\mu_{i,j+1} \tag{2.12}$$

        <u>else</u>

$$\mu_{i,j} = -g_j\mu_{i,j+1} , \quad \underline{\text{for}} \quad i\le\delta l+j-1 , \tag{2.13}$$

else

if $j \geq rl$ and $j \leq nmr$ then

if $i=j$ then

$$\mu_{i,j} = 1/d_j - g_j\mu_{i,j+1} \sum_{\lambda=0}^{nmr-j} h_{rl-1-\lambda,j+1-r+\lambda}\mu_{i,j+m-r+\lambda} \tag{2.14}$$

else

$$\mu_{i,j} = -g_j\mu_{i,j+1} - \sum_{\lambda=0}^{nmr-j} h_{rl-1-\lambda,j+1-r+\lambda}\mu_{i,j+m-r+\lambda}\ , \quad \text{for } i \leq \delta l+j-1\ , \tag{2.15}$$

else

if $j>nml+1$ and $j \leq rl-1$ then

if $i=j$ then

$$\mu_{i,j} = 1/d_j - g_j\mu_{i,j+1} - \sum_{\substack{\lambda=j+1-r \\ \lambda>0}}^{l} h_{j,\lambda}\mu_{i,m+\lambda-1} \sum_{\kappa=0}^{nml} h_{j-1-\kappa,l+1+\kappa}\mu_{i,ml+\kappa} \tag{2.16}$$

else

$$\mu_{i,j} = -g_j\mu_{i,j+1} - \sum_{\substack{\lambda=j+1-r \\ \lambda>0}}^{l} h_{j,\lambda}\mu_{i,m+\lambda-1} \sum_{\kappa=0}^{nml} h_{j-1-\kappa,l+1+\kappa}\mu_{i,ml+\kappa}$$

$$\text{for } i \leq \delta l+j-1\ , \tag{2.17}$$

else

if $j \leq nml+1$ then

if $i=j$ then

if $i=1$ then

$$\mu_{1,1} = 1/d_1 - g_1\mu_{1,2} - \sum_{\lambda=1}^{l} h_{1,\lambda}\mu_{1,m+\lambda-1} \tag{2.18}$$

else

$$\mu_{i,j} = 1/d_j \quad g_j\mu_{i,j+1} - \sum_{\substack{\lambda=j+1-r \\ \lambda>0}}^{l} h_{j,\lambda}\mu_{i,m+\lambda-1} - \sum_{\kappa=1}^{j-1} h_{j-\kappa,l+\kappa}\mu_{i,ml+\kappa-1} \tag{2.19}$$

else

$$\mu_{i,j} = -g_j\,\mu_{i,j+1} - \sum_{\substack{\lambda=j+1-r \\ \lambda>0}}^{l} h_{j,\lambda}\mu_{i,m+\lambda-1} - \sum_{\kappa=1}^{j-1} h_{j-\kappa,l+\kappa}\mu_{i,ml+\kappa-1}$$

$$\text{for } i \leq \delta l+j-1\ . \tag{2.20}$$

For $j \quad i, i-1,\ldots,1$

$$\mu_{j,i} \quad \mu_{i,j} \qquad \text{for } i \leq \delta l+j-1\ . \tag{2.21}$$

For $j = i-1, i-2,\ldots,1$ and $i>1$.

In the following, we present a simple (optimized) storage scheme for the banded approximate inverse, that enable us to keep in storage the $M_r^{\delta l}$ by using only $(2\delta l-1)$ vectors.

3. THE BAIM STORAGE SCHEME

Let us consider, without any loss of generality, the following banded form of an approximate inverse matrix (of low order dimension):

$$
M_r^{(I)} \equiv
\begin{array}{c}
\longleftarrow \delta l-1 \longrightarrow \\
\left[\begin{array}{cccccccc}
\mu_{1,1} & \mu_{1,2} & \mu_{1,3} & \mu_{1,4} & & & & \\
\mu_{2,1} & \mu_{2,2} & \mu_{2,3} & \mu_{2,4} & \mu_{2,5} & & \phi & \\
\mu_{3,1} & \mu_{3,2} & \mu_{3,3} & \mu_{3,4} & \mu_{3,5} & \mu_{3,6} & & \\
\mu_{4,1} & \mu_{4,2} & \mu_{4,3} & \mu_{4,4} & \mu_{4,5} & \mu_{4,6} & \mu_{4,7} & \\
 & \mu_{5,2} & \mu_{5,3} & \mu_{5,4} & \mu_{5,5} & \mu_{5,6} & \mu_{5,7} & \mu_{5,8} \\
 & & \mu_{6,3} & \mu_{6,4} & \mu_{6,5} & \mu_{6,6} & \mu_{6,7} & \mu_{6,8} \\
 & \phi & & \mu_{7,4} & \mu_{7,5} & \mu_{7,6} & \mu_{7,7} & \mu_{7,8} \\
 & & & & \mu_{8,5} & \mu_{8,6} & \mu_{8,7} & \mu_{8,8}
\end{array}\right]
\end{array}
\qquad (\delta l \text{ rows indicated at left: rows 1--4})
\tag{3.1}
$$

Then, by rotating the $M_r^{(I)}$ matrix through 180°, with respect to the to the main diagonal, we obtain the following matrix scheme:

$$
M_r^{(II)} \equiv
\begin{array}{c}
\longleftarrow \delta l \longrightarrow \\
\left[\begin{array}{cccccccc}
\mu_{8,8} & \mu_{8,7} & \mu_{8,6} & \mu_{8,5} & & & & \\
\mu_{7,8} & \mu_{7,7} & \mu_{7,6} & \mu_{7,5} & \mu_{7,4} & & \phi & \\
\mu_{6,8} & \mu_{6,7} & \mu_{6,6} & \mu_{6,5} & \mu_{6,4} & \mu_{6,3} & & \\
\mu_{5,8} & \mu_{5,7} & \mu_{5,6} & \mu_{5,5} & \mu_{5,4} & \mu_{5,3} & \mu_{5,2} & \\
 & \mu_{4,7} & \mu_{4,6} & \mu_{4,5} & \mu_{4,4} & \mu_{4,3} & \mu_{4,2} & \mu_{4,1} \\
 & & \mu_{3,6} & \mu_{3,5} & \mu_{3,4} & \mu_{3,3} & \mu_{3,2} & \mu_{3,1} \\
 & \phi & & \mu_{2,5} & \mu_{2,4} & \mu_{2,3} & \mu_{2,2} & \mu_{2,1} \\
 & & & & \mu_{1,4} & \mu_{1,3} & \mu_{1,2} & \mu_{1,1}
\end{array}\right]
\end{array} .
\tag{3.2}
$$

Finally, by shifting the first δl upper diagonals of $M_r^{(II)}$ into the δl first columns of a new matrix scheme $M_r^{(III)}$, cf.(3.3) and, consequently the remaining $(\delta l-1)$ lower diagonals of $M_r^{(II)}$ into the next $(\delta l-1)$ columns of $M_r^{(III)}$, we obtain an optimized compact storage scheme, which is particularly suited for the storage of (very) large order banded approximate inverse matrices, by reducing considerably the high computer fast storage requirements, viz.

$$M_r^{(III)} \equiv \begin{bmatrix} \mu_{8,8} & \mu_{8,7} & \mu_{8,6} & \mu_{8,5} & \mu_{7,8} & \mu_{6,8} & \mu_{5,8} \\ \mu_{7,7} & \mu_{7,6} & \mu_{7,5} & \mu_{7,4} & \mu_{6,7} & \mu_{5,7} & \mu_{4,7} \\ \mu_{6,6} & \mu_{6,5} & \mu_{6,4} & \mu_{6,3} & \mu_{5,6} & \mu_{4,6} & \mu_{3,6} \\ \mu_{5,5} & \mu_{5,4} & \mu_{5,3} & \mu_{5,2} & \mu_{4,5} & \mu_{3,5} & \mu_{2,5} \\ \mu_{4,4} & \mu_{4,3} & \mu_{4,2} & \mu_{4,1} & \mu_{3,4} & \mu_{2,4} & \mu_{1,4} \\ \mu_{3,3} & \mu_{3,2} & \mu_{3,1} & x & \mu_{2,3} & \mu_{1,3} & x \\ \mu_{2,2} & \mu_{2,1} & x & x & \mu_{1,2} & x & x \\ \mu_{1,1} & x & x & x & x & x & x \end{bmatrix} . \quad (3.3)$$

This three-stage matrix transformation technique, based on simple "rotation-and-shift" operations, reduces the storage requirements of an approximate (nxn) inverse matrix into (nx(2δl-1)) vector spaces and is particularly effective when $\delta l \ll n/2$. Let us denote by I and J the subscripts of an element of $M_r^{(I)}$ and by IT and JT the so-called transformed subscripts of the same element in $M_r^{(III)}$. Then, the relationships between these subscripts can be described by the following algorithmic scheme:

<u>if</u> (I.GE.J) <u>then</u>

$$IT = n + 1 - I \quad (3.4)$$

$$JT = I \quad J + 1 \quad (3.5)$$

<u>else</u>

$$IT = n + 1 - J \quad (3.6)$$

$$JT = \delta l + J \quad I \quad (3.7)$$

<u>endif</u> .

By examining the transformed subscripts IT and JT of the elements of $M_r^{(III)}$, which are dependent on the initial subscripts I and J of the corresponding elements of $M_r^{(I)}$, it is obvious that an explicit reference can be made for elements strictly on, below and above the diagonal of the initial matrix $M_r^{(I)}$.

The optimized banded Approximate Inverse Matrix (BAIFEMO) algorithm can then be expressed in the following compact form:

It should be noted that the variable indices x, y of the AIM algorithmic elements $\mu_{i,j} \equiv \mu^r_{i,j}$, are functions of the variables indices p, q respectively and can be determined by a test-routine (namely rstest), which is given at the end of the following algorithm.

Let us assume that

$$rl = r+1 \; ; \qquad ml = m+1 \; ; \qquad mr = m-r \quad (3.8)$$

and

$$nml \quad n-ml \; ; \qquad nmr = n-mr \, . \quad (3.9)$$

Then, the elements of the first-row and column of $M_r^{(III)}$ can be computed as follows:

$$\mu_{1,1} = 1/d_n \quad , \tag{3.10}$$

$$\mu_{1,n+1-j} = -g_j\mu_{1,n-j} - \sum_{\lambda=0}^{nmr-j} h_{rl-1-\lambda,j+1-r+\lambda}\mu_{x,y} \; ,$$

$\underline{\text{for}}$ $j=n-1,\dots,rl$ and $j>n-\delta l$, (3.11)

| p = n
| q = j+mr+λ <u>for</u> q≤j+δl-1 and q≥j-δl+1 ,
| call rstest(n,δl,p,q,x,y) ,

$$\mu_{1,n+1-j} = -g_j\mu_{1,n-j} - \sum_{\substack{\lambda=j+1-r\\ \lambda>0}}^{1} h_{j,\lambda}\mu_{x,y} \sum_{\kappa=0}^{n-ml} h_{j-1-\kappa,1+1+\kappa}\mu_{x1,y1} \quad ,$$

$\underline{\text{for}}$ $j=rl-1,\dots,nml+2$ and $j>n-\delta l$, (3.12)

| p = n | p = n
| q = m+λ-1 | q = ml+κ <u>for</u> q≤j+δl-1 and q≥j-δl+1 ,
| call rstest(n,δl,p,q,x,y) | call rstest(n,δl,p,q,x1,y1)

$$\mu_{1,n+1-j} = -g_j\mu_{1,n-j} - \sum_{\substack{\lambda=j+1-r\\ \lambda>0}}^{1} h_{j,\lambda}\mu_{x,y} - \sum_{\kappa=0}^{j-1} h_{j-\kappa,1+\kappa}\mu_{x1,y1} \quad ,$$

$\underline{\text{for}}$ $j=nml+1,\dots,n-\delta l$, (3.13)

| p = n | p = n
| q = m+λ-1 | q = ml+κ-λ <u>for</u> q≤j+δl-1 and q≥j-δl+1 ,
| call rstest(n,δl,p,q,x,y) | call rstest(n,δl,p,q,x1,y1)

$\mu_{z,w} = \mu_{x,y}$, $\underline{\text{for}}$ $j=n-1,\dots,n-\delta l+1$, (3.14)

| p = j | p = n
| q = n | q = j
| call rstest(n,δl,p,q,z,w) | call rstest(n,δl,p,q,x,y) .

<u>For</u> i = n-1,n-2,...,1

<u>if</u> j>nmr <u>then</u>

<u>if</u> i=j <u>then</u>

$$\mu_{n-i+1,i-j+1} = \frac{1}{d_j} \quad g_j\,\mu_{n-j,\delta l-i+j+1} \tag{3.15}$$

<u>else</u>

$$\mu_{n-i+1,i-j+1} \quad -g_j\,\mu_{n-i+1,i-j} \quad , \qquad \underline{\text{for}} \;\; i\le j+\delta l-1 \; , \tag{3.16}$$

<u>else</u>

<u>if</u> j≥rl and j≤nmr <u>then</u>

<u>if</u> i=j <u>then</u>

$$\mu_{n-i+1,i-j+1} = \frac{1}{d_j} - g_j\,\mu_{n-j,\delta l-i+j+1} - \sum_{\lambda=0}^{nmr-j} h_{rl-1-\lambda,j+1-r+\lambda}\mu_{x,y} \; , \tag{3.17}$$

p = i
q = j+mr+λ for q≤j+δl-1 and q≥j-δl+1 ,
call rstest(n,δl,p,q,x,y)

else

$$\mu_{n-i+1,i-j+1} = - g_j \mu_{n-i+1,i-j} - \sum_{\lambda=0}^{nmr-j} h_{rl-1-\lambda,j+1-r+\lambda}\mu_{x,y} , \quad \text{for} \quad i\le\delta l+j-1 , \tag{3.18}$$

p = i
q = j+mr+λ for q≤j+δl-1 and q≥j-δl+1 ,
call rstest(n,δl,p,q,x,y)

else

if j>nml+1 and j≤rl-1 then

if i=j then

$$\mu_{n-i+1,i-j+1} = \frac{1}{d_j} - g_j\mu_{n-j,\delta l-i+j+1} - \sum_{\substack{\lambda=j+1-r\\ \lambda>0}}^{1} h_{j,\lambda}\mu_{x,y} \quad \sum_{\kappa=0}^{nml} h_{j-1-\kappa,1+1+\kappa}\mu_{x1,y1} \tag{3.19}$$

p i
q = m+λ-1
call rstest(n,δl,p,q,x,y)

p = i
q = ml+κ for q≤j+δl-1 and q≥j-δl+1 ,
call rstest(n,δl,p,q,x1,y1)

else

$$\mu_{n-i+1,i-j+1} = \quad g_j\mu_{n-i+1,i-j} \quad \sum_{\substack{\lambda=j+1-r\\ \lambda>0}}^{1} h_{j,\lambda}\mu_{x,y} - \sum_{\kappa=0}^{nml} h_{j-1-\kappa,1+1+\kappa}\mu_{x1,y1} ,$$

$$\text{for} \quad i\le\delta l+j-1 \tag{3.20}$$

p i
q = m+λ-1
call rstest(n,δl,p,q,x,y)

p = i
q = ml+κ for q≤j+δl-1 and q≥j-δl+1 ,
call rstest(n,δl,p,q,x1,y1)

else

if j≤nml+1 then

if i=j then

if i=1 then

$$\mu_{n-i+1,i-j+1} = \frac{1}{d_j} \quad g_j \mu_{n-j,\delta l-i+j+1} \quad \sum_{\lambda=1}^{1} h_{1,\lambda}\mu_{x,y} , \tag{3.21}$$

p 1
q = m+λ-1 for q≤j+δl-1 and q≥j-δl+1 ,
call rstest(n,δl,p,q,x,y)

else

$$\mu_{n-i+1,i-j+1} = \frac{1}{d_j} - g_j\mu_{n-j,\delta l-i+j+1} \quad \sum_{\lambda=j+1-r}^{1} h_{j,\lambda}\mu_{x,y} - \sum_{\kappa=1}^{j-1} h_{j-\kappa,1+\kappa}\mu_{x1,y1} \tag{3.22}$$

| p = i
| q = m+λ-1
| call rstest(n,δl,p,q,x,y)

| p = i
| q = ml+κ-1 *for* q≤j+δl-1 and q≥j-δl+1 ,
| call rstest(n,δl,p,q,xl,yl)

else

$$\mu_{n-i+1,i-j+1} \quad - g_j\mu_{n-i+1,i-j} - \sum_{\substack{\lambda=j+1-r \\ \lambda>0}}^{l} h_{j,\lambda}\mu_{x,y} \cdot \sum_{\kappa=1}^{j-1} h_{j-\kappa,l+\kappa}\mu_{xl,yl} \quad ,$$

for i≤j+δl-1 , (3.23)

| p = i
| q = m+λ-1
| call rstest(n,δl,p,q,x,y)

| p = i
| q = ml+κ-1 *for* q≤j+δl-1 and q≥j-δl+1 ,
| call rstest(n,δl,p,q,xl,yl) .

For j = i,i-1,...,l-δl+1

$\mu_{z,w} = \mu_{x,y}$, *for* j≤i+δl-1 and j≥i-δl+1 , (3.24)

| p = i
| q = j
| call rstest(n,δl,p,q,z,w)

| p = j
| q = i
| call rstest(n,δl,p,q,x,y)

For j = i-1, i-2,..., 1 and i>1 .

Note that each successive call of the "rotation-and-shift" transformation routine rstest in the BAIFEMO algorithm, refers respectively to each corresponding algebraic sum, in the order that they appear in several relationships of equations (3.10)-(3.24). The routine rstest can be described by the following "pseudocode":

```
Procedure rstest(n,δl,p,q,x,y)
              if   p≥q   then
                      x  =  n + 1   p  ,
                      y  =  p - q + 1
              else
                      x  =  n + 1   q
                      y  =  δl + q - p                    (3.25)
```

The storage requirements of the BAIFEMO algorithm are [nx(2δl-1)] vectors and the computational work involved is

(l-1) [(δl-1)(2n-δl)/2 - (m+l-2)(m+l-1)/2 + m-1] +

+ r [n(2δl-m+4)/2   δl(δl+1)/2   (m-1)] + n(δl+1)   δl(δl-1)-1

multiplicative operations.

It is known, cf.[13], that the larger in magnitude elements of the inverse matrices resulting from the discretization of parabolic or elliptic PDE's, in almost every case, are clustered around the diagonals at distances ρ.m (where m is the semi-bandwidth of the original coefficient matrix, with $\rho=1,2,\ldots,m-1$) from the main diagonal in a "recurring wave" like pattern. Therefore, by considering several "dropping fill-in" strategies, cf.[11], it is reasonable to assume that the value of parameter δl can be chosen as multiple of m. Alternative strategies, still under investigation, retain only the large elements of the inverse, which occur along (or near) the diagonals $\rho.m$, where $\rho=1,2,\ldots m-1$.

4. EXPLICIT PRECONDITIONING SEMI-DIRECT METHODS

In this section we present a class of explicit preconditioning semi-direct methods based on the derived AIM techniques for solving the linear system (2.1).

The Explicit Simultaneous Displacement (ESD) and the Explicit first-order Richardson (ER1) methods are defined as

$$\delta u_{i+1} = \alpha.M_r^{\delta l}\, r_i\,, \qquad i\geq 0\,, \tag{4.1}$$

and

$$\delta u_{i+1} = \alpha_i.M_r^{\delta l}\, r_i\,, \qquad i\geq 0\,, \tag{4.2}$$

where α, α_i are respectively preconditioned acceleration parameter and sequence of parameter, $\delta u_{i+1} = u_{i+1}-u_i$, $r_i = s-Au_i$ and u_0 is an arbitrary initial approximation of the solution u. The parameters r and δl must be determined in order to produce a pseudoinverse $M_r^{\delta l}$ such that $\rho(I-M_r^{\delta l}.A)<1$, where $\rho(H)$ denotes here the spectral radius of matrix H, and the Explicit Preconditioning method, converge to the solution of the FE linear system. The explicit second-order semi-direct methods, i.e. Richardson's (ER2) and Chebychev (ECh) methods, can be respectively defined by

$$\delta u_{i+1} = \alpha\, M_r^{\delta l}\, r_i + \beta\, \delta u_i\,, \qquad i\geq 0\,, \tag{4.3}$$

$$\delta u_{i+1} = \alpha_v\, M_r^{\delta l}\, r_i + \beta_v\, \delta u_i\,, \qquad i\geq 0\,, \tag{4.4}$$

where α,β and α_v,β_v are preconditioned acceleration parameters and sequences of parameters respectively.

Let us assume now that A and $M_r^{\delta l}$ are real, symmetric, positive-definite (nxn)-matrices. Then, the Explicit Preconditioning Conjugate Gradient (EPCG) method can be stated as follows:

Let u_0 be an arbitrary initial approximation to the solution u. Then, form the residual

$$r_0 = s \quad Au_0\,, \tag{4.5}$$

compute

$$r_0^* = M_r^{\delta l}\, r_0\,, \tag{4.6}$$

and set $\quad \sigma_0 \quad r_0^* \quad .$ (4.7)

Then, for $i=0,1,2,\ldots,$(until convergence) calculate the vectors u_{i+1}, r_{i+1}, σ_{i+1} and the scalar quantities α_i, β_{i+1} as follows:

compute $\quad q_i = A\,\sigma_i$ (4.8)

form $\quad p_i = (r_i, r_i^*)$ (only when i=0) (4.9)

evaluate $\quad \alpha_i = p_i/(\sigma_i, q_i)$, (4.10)

$\quad u_{i+1} = u_i + \alpha_i \sigma_i$, (4.11)

and $\quad r_{i+1} = r_i - \alpha_i q_i$, (4.12)

Then, compute $\quad r_{i+1}^* = M_r^{\delta l} . r_{i+1}$ (4.13)

form $\quad p_{i+1} = (r_{i+1}, r_{i+1}^*)$ (4.14)

evaluate $\quad \beta_{i+1} = p_{i+1}/p_i$ (4.15)

and set $\quad \sigma_{i+1} = r_{i+1}^* + \beta_{i+1}\,\sigma_i$. (4.16)

The EPCG method, assuming that $M_r^{\delta l}$ is a $(2\delta l-1)$-diagonal matrix, requires a computational work of

$$[(2\delta l-10)n \text{ mults} + 3n \text{ adds}].\eta \ ,$$

where η is the number of iterations required for the convergence of EPCG to a certain level of accuracy.

The explicit preconditioning semi-direct methods, based on the BAIFEMO algorighmic techniques, are suitable for use on multiprocessor systems with a resulting factor of n in "speed-up", assuming that synchronization time and time to pass information between processors are negligible. Note also that both synchronous and asynchronous versions of the derived explicit semi-direct methods can be efficiently implemented on SIMD and MIMD machines.

5. NUMERICAL EXPERIMENTS

Let us consider a class of problems defined by the following 2D elliptic boundary-value problem with Dirichlet boundary conditions:

$$-\Delta u(x,y) + u(x,y) \quad f(x,y)\ , \quad (x,y)\in R\ , \tag{5.1}$$

$$u(x,y) = 0\ , \quad (x,y)\in\partial R\ , \tag{5.1a}$$

where Δ is the Laplacian operator, R is the unity square and ∂R the boundary of R.

The domain R was covered by a non-overlapping regular triangular network. For an efficient numerical solution of the 2D model problem, the linear system (2.1) resulting from the FE discretization of equation (5.1), was solved by the explicit semi-direct methods based on the BAIFEMO techniques. The non-zero elements of the coefficient matrix A were chosen as

$$b_i = 8.0\ ,\quad i=1,\dots,n\ ;\qquad a_i = -1.0\ ,\quad i=1,\dots,n-1\ ,$$

$$c_{i,j} = -1.0\ ,\quad i=1,\dots,n-m+1\ ,\quad j=1,\dots,l\ . \tag{5.2}$$

and the right hand side vector s was computed as the product of matrix A by the solution vector u (arbitrarily set to 1.0). The width parameter l in semi-bandwidth m was set to l=3. The iterative process was terminated when the maximum norm of the residual was less than the relative precision of the arithmetic, which has been chosen to be $\varepsilon=10^{-5}$.

Numerical results for several values of the parameter δl and a (near) optimum value of the fill-in parameter r=2, cf.[4], are presented in Tables 1 and 2. The values of "diagonal-retention" parameter δl, which controls the formation of a class of pseudoinverses $M_r^{\delta l}$, were chosen as multiples of semi-bandwidth m, cf.[13].

The behaviour of the error norms $\|r_i\|_2$ and $\|u_k-u_E\|_2$, where u_k is the k-approximation to the exact solution u_E, for the ESD method is given in Figure 1, while the behaviour of the error norms $\|r_i\|_2$ and $(r_i,M^{\delta l}r_i)^{1/2}$ for the EPCG method is given in Figure 2. Experimental results, for the EPCG method, by varying the magnitude of elements c_{ij} and the width-paremeter l, are also presented in Table 3 and the corresponding behaviour of the error norms $\|r_i\|_2$ and $(r_i,M^{\delta l}r_i)^{1/2}$ is given in Figure 3. Two cases concerning the magnitude of the elements c_{ij} have been considered, i.e. case I: c_{ij}=-1 and case II: c_{ij}=-3, cf.(2.1).

GRID	α	No. of iters	α	No. of iters	α	No. of iters	α	No. of iters	α	No. of iters	α	No. of iters
	$\delta l=2$		$\delta l=[m/2]=5$		$\delta l=m=10$		$\delta l=2m=20$		$\delta l=3m=30$			
(9x9) m=10	0.85	39	0.85	38	0.85	23	0.85	15	0.85	11		
	0.90	37	0.90	36	0.90	21	0.90	14	0.90	10		
	0.95	35	0.95	34	0.95	20	0.95	13	0.95	9		
	1.00	36	1.00	44	1.00	24	1.00	12	1.00	8		
	1.05	56	1.05	78	1.05	33	1.05	12	1.05	9		
	1.10	123	1.10	>200	1.10	48	1.10	14	1.10	10		
					1.15	85	1.15	16	1.15	11		
					1.20	>200	1.20	19	1.20	13		
	$\delta l=2$		$\delta l=[m/2]=7$		$\delta l=m=15$		$\delta l=2m=30$		$\delta l=3m=45$			
(14x14) m=15	0.85	47	0.85	46	0.85	25	0.85	17	0.85	12		
	0.90	44	0.90	43	0.90	24	0.90	15	0.90	10		
	0.95	42	0.95	41	0.95	23	0.95	14	0.95	9		
	1.00	46	1.00	61	1.00	28	1.00	13	1.00	9		
	1.05	85	1.05	157	1.05	41	1.05	13	1.05	9		
	1.10	>200			1.10	76	1.10	15	1.10	10		
					1.15	>200	1.15	18	1.15	11		
							1.20	21	1.20	13		
	$\delta l=2$		$\delta l=[m/2]=10$		$\delta l=m=20$		$\delta l=2m=40$		$\delta l=3m=60$		$\delta l=6m=120$	
(19x19) m=20	0.85	49	0.85	47	0.85	26	0.85	16	0.85	12	0.85	11
	0.90	46	0.90	44	0.90	24	0.90	15	0.90	11	0.90	10
	0.95	43	0.95	42	0.95	23	0.95	14	0.95	9	0.95	8
	1.00	46	1.00	67	1.00	28	1.00	13	1.00	8	1.00	7
	1.05	91	1.05	>200	1.05	42	1.05	13	1.05	9	1.05	8
	1.10	>200			1.10	85	1.10	15	1.10	11	1.10	10
					1.15	>200	1.15	18	1.15	12	1.15	11
							1.20	22	1.20	14	1.20	13
	$\delta l=2$		$\delta l=[m/2]=30$		$\delta l=m=60$		$\delta l=2m=120$		$\delta l=3m=180$		$\delta l=6m=360$	
(59x59) m=60	0.85	51	0.85	49	0.85	28	0.85	16	0.85	12	0.85	11
	0.90	48	0.90	46	0.90	26	0.90	15	0.90	11	0.90	10
	0.95	45	0.95	45	0.95	24	0.95	14	0.95	10	0.95	9
	1.00	50	1.00	77	1.00	28	1.00	13	1.00	10	1.00	10
	1.05	106	1.05	>200	1.05	43	1.05	13	1.05	12	1.05	12
					1.10	94	1.10	15	1.10	14	1.10	14
							1.15	18	1.15	16	1.15	16
							1.20	22	1.20	18	1.20	18

Table 1: The convergence behaviour of ESD method, incorporating BAIFEMO algorithm, for the 2D-model problem with "fill-in" parameter r=2.

GRID	Storage for $M_r^{\delta l}$	No.of iters	Storage for $M_r^{\delta l}$	No.of iters	Storage for $M_r^{\delta l}$	No.of iters	Storage for $M_r^{\delta l}$	No.of iters	Storage for $M_r^{\delta l}$	No.of iters	Storage for $M_r^{\delta l}$	No.of iters
	δl=2		δl=[m/2]=10		δl=m=20		δl=2m=40		δl=3m=60		δl=6m=120	
(19x19) m=20	3n	16	19n	17	39n	12	79n	9	119n	6	239n	6
	δl=2		δl=[m/2]=15		δl=m=30		δl=2m=60		δl=3m=90		δl=6m=180	
(29x29) m=30	3n	17	29n	17	59n	13	119n	9	179n	7	359n	6
	δl=2		δl=[m/2]=20		δl=m=40		δl=2m=80		δl=3m=120		δl=6m=240	
(39x39) m=40	3n	18	39n	18	79n	13	159n	9	239n	6	479n	6
	δl=2		δl=[m/2]=25		δl=m=50		δl=2m=100		δl=3m=150		δl=6m=300	
(49x49) m=50	3n	18	49n	18	99n	13	199n	9	299n	6	599n	6
	δl=2		δl=[m/2]=30		δl=m=60		δl=2m=120		δl=3m=180		δl=6m=360	
(59x59) m=60	3n	18	59n	18	119n	13	239n	9	359n	6	719n	6

Table 2: The convergence behaviour of EPCG method, incorporating BAIFEMO algorithm, for the 2D-model problem with "fill-in" parameter r=2.

Case	Width parameter l	δl m	δl = 2m
I	3	13	9
II	3	14	11
I	6	8	5
II	6	9	5

Table 3: The behaviour of EPCG method, incorporating BAIFEMO algorithm, for the 2D-model problem on a grid (29x29) with semi-bandwidth m=30, "fill-in" parameter r=2, for several values of width-parameter l and magnitude of elements c_{ij} of the coefficient matrix A.

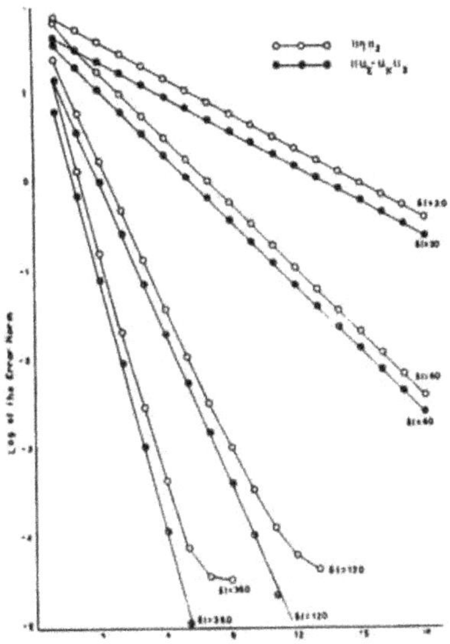

Figure 1: Behaviour of error measures $\|r_i\|_2$ and $\|u_u - u_c\|_2$ of ESD method applied to the 2D-model problem on a grid (59x59) with semi-bandwidth m=60, width-parameter l=3, fill-in parameter r=2.

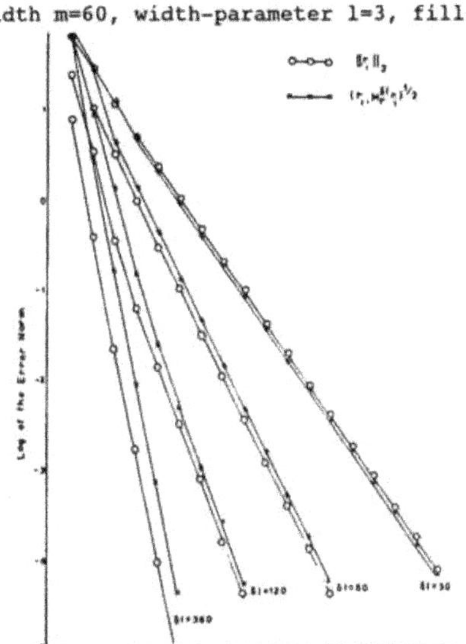

Figure 2: Behaviour of error measures $\|r_i\|_2$ and $(r_i, M_r^{\delta l} r_i)^{1/2}$ of EPCG method applied to the 2D-model problem on a grid (59x59), i.e. n=3481.

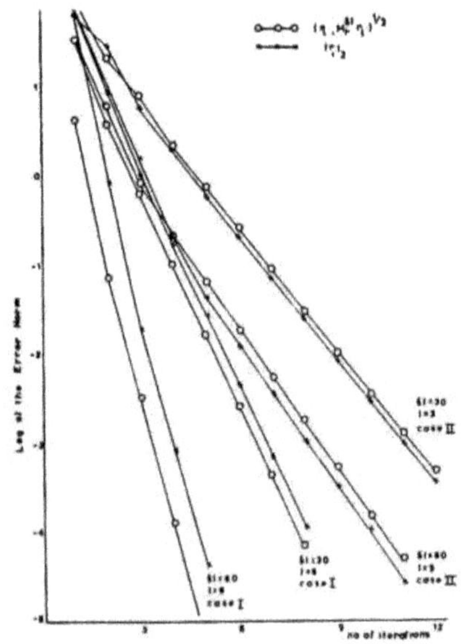

Figure 3: Behaviour of error measures $\|r_i\|_2$ and $(r_i, M_r^{\delta l} r_i)^{1/2}$ of EPCG method applied to the 2D-model problem on a grid (29x29), i.e. n=841, by varying the width parameter l and magnitude of elements of $c_{i,j}$.

Finally, we state that the new methods can be efficiently applied for solving linear and non-linear 2D and 3D elliptic boundary-value problems.

REFERENCES

[1] DANTZIG G.B., ORCHARD-HAYS W. (1954): The product form of the inverse in the Simplex method, Math.Comp., 64-67.

[2] DUBOIS P.F., GREENBAUM A., RODRIGUE G.H. (1979): Approximating the inverse of a matrix for use in iterative algorithms on vector processors, Computing 22, 257-268.

[3] EVANS D.J. (1984): Parallel SOR iterative methods, Parallel Computing 1, 3-18.

[4] EVANS D.J., LIPITAKIS E.A. (1980): A normalized implicit Conjugate Gradient method for the solution of large sparse systems of linear equations, Comp.Math.Appl.Mech.Engng. 23, 1-19.

[5] EVANS D.J., LIPITAKIS E.A. (1983): An approximate QIF method for parallel computers, In "Preconditioning Methods: Theory and Applications", Gordon & Breach Science Publishers, 537-552.

[6] EVANS D.J., LIPITAKIS E.A. (1983): Implicit semi-direct methods based on root-free sparse factorization procedures, BIT 24, 194-208.

[7] FADEEVA V.N.: Computational Methods of Linear Algebra, transl. by C.D. Benster, Dover, N.Y. 1959.

[8] GUSTAFSSON J. (1978): A class of first order factorization methods, BIT 18, 142-156.

[9] JOHNSON O.G., PAUL G. (1981): Vector algorithms for elliptic PDE's based on the Jacobi method, Proc. Elliptic Problem Solvers Conference, Los Alamos, Acad.Press, 345-351.

[10] LIPITAKIS E.A. (1984): Solving elliptic BV problems on parallel processors by Approximate Inverse Matrix semi-direct methods based on the multiple explicit Jacobi iteration, Comp.Math. with Applics 10, 171-184.

[11] LIPITAKIS E.A. (1984): Generalized extended to the limit sparse factorization techniques for solving unsymmetric FE systems, Computing 32, 255-270.

[12] LIPITAKIS E.A., EVANS D.J. (1980): Solving non-linear elliptic difference equations by extendable sparse factorization procedures, Computing 24, 325-339.

[13] LIPITAKIS E.A., EVANS D.J. (1987): Explicit semi-direct methods based on approximate inverse matrix techniques for solving BV problems on parallel processors, Math. and Comp. in Simulation 29, 1-17.

[14] LIPITAKIS E.A., EVANS D.J. (1981): The rate of convergence of an approximate matrix factorization semi-direct method, Numer.Math. 36, 237-251.

[15] LIPITAKIS E.A., GRAVVANIS G.A. (1990): A fast direct method for solving elliptic boundary-value problems on multiprocessor systems, Proc.Intern.Conf. on "Numerical Methods in Engineering: Theory and Applications", Swansea, UK, 7-11 Jan. 1990, 622-632.

[16] LIPITAKIS E.A., GRAVVANIS G.A. (1990): Implicit preconditioning methods based on root-free sparse finite-elements factorization techniques, T.R. ASOE/INFOS/ΠΥΜ^2,-90-36, Department of Informatics, Athens University of Economics.

[17] MEIJERINK J.A., VAN DER VORST H.A. (1977): An iterative solution method of linear systems of which the coefficient matrix is symmetric M-matrix, Math.Comp. 31, 148-162.

[18] MUNKSGAARD N. (1980): Solving sparse symmetric sets of linear equations by preconditioned Conjugate Gradients, ACM TOMS 6, 206-210.

[19] RODRIGUE G., HENDRICKSON C., PRATT M. (1982): An implicit numerical solution of the 2D-diffusion equation and vectorization experiments. In "Parallel Computations", Academic Press, NY, 101-128.

[20] STONE H.L. (1968): Iterative Solution of implicit approximation of multidimensional partial differential equations, SIAM J.Numer.Anal. 5, 530-558.

[21] TEWARSON R.P. (1966): On the product form of inverses of sparse matrices, SIAM Rev.8, 336-342.

[22] YOUNG. D.: Iterative solution of linear systems, Academic Press, 1971.

[23] VARGA R.S. (1962): Matrix iterative analysis, Prentice-Hall, Eastewood Cliffs, N.J.

[24] WAUGH F.V., DWYER P.S. (1945): Compact computation of the inverse of a matrix, Ann.Math.Stat.16, 259-271.

Professor E.A. Lipitakis
Department of Informatics
Athens University of Economics
76, Patission Street
Athens 104 34, Greece.

Integrated Portable Problem Solving Environments for the Parallel Solution of Elliptic PDE Problems and the case of the EUCLID system

T. S. Papatheodorou[1], G.E. Kalamaras[1,2], D.K. Kehagias[1],
P. I. Kourouniotis[1], K. N. Pantazopoulos[1,2] and E. D. Polychronopoulos[1].

Abstract

We present a fundamental set of requirements for Problem Solving Environments (PSE) for solving elliptic PDE problems on parallel computers. We propose a design that meets these requirements and defines the architecture of a full PSE, a system that contains most of the necessary components. The system is designed so that it interacts intelligently with both its environment and the user and so that it finds and uses the "best" available solution, for the specific computational resources. As a particular reduced version we obtain "EUCLID", a small, economic, flexible and efficient PSE, that can operate on small platforms and can be used for the study of parallel PDE algorithms as well as for research on PDEs. We also discuss the use of EUCLID with virtual parallel environments and we give details on the present state of implementation of the EUCLID system.

1. Introduction

In spite of their necessity and great potential for solving important problems in Science and Engineering, parallel computers have not yet been widely accepted and used. One of the main reasons for this, is the complexity of parallel programming and the luck of appropriate tools necessary for supporting this type of programming. Another important reason is the restricted availability of efficient application libraries, which severely limits the exploitation of the high computational power of these machines. Programming conventional computers (e.g. using FORTRAN) has been already a problem for many users. Programming parallel machines efficiently is almost impossible for most of them, as it requires expert knowledge of several architectural details.

1. Laboratory for High Performance Architectures And Applications, Dept. of Computer Engineering and Computer Science, University of Patras, P.O. Box 1045, 26 110, Patras, Greece.
Supported in part by Grant GSRT#89ED345, of the General Secreteriat of Research and Technology and by ESPRIT Basic Research Action project 6634 (APPARC).

2. FIRST Informatics S.A., HPC Sector, R&D Dept, P.O. Box 1340, 26 110, Patras, Greece.
Work partially supported by FIRST Informatics S.A and by ESPRIT IPSS project 6643 (PPPE).

These problems have led to intensive efforts for the development of parallel **Problem Solving Environments (PSE)**. Briefly, a PSE is an integrated system that does the work on behalf of the user, in the sense that it integrates all the necessary software and hardware technologies and takes care of the programming details, in order to solve the problem on user's behalf. More details are provided in Section 2. It should be noted that the development of efficient PSE requires collaborative effort among many research disciplines (numerical analysis of PDE, computational linear algebra, physics, engineering, computer sciences, ...) as well as the integration of several technologies. These include human-machine interfaces, visualization, interactive color graphics and related peripherals, software development tools, expert systems, application libraries, system software for parallel processing, modern computer architectures and, generally, what is normally meant with the term "High Performance Computing".

For our purposes, the user is a scientist or engineer who needs to solve a PDE problem on a parallel computer. PDE are both important for "real-world" applications and representative of what is involved in large scale computing. Although it is possible to include other types of PDE, we restrict the discussion to two-dimensional elliptic boundary value problems. The critical importance of PSE has been recognised even for the (much simpler but still very complicated) case of sequential computers. Thus, the first significant efforts are dated back in the 1960's. However, they were mostly abandoned during the 1970's, due to the inability of the existing technological infrastructure to support the development of efficient PSE. The recent advances in both, the computer technologies and the computational science, make it realistic to develop efficient PSE in the 1990's. In fact, research in this direction is recently intensified and some systems have energed which, though far from what is desired, can be considered as modern operational PSE. Most noticable among them are ELLPACK ([11]), operating since the early 1980's in more than 100 academic and industrial facilities worldwide, a parallel version of it, //ELLPACK, designed mainly for hypercube architectures ([5], [7]), DEQSOL ([8]) which is similar in function and scope (see [1] for comparisons), the early efforts of the ESPRIT projects SPAN ([10]) and PADMAVATI ([15]) and, finally, FIRST's PARDIF prototype, projected to become a finished product by the end of 1994. Other enviroments of similar nature include AGNES [9] and ALPAL[2]. Further work is being done in recent european projects such as PPPE. For further discussion see also [4] and [3].

In this paper we report on our group's recent work on PSE for solving elliptic PDE problems on parallel machines. Specifically, we define the necessary functions and corresponding requirements expected from a PSE and, based on these, we introduce the design of a "full" PSE. As a reduced version of it we obtain the EUCLID system. This system is not meant to become a full PSE and in some respects can be considered as a subject of large PSE, such as //ELLPACK and PARDIF. Nevertheless, EUCLID is expected to serve different purposes. One of them is the development of know-how in building and studying this type of complicated systems. More importantly however, EUCLID is not intended to become another PSE for broad use by inexperts, or a product for plain PSE users. It is primarily designed as a tool for research on PSEs themselves, a task that cannot be equally fullfilled by large and "canned" systems. In this direction, EUCLID is expected to be a small and flexible system, that can operate on smaller platforms, can be easily modified, and can be easily used to incorporate and test new ideas, or include new modules and study their effect on the system and its performance. Still, EUCLID will be capable to serve as a facility for studying the performance of parallel algorithms.

In Section 2 we provide some necessary details on PSE, with emphasis on five classes of functions and corresponding requirements that they should satisfy. In Section 3 we introduce a design for full PSE, i.e. systems that contain "all" necessary components which are also fully developed, so that the PSE can satisfactorily meet "all" the requirements. Then, as a reduced version of it, we obtain the design of the EUCLID system.

Issues such as modularity, flexibility, scalability and "best" exploitation of the available networked heterogeneous resources, are most important in the design and development of PSE. In order to resolve these issues, we propose in Section 4 the use of **Virtual Parallel Environments (VPE)**, which we analyze in relation to the EUCLID system. Finally, Section 5 is used for discussion and conclusions.

2. Problem Solving Environments for PDE

The necessity and role of PSE was briefly outlined above. In this section we provide some details that are necessary for determining the functions required from efficient PSE and understand the technical areas and technologies that must be integrated for their development. Emphasis is placed on the main requirements that a PSE should satisfy, five sets of which are determined in this section.

Recall that in our case, the user is normally a scientist or engineer who, though an expert in his/her own field, does not (and does not wish to) have sufficient knowledge of numerical methods, or computer architectures (especially parallel ones), or other related areas. To fix the ideas, it may be useful to think of a scientist in need of solving a problem such as the one described in Figure 1. The user knows all problem attributes, i.e. all the coefficient functions and the right side function f in the PDE (1), the equations of all four boundary curves for the domain in (2), and all functions involved in the boundary conditions (3). The user does not wish to know anything more, except perhaps the name of the parallel machine to be used. In some cases the PSE should be able to also make a "best" platform selection for the user, among a set of several systems available at the specific computing facility. The user must enter the information that defines the problem together with the desired accuracy ε, within which he/she wishes to obtain an approximate solution. In doing so, the user should be allowed to use his/her "own terms", i.e. descriptions of the equations and the domain in the same form as in Figure 1.

With this in mind, we can discover most of the important functions and corresponding requirements that an efficient PSE should satisfy.

Main PSE requirements

R1. INPUT. The system must assist the user in correctly entering the problem attributes, in a way that is friendly, easy and natural for the user. It must also check the validity and consistency of the information entered, guide the user through possible modifications and discover other properties that are crucial for further actions, e.g. that the specific problem is self-adjoint.

R2. METHOD SELECTION AND CODE GENERATION. The system must discover "all" applicable methods for solving the specific PDE problem and eventually select the "best" one on behalf of the user, or help the user make this selection if he/she so desires. It should also offer explanations for various suggestions or discussions. The system should always offer state-of-the-art methods. Finally, the system must generate the code necessary to link together appropriate modules of the parallel numerical library, for the execution of the selected method.

R3. EXECUTION. The code is executed using the parallel library and underlying architecture, or even a network of several computing systems. The PSE must make "best" exploitation of the computing power of the underlying resources. Critical for this is the automatic partition and allocation of workload to processors.

R4. ENVIRONMENT KNOWLEDGE AND ADAPTABILITY. This set of requirements refers to the ability of the system to adapt to changes in its own (already existing) environment and to be easily interfaced with, or ported to, other environments. As such, it does not directly refer to the interaction with the user or to a specific problem solution, as it is mainly the case with the other sets of requirements. Instead, the interaction addressed is with systems and modules other than its own, as well as with PSE engineers and High Performance Computing (HPC) experts. This set mainly refers to technical requirements, such as portability and scalability, and offers the support for efficiently performing the functions described by other sets of requirements, in particular R2 and R4. Examples include the efficient responce to addition or removal of resources in the already existing environment, adaptability to changes in existing parameters, such as the number of available processors in a specific parallel system, facilities for authoring of expert system extensions in order to include new architectures, provisions for easy modifications of the underlying parallel library and other system components, as well as the efficient interfacing with Virtual Parallel Programming Environments (VPE).

R5. RESULTS AND OUTPUT. The system must analyze and present the results to the user in a meaningful and easily understandable way. It should offer alternatives for the presentation of the results and should interact with the user in order to assist in their interpretation. It should also be able to use the analysis of the results for self-learning and proper updating of knowledge bases.

One immediate conclusion is that the development of efficient PSEs requires the collaborative effort among many research disciplines (such as numerical analysis of PDE, numerical linear algebra, physics, engineering, computer science, ...) and the integration of several software and hardware technologies. These include human-machine interfaces, visualization and related peripherals, integrated symbolic-numerical processing, software development tools, expert systems, computational geometry, graph algorithms application algorithms and libraries, system software for parallel processing and computer architectures in their broad sense, powerful processors, and more.

3. Design of full PSE and the case of the EUCLID system.

In this section we first propose a design of a full PSE for solving PDE on parallel machines and virtual parallel environments. The term "full" is used to denote that such a PSE, although not complete or ideal, contains most of the necessary components and each component is fully developed, so that the system can satisfactorily meet the functions/requirements specified in the proceeding section. The design of EUCLID is then obtained as a reduced version of the full system, in the sense that several components are eliminated and others are reduced in size and scope. The design of the full PSE is presented in Figure 2, in correspodence to requirements R1-R5 of Section 2. Next, we discuss its main components.

The systems for general support are meant to be used with front-end sequential machines and powerful graphics peripherals. Example routines in the small (and sequential) numerical-statistical library include root finding for the determination of intersection points between boundary curves and vertical/horizontal grid lines as well as systems for analysis and statistics on newly obtained results, to be used for self-learning and updating of appropriate knowledge bases. Examples for the symbolic utility include several actions regarding input tests, interaction with the user, code generation and symbolic differentiation. For intance, the user may have to define the right side function *f* implicitly. In such cases, the system must perform symbolic differentiation, in order to obtain the necessary code for *f*. Mixed symbolic-numeric processing is considered essential for PDE (see also [10] and [6]).

The entire system is controlled by the Intelligent Graphical User Interface (IGUI). This module also interacts with the user, mainly in order to meet requirements R1 and R5. Appendix A provides some examples (printed screens), regarding the present state of development of the IGUI-module for the EUCLID system. Here, considering that the role of the expert system in R5 is rather evident, we summarize some key points regarding R1. After extensive and friendly interaction with the user, all problem attributes are correctly defined and understood by the system, intermediate code has been generated for the various functions and boundary curves involved and important properties have been derived, e.g. that the problem is self-adjoint or that the coefficients and right side is the differental equation are all continuous functions. This information is used for the part in R2-R4, which we now describe. Regarding R2, first note that for a given problem the number of applicable methods may be very large, even when a relatively small number of routines is to be used. Consider for instance a very small parallel library capable for running only six discretization methods, namely two finite difference variants (5-point, 9-point star), two Galerkin variants (linear triangular elements, cubic splines) and two Hermite-cubic collocation variants (interior, general) and those only for solving Poisson's equation with Dirichlet Boundary conditions on the unit square. In addition, ignore all other steps in the procedure and consider only two, namely equation formation and equation solution. Furthermore, assume that the library contains two variants for equation formation for each method and 12 linear system solvers, for a total of 24 routines. Assume that half of the solvers apply only to symmetric and positive definite matrices. Hence, all solvers apply to the finite difference and Galerkin variants, while only (the nonsymmetric) half apply to the two collocation variants. Combining each of the two applicable routines for equation formation with each of the 12 (finite difference and Galerkin variants) or 6 (collocation variants) solvers, we find a total of 120 applicable combinations in the library. The role of the various components in R2 is to determine all applicable combinations and extract one which is "best" for the specific problem and the specific resources. This is a complicated task, having in mind only the large variety of possible equations, domains and problem properties. Nevertheless, some techniques have been suggested ([12], [13]) which provide an acceptable solution. These are implemented with the component corresponding to R2, which are controlled by the Expert System Application Generator (ESAG). The technique is based on a partition of possible problems into several classes. For instance, Class 1 contains the problems with the property "self-adjoint, constant coefficient smooth right side, rectangular domain, Dirichlet boundary conditions". Variations in one or more of these properties define different classes. Variants may include discontinuous functions for coefficients or the right side, different domains that are standard and are immediately recognized by the system (e.g. L-shaped, circle, ellipse) or general domains that are simple or may even contain wholes (cf. Figure 1). Three databases are maintained by the system which are indepenedent of the environment or the parallel machines on which the problem is to be solved. The Application Classes DB contains all seperate classes with their properties and possibly some example problems. The Applicable Method DB contains a list of all possible methods applicable for each problem class. Examples could be, for a given class.

* 5-point star, Cholesky decomposition variant 3,...
* interior collocation, bounded Gauss partial pivoting version 2,...
* 4-point star, SOR, ω=1.4, ...
* Galerkin, linear elements, conjugate gradient, version 1,...

The third is the "ε vs size database" which provides the minimum grid size necessary to achieve the accuracy ε prescribed by the user, for each applicable method and for the specific problem class. It is understood that all three DB, are updated accordingly, each time the analysis of the output requires it. The Application Analyser-Classifier discovers the class in which the given problem belongs, by search and comparison in the Application Classes DB. Next, the All Applicable Methods Generator produces a list of all application methods, using the Applicable Method DB, and passes it on to the Best Method Generator (BMG). As a first step, BMG uses the "ε vs size DB" and supplements each applicable method with the necessary minimum grid size. This procedure is architecture indepedent, in line with the design proposed in [13].

At this point the system requires knowledge of some architectural details, eg. that the problem is to be solved on the Intel iPSC/2 with 16 processors. The "size-time-machine DB" provides statistics on the performance of each method on this machine. For instance, for the 24-routine library and the problem class of the preceding example, the DB will contain the execution time required to execute each of the 60 combinations for several grid sizes. If a size is not present in the DB the corresponding execution time can be easily estimated. A conversion factor ([5]) may also be used for moving on to different machines. At this stage the BMG can easily select as best method one which requires the smallest execution time for the great problem class and accuracy (grid size) and for the specific machine.

Finally, the Parallel Program DB provides the string of library routines that must be executed, which allows the Code Generator to "synthesize" the necessary driver code (an analogue of a FORTRAN main program). This is passed on for execution to the R3-components, ie. the allocator and the parallel library. The former solves the important problem of partition and allocation of the computational load to processors. This corresponds to the requirement that the PSE must manage the existing resources in the "best" possible way. The problem constitutes by itself a subject of active research in High Performance Computing and we will not pursue it further in this paper. The parallel library in R3 may be specific to the architecture (in which case it is provided through the RSAS discussed below) or it can be portable, as provided by the Virtual Parallel Environments (Section 5).

It remains to discuss the system components corresponding to R4. The Resource Specific Authoring System (RSAS) is in line with the design proposed in [13]. It is by now evident that a good part of a full PSE is indepedent of the architecture and the environment in which the system is to be ported. This suggests that an environment or architecture indepedent part (or kernel) is built. Then the RSAS with the "kernel" will assist the PSE engineers and High Performance Computing experts to provide all additional information that the PSE requires about the specific environment. Finally the VPE Interface provides for the integration of the system into a Virtual Parallel Environment, discussed in Section 4.

As stated earlier, EUCLID is defined as a reduced version of the full PSE, taking into account that the task is to build a small system and that, for the present, the system is designed for the Intel iPSC/2 hypercube with a Sun Workstation as a front-end. It should be noted that the parallel library is intended to have a relatively small number of well selected routines, which will result to a considerable reduction on the size of the DBs in R2. The careful selection of problem classes and library modules is essential for obtaining a high quality system.

4. Virtual Parallel Environments (VPE)

In general, a VPE is a software system that makes it possible to "run" parallel software modules on heterogeneous components, which are interconnected through some type of a network (eg. LAN, MAN, WAN). For a conceptual model see Figure 3.

The VPE is built as a set of control daemons and an interface library. The system to be implemented on the VPE consists of a set of software modules that have varying needs of hardware and software resources. The hardware pool is offered in a transparent way as a single multi-character resource, through the support of cooperating distributed control daemons. The software resources are linked with the VPE via interfaces to the control daemons. The resulting situation is a transparent environment which offers various software and hardware resources through interface libraries. In order to port the targeted system on the VPE, the various components must indicate their h/w and s/w needs and interface with the appropriate control daemon. The interconnection and cooperation of the modules comprising the targeted system is designed with the appropriate programming model, that is, message passing or shared-memory.

Several prototypes implementing the VPE approach exist, although research in this direction is open. Such prototypes are mainly judged by the number of platforms they support, the efficiency of the control daemons and the type of the interface library they provide. Examples of existing prototype VPEs include PVM, P4, APPL, TCSMSG, VMMP and LINDA. We have selected two of them, PVM and P4, for interfacing and experimenting with the EUCLID system. We restrict attention to the PVM (Parallel Virtual Machine) ([14]) case, although most of the discussion applies to other VPEs as well.

Designing the architecture of EUCLID system using VPE approach, modularity is secured due to the nature of the approach and the notion of the interface library and programming model and portability is envisaged because it is only a matter of implementing the control daemons to the intended h/w platforms using an ad hoc approach. For the prototype EUCLID system, the selection of PVM ensures portability to almost all Unix-based workstations and to the majority of commercially available distributed memory MIMD and shared memory machines.

5. Conclusions

We have presented a set of important requirements for Problem Solving Environments for PDE and have proposed a design for full PSEs. As a reduced version of a full system we have presented the design of EUCLID, which is a small, flexible and easily modifiable PSE that can operate on small platforms and can be used for the study of parallel algorithms for PDE and for research on PSE development. We have also discussed the use of EUCLID in conjuction with Virtual Parallel Environments. Finally, we have given details on the present state of the implementation of the EUCLID system and, in particular, its Intelligent Graphical User Interface which controls the entire system.

6. References

[1] R.F. Boisvert and D.K. Kahaner, DEQSOL and ELLPACK: PSEs for PDEs, office of Naval Research, Asia Office, Scientific Information Bulletin (NAVSO P-3580), 16(1), pp7-19, 1991

[2] G.O. Cook, Jr., ALPAL: A program to generate physics simulation codes from natural descriptions, Int. J. Modern Physics, 1(1), pp1-51, 1990

[3] S. Galopoulos, E.N. Houstis, J.R. Rice, eds, Future Research Directions in PSE for Computational Science, Report of a Workshop on Research Directions in Intergrating Numerical Analysis, Symbolic Computing, Computational Geometry and Artificial Intelli-

gence for Computational Science, Washington D.C., April 11-12, 1991

[4] J.Fr. Hake and W. Homberg, Management of PDE Software for Supercomputing, Supercomputing Lecture Notes in Computer Science 297, E.N. Houstis, T.S. Papatheodorou and C.D. Polychronopoulos eds, Springer Verlag, Berlin, New York, 1988

[5] C.E. Houstis, E.N. Houstis, J.R. Rice, P. Varodoglou, T.S. Papatheodorou, Athena: A Knowledge Base System for // ELLPACK, Symbolic-Numeric Data Analysis and Learning, E. Diday and Y. Lechevalher Eds, Nova Science Publishers Inc., New York 1992

[6] E.N. Houstis, M. Katzouraki, T.S. Papatheodorou and V. Sotiropoulou, Logic Parallelism in an Expert System for Solving PDEs, Proceedings of the Int. Conference on Expert Systems for Scientific Computing, 1990

[7] E.N. Houstis, J.R. Rice, T.S. Papatheodorou, //ELLPACK: An Expert System for Parallel Processing of PDE, Supercomputing Lecture Notes in Computer Science 297, E.N. Houstis, T.S. Papatheodorou and C.D. Polychronopoulos eds, pp497-507, Springer Verlag, Berlin, New York, 1988

[8] C. Konno, Y. Umetani, M. Igai and T. Ohta, Interactive/Visual DEQSOL: Inetractive Creation, Debbuging, Diagnosis and Visualization of Numerical Simulation, Mathematics and Computers in Simulation 31(1989), pp353-369

[9] A.D. Kowalski, R.L. Peskin, M.F.Russo, Anatomy of AGNES: An automatic generator of numerical equation solutions, Mathematics and Computers in Simulation 31(1989), pp343-352

[10] T.S. Papatheodorou, E.N. Houstis, C.E. Houstis, M. Kalogerakis, M. Katzouraki, M. Tsoukarellas, V. Sotiropoulou, Y. Ramfos and E. Polychronopoulos, The design and implementation of a parallel PDE Expert Solving System, using languages that support numeric and symbolic processing, ESPRIT project 1588 (SPAN) report Jan. 1990

[11] J.R. Rice and R.F. Boisvert, Solving Elliptic Problems using ELLPACK, Springer Verlag series in Computational Mathematics 2, 1984

[12] V. Sotiropoulou and T.S. Papatheodorou, A Case Study of Knowledge Engineering in the Application Area of elliptic PDE, Proc. of the fourth Int. Conference on Software Engineering and Knowledge Engineering, Capri, Italy, June 1992, pp261-268

[13] V. Sotiropoulou and T.S. Papatheodorou, A Task-level Architecture for Expert Systems for Scientific Computing, submitted IEEE Transactions on Systems, Man and Cybernetics

[14] V.S. Sunderam, PVM: A Framework for Parallel Distributed Computing, Concurrency: Practice and Experience, vol 2, no 4, Dec 1992

[15] PADMAVATI, Esprit project 1219 (967), Analysis and Design of the Parallel Epxert System, report, Oct 1990

Appendix A

PDE Form

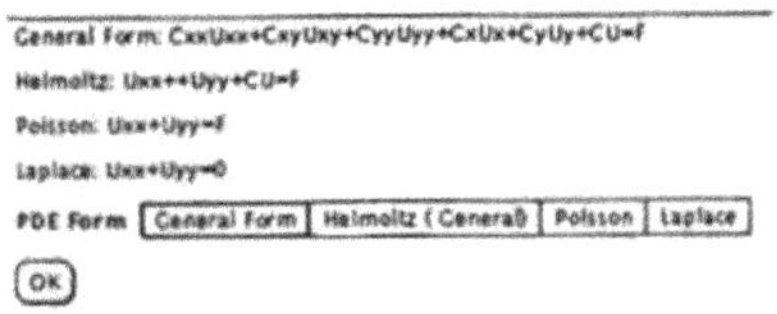

PDE Description (General Form)

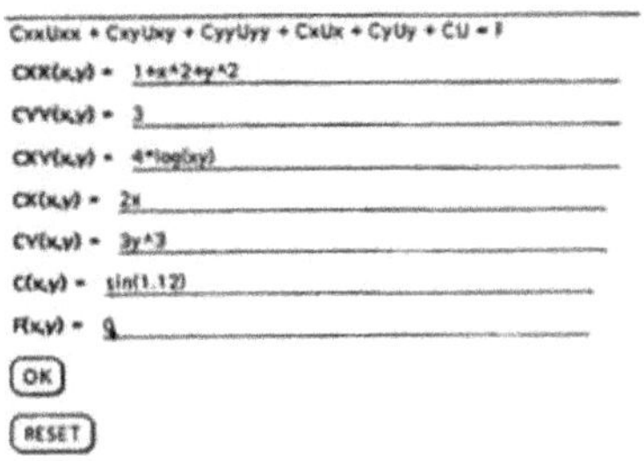

Domain-Grid Visualization

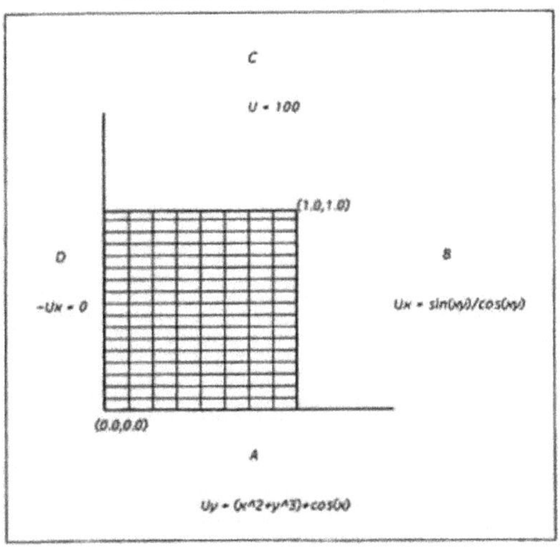

Domain Description

LowerLeft(x) = 0.0

LowerLeft(y) = .0

UpperRight(x) = 1

UpperRight(y) = 1.

A: Cy = 1

A: C = 0

A: F = PE2+cos(x)

B: Cx = 1

B: C = 0

B: F = PE1

C: Cy = 0

C: C = 1

C: F = 100

D: Cx = -1

D: C = 0

D: F = 0

OK

RESET

Grid Description

Ax = 0.0

Ay = 0.0

Bx = 1

By = 1

Grid points on X-axis 1 1024 8

Grid points on Y-axis 1 1024 16

OK

RESET

Example

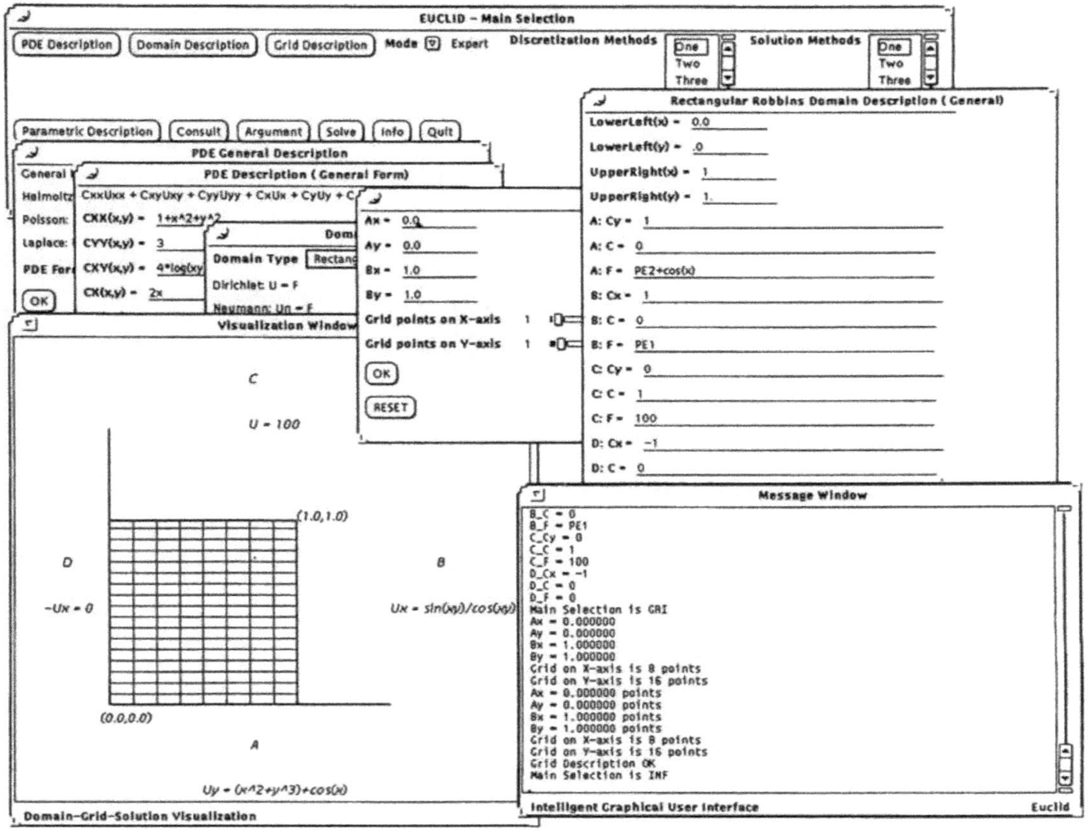
EUCLID – Main Selection
PDE Description
Domain Description
Grid Description
Mode
Expert
Discretization Methods
One
Two
Three
Solution Methods
One
Two
Three
Parametric Description
Consult
Argument
Solve
Info
Quit
PDE General Description
PDE Description (General Form)
CxxUxx + CxyUxy + CyyUyy + CxUx + CyUy + C
CXX(x,y) = 1+x^2+y^2
CYY(x,y) = 3
CXY(x,y) = 4*log(xy
CX(x,y) = 2x
OK
Domain Type
Dirichlet: U = F
Ax = 0.0
Ay = 0.0
Bx = 1.0
By = 1.0
Grid points on X-axis 1
Grid points on Y-axis 1
OK
RESET
Rectangular Robbins Domain Description (General)
LowerLeft(x) = 0.0
LowerLeft(y) = .0
UpperRight(x) = 1
UpperRight(y) = 1.
A: Cy = 1
A: C = 0
A: F = PE2+cos(x)
B: Cx = 1
B: C = 0
B: F = PE1
C: Cy = 0
C: C = 1
C: F = 100
D: Cx = -1
D: C = 0
Visualization Window
C
U = 100
(1.0,1.0)
D
B
-Ux = 0
Ux = sin(xy)/cos(xy)
(0.0,0.0)
A
Uy = (x^2+y^3)+cos(x)
Domain-Grid-Solution Visualization
Message Window
B_C = 0
B_F = PE1
C_Cy = 0
C_C = 1
C_F = 100
D_Cx = -1
D_C = 0
D_F = 0
Main Selection is GRI
Ax = 0.000000
Ay = 0.000000
Bx = 1.000000
By = 1.000000
Grid on X-axis is 8 points
Grid on Y-axis is 16 points
Ax = 0.000000 points
Ay = 0.000000 points
Bx = 1.000000 points
By = 1.000000 points
Grid on X-axis is 8 points
Grid on Y-axis is 16 points
Grid Description OK
Main Selection is INF
Intelligent Graphical User Interface
Euclid

Appendix B

PDE: $(1 + x^2 + y^2)U_{xx} + 3U_{yy} + \ldots = f(x,y)$
(1)

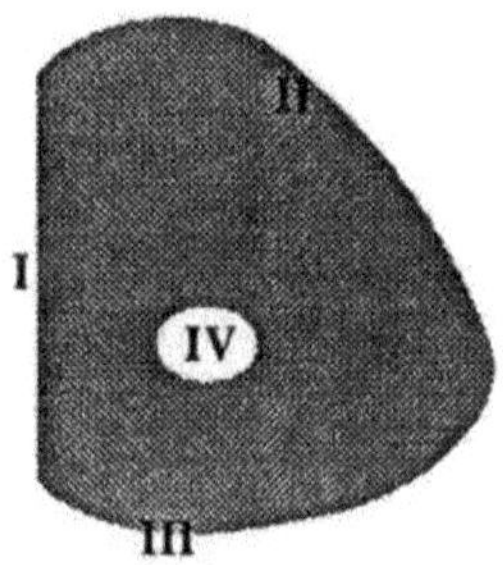

Domain (2)

Boundary Conditions (3)

$- U_x(x,y) = 0$ on I

$U_y(x,y) = g_{II}(x,y)$ on II

$U_y(x,y) = g_{III}(x,y)$ on III

$U(x,y) = 100$ on IV

(1)Differential Equation (2) Domain (3) Boundary Conditions

Figure 1.

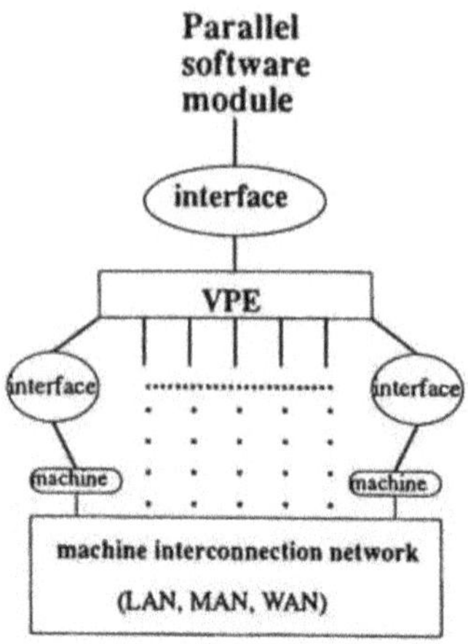

Figure 3. Conceptual Model of VPE.

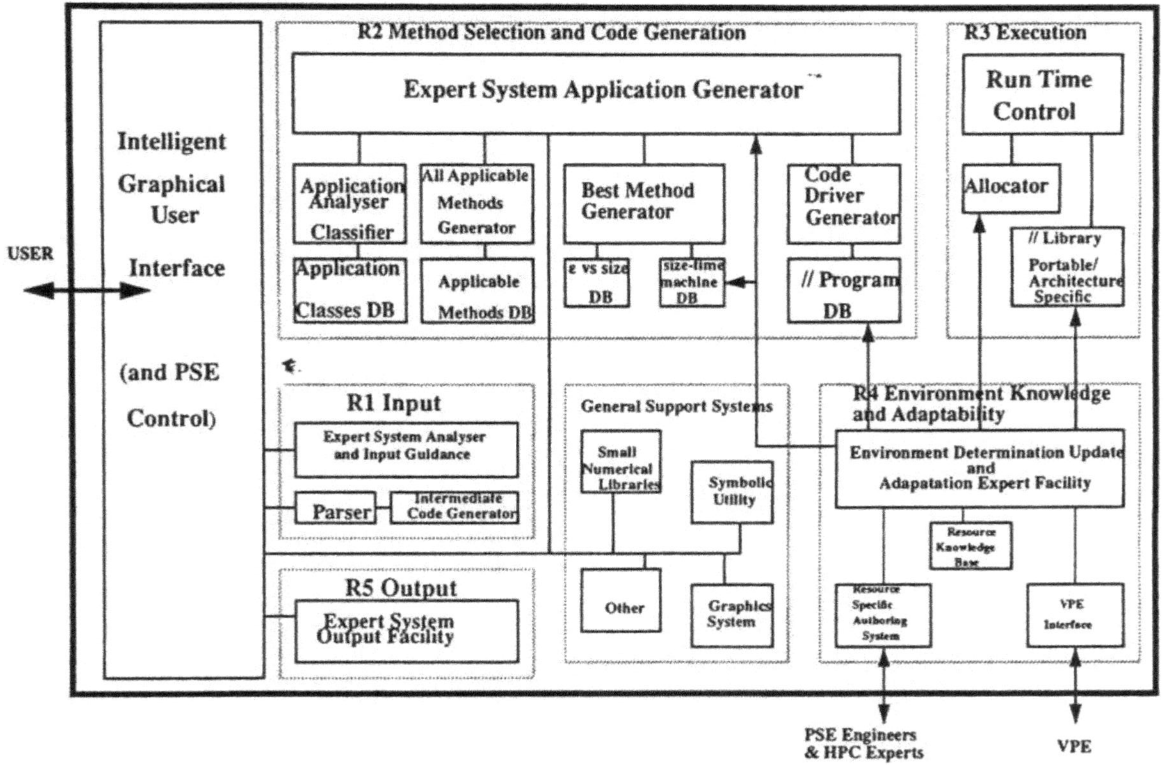

Figure 2. Architecture of "full" PSE

A SUBDOMAIN-BY-SUBDOMAIN CONJUGATE GRADIENT METHOD FOR PARALLEL PROCESSING

M. Papadrakakis, S. Bitzarakis, N. Bitoulas

Institute of Structural Analysis & Aseismic Research, National Technical University
Zografou Campus, Athens GR-15773, Greece.

ABSTRACT

In this study a subdomain-by-subdomain (SBS) preconditioner is developed in conjunction with the conjugate gradient method for the solution of large problems resulting from the application of the finite element method. The proposed technique is based on a multi-element group partitioning of the entire domain. The approximate inverse of the global coefficient matrix, which acts as a preconditioner, is expressed by a second-order truncated Neumann series where an additive composition of the matrix by its subdomain contributions is achieved. An important feature of this subdomain preconditioner is its local additive nature which is susceptible to parallelization.

INTRODUCTION

One of the most efficient iterative methods of solution is the preconditioned conjugate gradient (PCG). The incorporation of proper preconditioners improves the convergence properties of the method and combines some of the characteristics of the direct methods. Preconditioning techniques based on global implementation of the SSOR characteristic matrix and the incomplete Cholesky factorization have been proposed. Alternatively, preconditioners based on the element-by-element (EBE) concept seem to offer certain advantages on vector and parallel computers. This is achieved by utilizing local data structure, which is natural in the finite element framework and performing iterations on an EBE basis. Reference 1 reviews a wide range of preconditioning techniques for solving large-scale problems in computational mechanics.

Element-by-element methods were motivated by the intrinsic characteristic of finite element implementations in which global data structure can be stored and maintained at the local (element) level. Fox and Stanton[2] and Fried[3] were among the first to point out that the assembly of the coefficient matrix is not essential and that the matrix-vector product required for the CG iterations can be obtained by performing operations at the element level. Hughes and co-workers[4] and Nour-Omid and Parlett[5] were the first to address the problem of preconditioning for the solution of differential equations based on EBE implementations[6,7]. The primary purpose of the EBE applications was to keep storage requirements to a minimum, but the advent of vector and parallel computers has given to this approach a new interest especially in large-scale 3D problems[8,9].

In this work an SBS preconditioner is proposed based on a multi-element group partitioning of the entire domain. The approximate inverse of the global coefficient matrix which acts as a preconditioner is expressed by a second-order truncated Neumann series where an additive composition of the matrix by its subdomain contributions is achieved. In order to improve the efficiency of the preconditioner, in terms of number of operations, a rejection by magnitude criterion is adopted. Thus, the off-diagonal terms of the preconditioning matrix that influence to a lesser extend the outcome of its multiplication with the residual vector are rejected. A compact storage scheme is adopted which avoids storing unnecessary terms in the preconditioning matrix. An important feature of this subdomain preconditioner is its local additive nature which is inherently parallelizable without requiring complicated manipulations or any colouring schemes. Such implementation may be envisaged as mapping one or more subdomains onto each processor, while independent preconditioning matrices could be formed and operated on each processor.

DEFINITION OF ELEMENT AND GLOBAL QUANTITIES

The application of the displacement-based finite element method in linear problems results in sparse symmetric and positive definite systems

$$Ax = b \tag{1}$$

relating the global stiffness matrix, nodal displacements and nodal forces. The global nodal displacements can be related to the unassembled global nodal displacements x_e by the transformation

$$x_e = Px \tag{2}$$

where x_e is defined from the element nodal displacements in global degrees of freedom as

$$x_e = [x_{(1)}, x_{(2)}, \ldots, x_{(p)}] \tag{3}$$

and P is the global connectivity matrix which maps the global displacement vector to element displacement vectors and p is the total number of elements. Similarly the unassembled global force vector can be defined as

$$b_e = [b_1, b_2, \ldots, b_p] \tag{4}$$

where b_j ($j = 1, \ldots, p$) is related to element j force vector.

The application of the principal of virtual work gives the global stiffness and global force vector from element quantities

$$A = P^T A_e P \tag{5}$$

$$b = P^T b_e \tag{6}$$

where A_e is the unassembled block diagonal stiffness matrix

$$A_e = \lceil A_1, A_2, ..., A_p \rfloor \tag{7}$$

and A_j (j = 1, ..., p) is the element stiffness matrix. An alternative expression for A may be written in the form

$$A = \sum_{j=1}^{p} \underset{\sim}{A}_j \tag{8}$$

in which $\underset{\sim}{A}_j$ is basically the element stiffness matrix but of the same size as A with the appropriate correspondence between local and global degrees of freedom (d.o.f.). This mapping is expressed as

$$\underset{\sim}{A}_j = P^T A_j P \tag{9}$$

AN ELEMENT-BASED PRECONDITIONER

Iterative methods are widely used for the solution of linear systems where computer storage is limited. Besides, techniques based on the idea of preconditioning have significantly enhanced the efficiency of the methods and broadened their area of application. Under this concept Eq. (1) is transformed into

$$M A x = M b \tag{10}$$

where M is the transformation or preconditioning matrix.

The PCG algorithm for the solution of Eq. (10) may be expressed as follows

(1) Initialization
$x_0 = 0, \quad r_0 = b, \quad d_0 = Mr_0$

(2) $u_i = Ad_i$

$$\alpha_i = \frac{r_i^T M r_i}{d_i^t u_i}$$

$$x_{i+1} = x_i + \alpha_i d_i$$

$$r_{i+1} = r_i - \alpha_i u_i \tag{11}$$

(3) Check for convergence : $\|r_{i+1}\| < \varepsilon \|r_0\|$
else continue

(4) $\beta_i = \dfrac{r_{i+1}^T M r_{i+1}}{r_i^T M r_i}$

(5) $d_{i+1} = Mr_{i+1} + \beta_i d_i$

go to (2)

It is apparent that if $M^{-1}=A$ then the method converges under exact arithmetic in one iteration. Experience shows that a simple and efficient transformation is obtained when identifying M with the diagonal entries of A. This constitutes the diagonally scaled conjugate gradient (D-CG). Alternative transformation matrices may be constructed from approximate expressions of the inverse of A.

PRODUCT FORM EBE PRECONDITIONERS

The most widely used EBE preconditioners are based on approximate factorization techniques of element data structures. The so-called Crout EBE preconditioner is defined as follows ($B=M^{-1}$):

$$B = D^{1/2}\left(\prod_{j=1}^{p} P_j^T \widetilde{L}_j P_j\right)\left(\prod_{j=1}^{p} P_j^T \widetilde{D}_j P_j\right)\left(\prod_{j=p}^{1} P_j^T \widetilde{L}_j^T P_j\right)D^{1/2} \tag{12}$$

where $\widetilde{L}_j, \widetilde{D}_j, \widetilde{L}_j^T$ are the Crout factors of the regularized matrix $\widetilde{A}_j = I + (\overline{A}_j - \overline{D}_j)$ with $\overline{D}_j = \mathrm{diag}(\overline{A}_j)$ and $\overline{A}_j$ is the scaled element matrix

$$\overline{A}_j = P_j(D^{-1/2} A_j D^{-1/2})P_j^T \tag{13}$$

where D is the assembled global diagonal matrix D=diag(A). A drawback of the Crout preconditioning is the need to store twice as much data as for the simple CG.

An alternative preconditioner called symmetrized Gauss-Seidel EBE preconditioner

$$B = D^{1/2}\left(\prod_{j=1}^{p} P_j^T[I + L_j^s]P_j\right)\left(\prod_{j=p}^{1} P_j^T[I + (L_j^s)^T]P_j\right)D^{1/2} \tag{14}$$

with $\overline{A}_j = L_j^s + D_j^s + (L_j^s)^T$, uses the same element data as required to compute the matrix product Ad_i, thereby reducing storage requirements to 50 percent compared to Crout-EBE.

AN ADDITIVE EBE PRECONDITIONER

An EBE preconditioner based on polynomial preconditioning techniques is proposed by Papadrakakis et al[10]. The inverse of $\overline{A} = D^{-1/2} A D^{-1/2}$ is expressed as

$$\overline{A}^{-1} = (I - V)^{-1} \tag{15}$$

with

$$V = D^{-1/2}(D - A)D^{-1/2} \tag{16}$$

For $|V| < 1$ the inverse of (I-V) can be written by the polynomial expansion

$$(I - V)^{-1} = \eta_0 I + \eta_1 V + \eta_2 V^2 + \ldots + \eta_k V^k \tag{17}$$

where η_k are properly selected coefficients for accelerating convergence. Following the same definitions for the element quantities of V as for A with $V_j = D^{-1/2}(D_j - A_j)D^{-1/2}$ the inverse expression (7) may be written as

$$(I - V)^{-1} = \eta_0 I + \eta_1 \sum_{j=1}^{p} \underset{\sim}{V}_j + \eta_2 \left(\sum_{j=1}^{p} \underset{\sim}{V}_j \right)^2 + \ldots + \eta_k \left(\sum_{j=1}^{p} \underset{\sim}{V}_j \right)^k \tag{18}$$

An approximation to equation (18) is obtained by replacing $(\sum \underset{\sim}{V}_j)^k$ by $\sum (\underset{\sim}{V}_j)^k$ which is computationally more efficient in terms of operations and communication between elements. Thus, equation (18) becomes

$$(I - V)^{-1} \approx \eta_0 I + \eta_1 \sum_{j=1}^{p} \underset{\sim}{V}_j + \eta_2 \sum_{j=1}^{p} \underset{\sim}{V}_j^2 + \ldots + \eta_k \sum_{j=1}^{p} \underset{\sim}{V}_j^k \tag{19}$$

and the preconditioning matrix may now be written as

$$B^{-1} = M = D^{-1/2}(I-V)^{-1}D^{-1/2}$$

or

$$M = \eta_0 D^{-1} + \sum_{j=1}^{p} \underset{\sim}{R}_j \tag{20}$$

The inverse preconditioning matrix is formed from element quantities

$$M = \sum_{j=1}^{p} \underset{\sim}{M}_j \tag{21}$$

with $\underset{\sim}{M}_j = \eta_0 \underset{\sim}{D}_j^{-1} + \underset{\sim}{R}_j$ and $\underset{\sim}{D}_j$ the part of D which corresponds to the element d.o.f. The unassembled block diagonal approximation to the inverse of the coefficient matrix is given by

$$M_e = \lceil M_1, M_2, \ldots, M_p \rfloor \tag{22}$$

THE EBE-PCG ALGORITHM

The EBE-PCG algorithm following the definition of the previous sections may be stated as follows:

Initialize:

$x_0 = 0, \;\; r_0 = b$

$(r_e)_0 = Pr_0 \equiv \; [r_{(1)} \; r_{(2)} \cdots r_{(p)}]_0^T$

$$(z_j)_0 = M_j r_{(j)}\ ,\ j = 1,\ldots, p$$

$$(z_e)_0 = [z_1\ z_2\ \ldots\ z_p]_0^T$$

$$z_0 = P^T(z_e)_0$$

$$d_0 = z_0$$

for $i = 0,1,\ldots$

$$(u_j)_i = A_j(d_{(j)})_i\ ,\ j=1,\ldots, p$$

$$(u_e)_i = [u_1\ u_2\ \ldots\ u_p]_i^T$$

$$u_i = P^T(u_e)_i$$

$$\alpha_i = \frac{r_i^T z_i}{d_i^T u_i}$$

$$x_{i+1} = x_i + \alpha_i d_i$$

$$r_{i+1} = r_i - \alpha_i u_i$$

Convergence check:

if $||r_{i+1}|| \leq \varepsilon||r_0||$ stop

$$(r_e)_{i+1} = P r_{i+1}$$

$$(z_j)_{i+1} = M_j(r_{(j)})_{i+1}\ ,\ j = 1,\ldots, p$$

$$(z_e)_{i+1} = [z_1\ z_2\ \ldots\ z_p]_{i+1}^T$$

$$z_{i+1} = P^T(z_e)_{i+1}$$

$$\beta_i = \frac{r_{i+1}^T z_{i+1}}{r_i^T z_i}$$

$$d_{i+1} = z_{i+1} + \beta_i d_i$$

end for

The element contributions may also be considered to be contributions by groups of elements or subdomains depending on the partition of the structure into a number of subdomains. This concept is adopted in this study and is equivalent to an overlapping block partitioning in which each block corresponds to the assembled coefficient matrix of each subdomain.

POLYNOMIAL PRECONDITIONING

Polynomial preconditioning consists of choosing a polynomial $s(\overline{A})$ to replace $\overline{A}^{-1}$ The simplest polynomial expansion of equation (17) is given by the Neumann series[11] with $n_0=\ldots=n_k=1$: $s(\overline{A}) = I + V + V^2 + \ldots$. More efficient polynomials can be obtained by making the spectrum of the preconditioned matrix as close as possible to that of the identity. One way to achieve this is to minimize the residuals $\|I - s(\bar{A})\bar{A}\|$, where $\|.\|$ represents the L_2-norm.

According to Johnson et al.[12] and Saad[13] this can be achieved by defining the best polynomial which minimizes

$$\max_{\lambda \in [a,b]} |1 - \lambda s(\lambda)| \tag{24}$$

where [a,b] is some interval containing the eigenvalues of $\bar{A}$, with $0<a<b$. The Chebychev polynomials or alternatively a least squares minimization procedure may be implemented to solve this problem. The latter approach takes the form: Find s that minimizes $\|1 - \lambda s(\lambda)\|_w$, where $\|.\|_w$ is the L_2-norm on [a,b] with respect to some weight function $w(\lambda)$. It has been observed in Reference 13 that least squares polynomials tend to perform better than those based on the Chebychev ones, in that they lead to a better overall clustering of the spectrum.

Following the derivation of Saad[13] the parameters a and b are computed as the Gershgorin estimates of λ_{min} and λ_{max}. However, since the Gershgorin lower bound "a" may be nonpositive even when $\bar{A}$ is positive definite, the least squares polynomials are thus defined in the interval [0,b]. The computed first two polynomials take the form

$$S_1(\lambda) = -\frac{5}{4}b + \lambda \tag{25}$$

$$S_2(\lambda) = \frac{7}{8}b^2 - \frac{7}{4}b\lambda + \lambda^2 \tag{26}$$

and the coefficients in eq(17) correspond to the coefficients of λ in equations (25) and (26), respectively. Thus for the first order approximation $\eta_0 = -5b/4$, $\eta_1 = 1$ and for the second order approximation $\eta_0 = 7b^2/8$, $\eta_1 = -7b/4$, $\eta_2 = 1$.

The main attraction of polynomial preconditioning is that the only operations involving matrices are products with vectors. As a consequence of this, the performance of this preconditioning strategy in vector and parallel machines is facilitated and portability is achieved by optimizing matrix by vector multiplications only[14]. The performance of polynomial preconditioning is enhanced on parallel machines with large number of processors. On the other hand, global incomplete factorization preconditioning exhibits poor performance during forward and backward substitutions in vector and parallel machines. To alleviate this inefficiency, approximations to the inverse factors are obtained by also using polynomial expansions[15]. More general type polynomial preconditioners may be formed when combined with k steps of a relaxation type iterative method. Application of polynomial preconditioning in structural analysis problems has been recently presented in References 10,16.

THE HANDLING OF THE PRECONDITIONING MATRIX

A preconditioner has to combine the following characteristics in order to be effective: (i) to reduce as much as possible the ellipticity of the coefficient matrix; (ii) to be easily computed and handled by the computer; (iii) to be sparse in order to reduce storage demands and keep matrix-vector operations at low cost; (iv) to be susceptible to parallelization with reduced communication overheads between the processors. Since some of these requirements are conflicting a dynamic balance has to be kept among them according to the type of problem and the available hardware. The present SBS preconditioner has the following features: A second order approximation is adopted for the definition of the preconditioning matrix from eq. (20). The elemental contributions are considered to be contributions by groups of elements or subdomains depending on the partition of the structure into a number of subdomains. This is equivalent with an overlapping block partitioning in which each block corresponds to the assembled coefficient matrix of each subdomain without performing any condensation. Thus, operations are performed at the level of each subdomain j (j = 1, ..., p) and the formation of the global coefficient matrix is avoided. The replacement of $\left(\sum_{j=1}^{p} \underline{Y}_j\right)^k$ by $\sum_{j=1}^{p} (\underline{Y}_j)^k$ aims in reducing the multiplications between subdomains in order to keep operations and communication overheads to a minimum.

The handling of $\underline{R}_j$ in Eq. (20) is another important factor for the efficiency of the preconditioner. In order to avoid operations with the components of $\underline{R}_j$ in each PCG iteration, this matrix is formed only once and stored before the iterative procedure begins. This in fact is the only additional storage required for the preconditioning strategy proposed. In an effort for further reducing the operations involved in the multiplication $M_j\{r_{(j)}\}_i$, a rejection by magnitude criterion is adopted from the off-diagonal entries of M. Thus, the off-diagonal entries of $\underline{R}_j$ are obtained by adding those elements which influence to a lesser extend the outcome of the multiplication $M_j\{r_{(j)}\}_i$. The rejection criterion is stated as

$$(R_j)_{kl}^2 \leq \psi\, D_{kk}^{-1} D_{ll}^{-1} \qquad (27)$$

which controls the magnitude of the elements of $\underline{R}_j$ with respect to the magnitude of the diagonal elements of M. The control parameter ψ takes the values $0 \leq \psi \leq 1$. In this study the coefficient matrix is stored in skyline form, while the corresponding preconditioning matrix is stored in compact form.

PARALLEL COMPUTER IMPLEMENTATION

The numerical results corresponding to the parallel implementation of the proposed methods are performed on the Alliant FX/80 computer. The Alliant FX/80 is a mini-supercomputer consisting of two kind of processors: (i) Computational Elements (CEs) which have vector and concurrent capabilities and (ii) Interactive Processors (IPs) which serve as front-end processors running the tasks of the operating system (Concentrix) and performing all I/O operations. In order to enhance memory read-write operations, all processors have access to cache memory. A special hardware called cross-bar is used to allow access for all CEs to any cache port. All programs except operating system tasks are running onto the CEs which are connected by a concurrency bus. Each CE can run either in a detached mode or in complex mode with other CEs. The FORTRAN compiler of the Alliant automatically detects both vectorization and concurrency and is perhaps the most advanced one in the computer industry in this respect. This operation, however, is done in a conservative manner. The programmer can use directives to instruct the compiler to perform certain optimizations. The most common one is to inform the compiler that there is no data dependency in a loop.

The most consuming parts in terms of CPU time are the multiplications of both the coefficient matrix and the inverse of the preconditioner by a vector. The present implementation utilizes the SBS nature of both the coefficient matrix and the preconditioner. The subdomain coefficient matrix is stored by the skyline scheme. Thus , the multiplication by a vector involves addressings which inhibit the proper parallelization and vectorization. In order to circumvent this problem, a concurrently callable subroutine has been written to perform the multiplication of the subdomain coefficient matrix by the corresponding part of the global vector:

```
cvd$ cncall
     do 10 i=1,n
          call ebemultiply(...)
  10 continue
```

In this subroutine only vectorization is used. After all subdomain multiplications have been performed, a loop gathers the subdomain contributions to the global vector. Concurrency and vectorization can be employed only inside the subdomain loops to avoid write conflict on the same element of the global vector. The same approach is used for the multiplication of the inverse of the preconditioning matrix, stored by a compact storage scheme, by a vector. This scheme is used to reduce storage requirements and to avoid unnecessary operations.

NUMERICAL RESULTS

In order to investigate the behavior and efficiency of the additive SBS preconditioner, a series of structural mechanics test examples have been performed. A plane frame and a cantilever beam subjected to transverse end loads as well as a simply supported plate under a central point load are examined. The plane frame depicted in Figure 1 has 64 storeys, 7 bays, two shear walls with 1560 d.o.f. The shear walls are used to increase the ellipticity of the coefficient matrix. The kinematic boundary conditions of half the cantilever beam example are illustrated in Figure 2 for a mesh with 4x16 plane strain elements, however all numerical results reported are for a mesh containing 16x32 elements (1122 d.o.f.). The plate example is discretized with a mesh of 16x16 elements (855 d.o.f.). We will index the results of the last two test examples by the element aspect ratio (ratio of the largest dimension to the smallest) ignoring discretization errors. The length of both structures is held constant but the other dimension is varied to alter the condition number of the stiffness matrix. The ill-conditioning is increased with large element aspect ratios.

Figure 3 shows the number of iterations, the CPU time (sequential implementation) and the elements stored for A and M of SBS-CG ($\psi = 10^{-3}$) and SBS-D-CG, implemented on different number of subdomains for the plane frame example. In estimating the storage the elements of the stiffness matrix are stored in skyline form and the elements of the preconditioner in compact form. Table 1 depicts the behavior of SBS-CG for different values of ψ corresponding to a division into 32 subdomains. The iterative procedure is terminated when the residual vector divided by the norm of the force vector is less than or equal to 10^{-6}.

Table 2 shows the performance of the methods for the concurrent and vector-concurrent modes of the 8 subdomain subdivision. The first column refers to the automatic detection of vectorization and concurrency by the computer of the sequential version of algorithm, while the second column refers to the optimized algorithm specially tailored for parallel and vector implementation. Table 3 shows the performance of the methods for different number of subdivisions for the concurrent-vector mode of the computer. The results clearly indicate the superiority of the proposed SBS formulation both in terms of storage and computing time.

The results depicted in Figures 4 and 5 provide a comparison of the SBS(ψ) preconditioner with the diagonally scaled CG ("D") (convergence tolerance $\varepsilon=10^{-5}$) for examples 2 and 3 respectively. Different values of the rejection parameter ψ which controls the number of retained elements in Rj of equation (20) are used for the SBS preconditioner. The SBS formulation of algorithm (23) is implemented with 32 subdomains for the beam and 8 subdomains for the plate. Each subdomain contains 68 and 135 d.o.f. respectively. The results clearly indicate that the superiority of the SBS preconditioner is improved with the increase of the ellipticity of the problem.

ACKNOWLEDGEMENTS

The authors are grateful to Prof. K. Papailiou for providing the computing facilities of the Laboratory of Thermal Turbomachines, National Technical University of Athens and to its staff for their assistance.

REFERENCES

1. M. Papadrakakis, "Solving large-scale linear problems in solid and structural mechanics", in M. Papadrakakis (ed.) Solving Large Scale Problems in Mechanics, J. Wiley, 1993, pp. 1-37.

2. R.L. Fox and E.L. Stanton, "Developments in structural analysis by direct energy minimization", AIAA J., 6, 1036-1042 (1968).

3. I. Fried, "More on gradient iterative methods in finite element analysis", AIAA J., 7, 565-567 (1969).

4. T.J.R. Hughes, J. Winget, I. Levit and T. Tezduyar, "New alternating direction procedures in finite element analysis based upon EBE approximate factorizations", in S.N. Atluri and N. Perrone (eds.), Computer Methods for Nonlinear Solids and Structural Mechanics, AMD-Vol. 4, ASME, New York, 1983, pp. 75-109.

5. B. Nour-Omid and B.N. Parlett, "Element preconditioning using splitting techniques", SIAM J. Sci. Stat. Comput., 6, 761-770 (1985).

6. M. Ortiz, P.M. Pinsky and R.L. Taylor, "Unconditionally stable element-by-element algorithm for dynamic problems", Comp. Meth. Appl. Mech. Enging., 59, 327-362 (1986).

7. T.J.R. Hughes, I. Levit and J.M. Winget, "Implicit stable algorithms for heat conduction analysis", J. Enging. Mech. Div., ASCE, 109, 576-585 (1983).

8. M. Papadrakakis and M.C. Dracopoulos, "A global preconditioner for the element-by-element solution methods", Comp. Meth. Appl. Mech. Enging., 88, 275-286 (1991).

9. T.J.R. Hughes, R.M. Ferencz and J.O. Hallquist, "Large scale vectorized implicit calculations in solid mechanics on a Cray X-MP/48 utilizing EBE preconditioned conjugate gradients", Comp. Meth. Appl. Mech. Enging., 61, 215-248 (1987).

10. M. Papadrakakis, N. Bitoulas and K. Hatzikonstantinou, "An efficient superelement-by-superlement method for large finite element computations", Computing Systems in Enging., 2, 535-540 (1991).

11. P.F. Dubois, A. Greenbaum and G.H. Rodrigue, "Approximating the inverse of a matrix for use in iterative algorithms on vector processors", Computing, 22, 257-268 (1979).

12. O.G. Johnson, C.A. Michelli and G. Paul, "Polynomial preconditioners for conjugate gradient calculations", SIAM J. Numer. Anal., 20, 362-376 (1983).

13. Y. Saad, "Practical use of polynomial preconditionings for the conjugate gradient method", SIAM J. Sci. Stat. Comput., 6, 865-881 (1985).

14. Y. Saad, "Krylov subspace methods on supercomputers", SIAM J. Sci. Stat. Comput., 10, 1200-1232 (1989).

15. O. Axelsson, "A survey of vectorizable preconditioning methods for large scale finite element problems", Report CNA-190 University of Texas, Austin, Texas, 1984.

16. A.L.G.A. Coutinho, J.L.D. Alves, N.F.F. Ebecken and L.M. Troina, "Conjugate gradient solution of finite element equations on the IBM 3090 vector computer utilizing polynomial preconditionings", Comp. Meth. Appl. Mech. Enging., 84, 129-145 (1990).

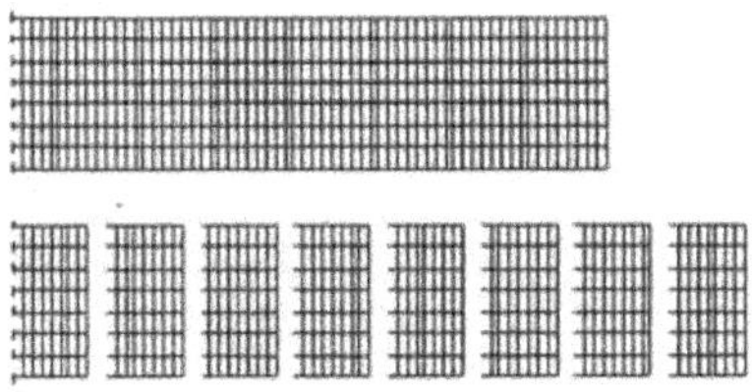

Figure 1. Example 1: Plane frame (1560 d.o.f.) and division to 8 subdomains

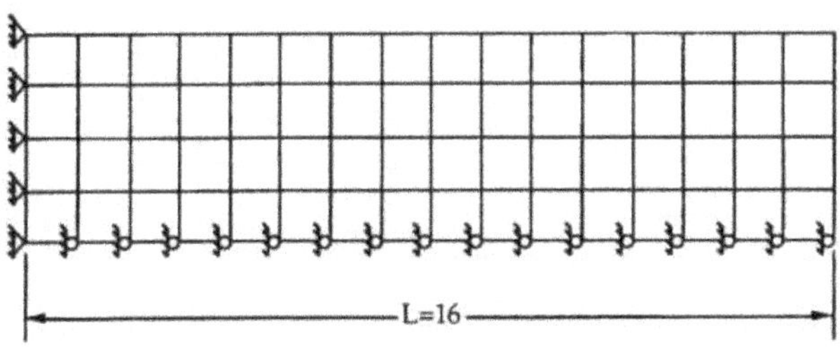

Figure 2. Example 1: Kinematic boundary conditions for the plane strain cantilever beam

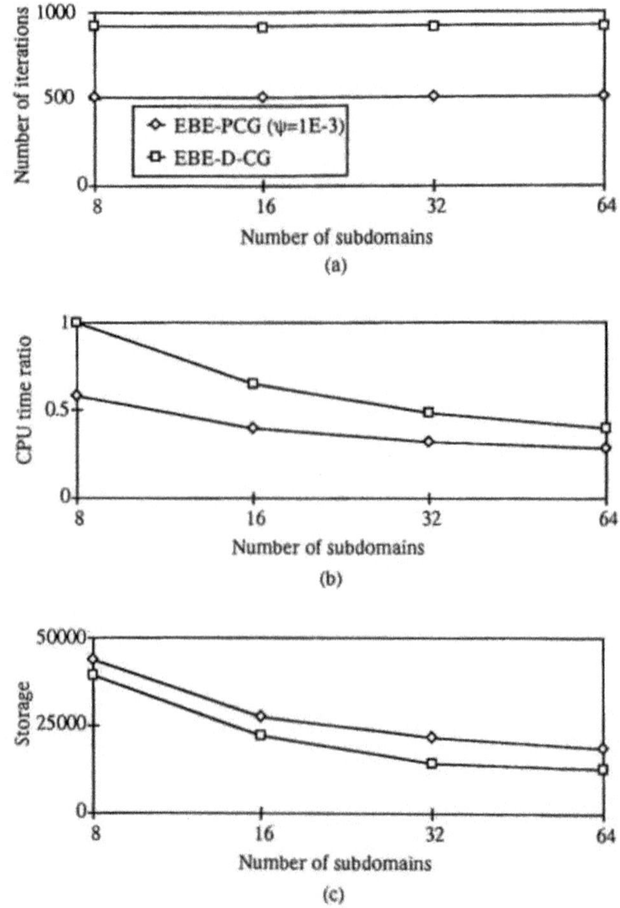

Figure 3. Example 1: (a) Number of iterations, (b) CPU time ratio (sequential mode), (c)Elements stored in A and M

ψ	0.0	10^{-4}	10^{-3}	10^{-2}	10^{-1}
Iterations	478	504	504	517	588
Time ratio	1	0.86	0.82	0.80	0.86
Storage	14976	5636	4169	3110	1634

Table 1. Example 1: Performance of SBS-CG for different values of ψ for 32 subdomain (sequential mode).

Number of processors	Concurrent mode		Concurrent-vector mode	
	1	2	1	2
1	672	911	446	228
2	505	323	291	79
4	331	173	257	49

Table 2. Example 1: Performance of SBS-CG ($\psi=10^{-3}$) for concurrent and concurrent-vector modes. Column 1: Automatic detection of vectorization and concurrency. Column 2: Optimized code.

Number of processors	Number of subdomains				Number of iterations
	8	16	32	64	
SBS-PCG ($\psi=10^{-3}$)					
1	142	110	97	114	504
2	79	63	55	63	504
4	49	38	33	37	504
SBS-D-CG					
1	170	129	114	136	927
2	92	71	65	74	927
4	56	43	40	45	927

Table 3. Example 1: Performance of SBS-PCG ($\psi = 10^{-3}$) and SBS-D-CG for different number of subdomains.

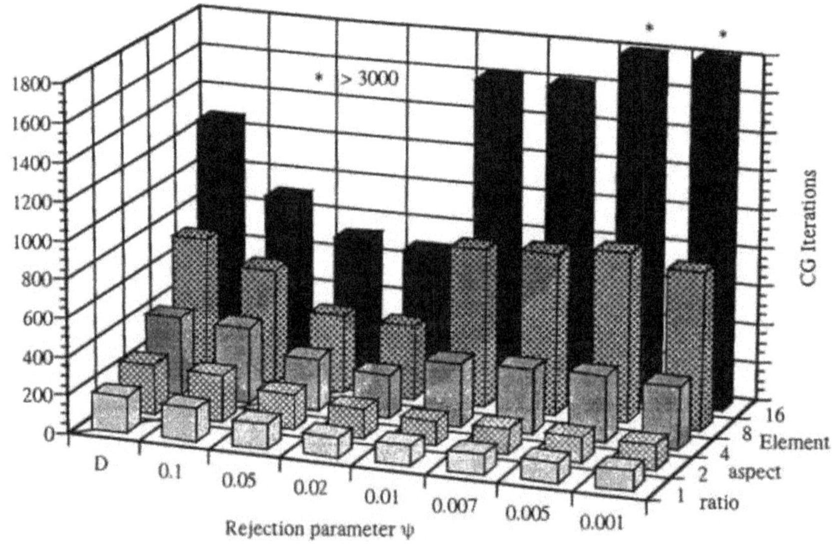

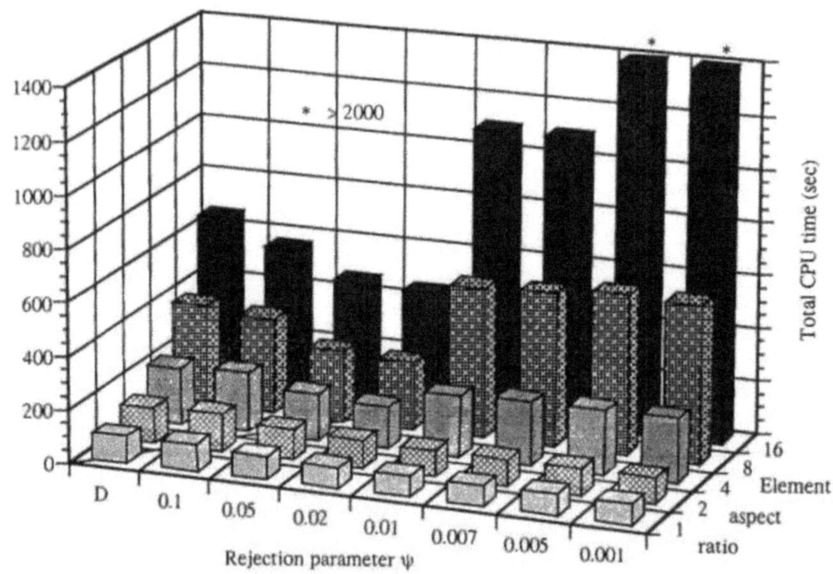

Figure 4. Example 2: Performance of diagonal and SBS(ψ) preconditioners versus element aspect ratio for the plane strain cantilever beam (16x32 elements) with 32 subdomains.

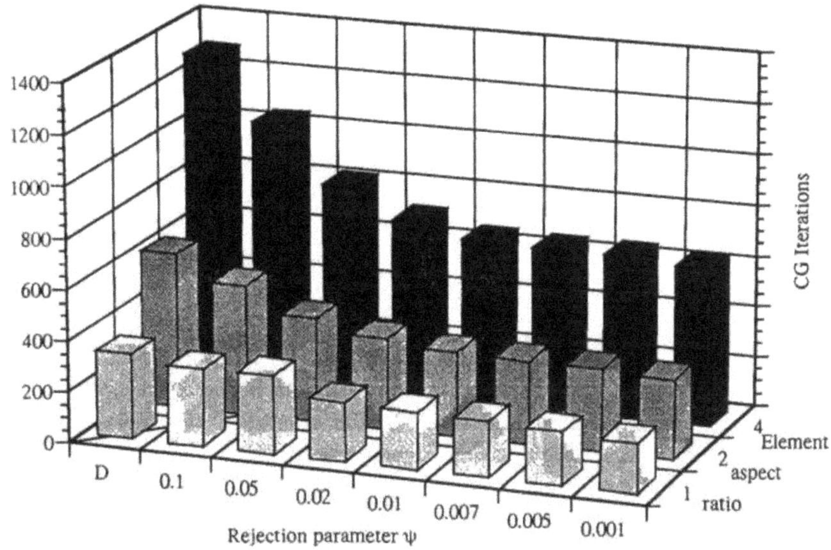

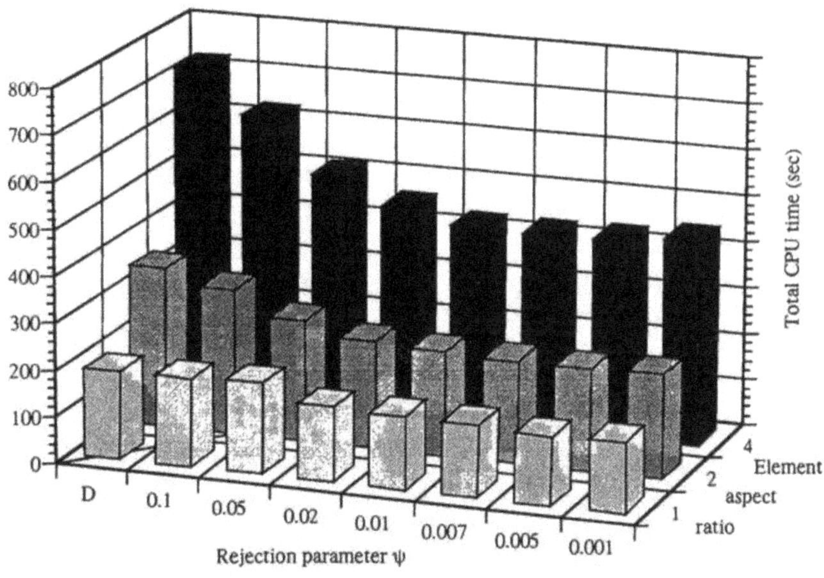

Figure 5. Example 3: Performance of diagonal and SBS(ψ) preconditioners versus element aspect ratio for the plate problem (16x16 elements) with 8 subdomains.

PARALLEL ALGORITHMS FOR INTEGER FACTORISATION

RIZOS SAKELLARIOU
Department of Computer Science, University of Manchester
Oxford Road, Manchester M13 9PL, U.K.

Abstract

Factorisation of long integers is a key issue in cryptographic applications and there has been much research in the field during recent years. In this paper we describe several of the most important integer factorisation algorithms; we start with the classical trial division and Fermat's algorithms, then we describe Pollard's and the Elliptic Curve probabilistic methods and, finally, the family of quadratic sieve algorithms. Possible uses of parallelism, both at the high and low level of these algorithms, are presented.

1 Introduction

The art of decomposing integers into their prime factors, known simply as *factorisation*, has a long tradition and many famous mathematicians have been involved with the subject. The advent of electronic computers permitted the factorisation of long integers but, even today, numbers around 100 digits or more are still considered difficult for factorisation and the problem remains "so elegant and so celebrated", as Gauss described it in his *Disquisitiones Arithmeticae*.

During the last 15 years interest in factoring algorithms has been renewed; public-key cryptography [2] and the RSA cryptosystem [3, 14], whose secrecy is based on the difficulty of factoring a large integer consisting of two almost equal primes, have stimulated much research in the field. The use of parallelism, i.e. the execution of more than one operation at a time, has been a practical way to improve the performance of these algorithms and some attempts have been made to parallelise parts of the existing sequential algorithms. Speedup, which is defined as the ratio between the time taken for the execution of the algorithm on one processor and the time taken for the execution on P processors, is the most common criterion for their performance [4].

Factorisation algorithms are usually divided into *special purpose factorisation algorithms*, which have a running time dependent mainly on the size of the factors of n [1], and *general purpose factorisation algorithms*, which have a running time dependent mainly on the size of n. Several factorisation algorithms are examined in the following sections; for more comprehensive surveys the reader is referred to [5, 13, 17].

2 Trial division

Trial division consists of making trial divisions of n by successive primes $p = 2, 3, 5, \ldots$ until discovering the smallest p for which $n \bmod p = 0$. Then p is the smallest prime factor of n, and the same process may be applied to n/p, dividing this by p and higher primes. Whenever we find that $n \bmod p \neq 0$ but $\lfloor n/p \rfloor \leq p$, we conclude that n is prime.

Since the number of primes less than x asymptotically tends to $x/\ln x$, trial division is totally impractical for factorising large integers; it has a worst-case time complexity of $O(n^{1/2})$ for the complete factorisation of n. However, the method should be preferred for finding small factors, up to 10^8 [13].

Trial division can be easily parallelised; using P processors, up to P trials can be performed in parallel and, in certain cases, an almost linear speedup can be achieved.

3 Fermat's algorithm

This algorithm, first used by Pierre de Fermat in 1643, tries to represent an odd composite integer n as the difference of two squares, say x^2, y^2; because $x^2 - y^2 = (x + y)(x - y)$, $x + y$, $x - y$ are factors of n. In its simplest form, the method examines all increasing values for x, starting with $\lfloor \sqrt{n} \rfloor + 1$, until there is one for which $x^2 - n$ is a perfect square. A modification, due to Knuth [5], removes the necessity of computing

[1]Throughout this paper n denotes a number to be factored.

the square root and the main loop of the algorithm can be performed using only additions and subtractions.

Fermat's algorithm is fast when n is the product of two almost equal factors but extremely slow in any other case. For instance, if n is composed of two factors $p_1 \simeq n^{1/3}$ and $p_2 \simeq n^{2/3}$, the complexity of the method is $O(n^{2/3})$, which makes it even worse than trial division. This result led Lehman [6] to look for the factorisation of $\sqrt{kn}$ where k is a random integer. Ideally, if $k = (p_1/p_2 + p_2/p_1 + 2)/4$, where $n = p_1 \cdot p_2$, Fermat's algorithm finds the factorisation immediately. However, since nothing is known in advance for the ratio of p_1, p_2, only random suggestions for the value of k can be made.

Lehman's variation is suitable for parallel implementation; each processor may work with a different value for k but, given the worst case complexity of Fermat's algorithm, the results are not impressive.

4 Pollard's and Elliptic Curve methods

The algorithms presented in this section are based on statistical techniques and they are powerful for discovering small factors.

Pollard's "$p-1$" method [10], can find, with little computing, a large factor, p, of n if $p-1$ splits entirely, or almost entirely, into a product of small primes. The method proceeds according to the following steps:

1. Choose an arbitrary integer a, where $2 \leq a \leq n-2$, and an integer k which is divisible by many small primes.
2. Compute $a^k \bmod n$ and then $d = \gcd(a^k - 1, n)$.
3. If d is not a factor of n, repeat with different a, k.

Pollard's second algorithm [11] is known as the "rho" or "Monte-Carlo" method of factorisation. Using an iteration of the form $x_{i+1} = f(x_i) \bmod n$, $i \geq 0$, one expects to find the smallest factor p of n after $O(p^{1/2})$ operations. The function $f(x)$ can be any nonlinear function; in practice, a quadratic polynomial $f(x) = x^2 + a$ ($a \not\equiv 0, -2 \bmod n$) is often used. The algorithm proceeds as follows:

1. Choose x_0 and a ($a \not\equiv 0, -2 \bmod n$) in the recursive formula $x_{i+1} \equiv x_i^2 + a \bmod n$, $i \geq 0$.

2. According to the previous formula, generate triples (x_i, x_{2i}, Q_i), $i = 1, 2, \ldots$, where $Q_i = \prod_{j=1}^{i}(x_{2j} - x_j) \pmod{n}$.
3. In short periods compute $d = \gcd(n, Q)$.
4. The algorithm stops after a predefined number of iterations or when $x_{2k} = x_k$.

In both Pollard's methods the internal loops are serial. At best, parallelism can be exploited by running the algorithms using different initial values; even in this case the gains are not considerable. For example, in the "rho" method, using P processors working on P different sequences, there is a sublinear speedup of $O(P^{1/2})$.

The recently invented Elliptic Curve Method (ECM) is regarded as a generalisation of Pollard's "$p-1$" algorithm [7]. An elliptic curve is a pair of elements (a, b) for which $4a^3 + 27b^2 \neq 0$. A point on an elliptic curve is a pair of elements (x, y) which satisfy the equation $y^2z = x^3 + axz^2 + bz^3$. One performs a computation similar to Pollard's "$p-1$" method by selecting an elliptic curve and a point x in the group of points of this elliptic curve and then compute x^k where k is the product of all primes less than a selected bound [8].

While the method has a good expected running time, it has been conjectured that it is unlikely to find factors larger than 40 digits [2]. In common with Pollard's methods, ECM can be performed in parallel by running the algorithm on different processors using different elliptic curves. Although a linear speedup can be achieved theoretically, storage requirements limit the speedup in practice.

5 Quadratic Sieve Algorithms

A generalisation of the idea behind Fermat's algorithm, originally due to Legendre, leads to much more efficient factoring methods. Legendre observed that finding x and y satisfying the congruence $x^2 \equiv y^2 \bmod n$, where $x \not\equiv \pm y \bmod n$, gives a factor of n by computing $\gcd(x + y, n)$ or $\gcd(x - y, n)$. This is because n divides $(x^2 - y^2) = (x + y)(x - y)$, whilst n does not divide $x + y$ or $x - y$; thus $\gcd(x + y, n)$ must be a proper factor, p, of n, and then n/p divides $\gcd(x - y, n)$.

In 1926, M. Kraitchik published a general method for factoring using Legendre's congruence [12]. The steps of his method are :

1. Generate a number of congruences $u \equiv v \bmod n$.
2. Factorise u and v for the above congruences.
3. Determine a subset of the factored congruences which can be multiplied to produce a congruence of the form $x^2 \equiv y^2 \bmod n$.
4. Obtain a factor of n from the computation $\gcd(x \pm y, n)$.

Since the running time of quadratic sieve algorithms depends mainly on the size of n, they have been extensively used for factoring integers used in cryptographic applications. The most common variations are presented in the following paragraphs.

The Continued Fraction Factoring Algorithm : This was the first algorithm in this category which was successfully implemented on a computer, leading to the complete factorisation of F_7 [9]. Using the continued fraction expansion of $\sqrt{n}$ one can produce congruences of the form $x^2 \equiv \pm v \bmod n$ which can be combined to solve Legendre's congruence. Although it has been suggested that the computation of these congruences can be done in parallel [18, 19], no implementations have been reported and it seems that the overall gain using parallelism is small.

The Quadratic Sieve Factoring Algorithm : The Quadratic Sieve resembles the Continued Fraction Algorithm in that both algorithms produce pairs x,v with $x^2 \equiv \pm v \bmod n$. The methods differ in the generation of these pairs and the method of factoring the v's. The Quadratic Sieve uses a polynomial $Q(x) = (\lfloor\sqrt{n}\rfloor + x)^2 - n$ from which results the congruence $(A + \lfloor\sqrt{n}\rfloor)^2 \equiv Q(A) \bmod n$, where $Q(A) < 2\sqrt{n}|A| + A^2$ and $Q(A) > 0$ for $A \neq 0$. The numbers $Q(A)$ are factored and, applying Gaussian elimination, one can produce a solution to Legendre's congruence [12].

Some variations of the algorithm have been developed, the most important of which is the multiple polynomial quadratic sieve [16]. This uses different polynomials, of the form $Q(x) = ax^2 + bx + c$, where $b^2 - 4ac = kn$, for the first, time-consuming stage of the algorithm, thus permitting parallel calculation of the congruences. This version has been successfully implemented in parallel by Lenstra and Manasse [8]. They distributed the task of computing the congruences for the first part of the algorithm among various computers around the world which were communicating via electronic mail; using cycles of the machines that would otherwise have been idle, they factorised a 100-digit integer within a month with almost free computational cost. However, the parallel part of the algorithm contributed only 15% of the overall work.

The Number Field Sieve has been proposed recently and it has been conjectured that this is the fastest general purpose factoring algorithm [2]. The collection of the residues is performed using properly selected polynomials. Although this stage is parallelisable, the final stage of Gaussian elimination remains sequential. It is worth noting that the algorithm is currently being used in an attempt to factor the 129-digit integer which was proposed in 1977 as being "safe" for the RSA cryptosystem [3].

6 Exploiting Low Level Parallelism

Implementations of factorisation algorithms on computers are based on especially designed software routines for handling large integers. Routines for the implementation of the four basic arithmetic operations have been analysed in [1, 5] and it has been recognised that the optimisation of any factorisation algorithm relies mainly on the optimisation of these operations. While it is accepted that these routines are parallelisable (i.e they present *fine-grain parallelism*), only a few attempts have been reported.

Experiments by the author on the KSR1, a virtual memory MIMD parallel machine, have shown that parallelism can result in a small speedup of the basic arithmetic operations for integers larger than around 100 digits [15]. For shorter integers there is no speedup, the main reason being the need for frequent communication and synchronisation between processors, leading to overheads. The use of modular arithmetic for the representation of integers is more amenable to parallelism but modular representation presents other disadvantages (e.g. it is difficult to compare integers and to perform division; for an analysis see [1, 5]).

7 Conclusions

While most of the factorisation algorithms presented in the previous sections are parallelisable, the practical gains are not significant. Trial division and Fermat's method can achieve a linear speedup but their high worst case complexity renders them unusable. Parallelism in probabilistic methods is exploited by running the algorithms on different machines using different initial values; although the theoretical speedup is linear, the probability of finding a factor does not increase proportionally. Quadratic sieve algorithms have an initial stage where parallelism can be exploited effectively, but the final stage is highly sequential. On the other hand, low level parallelism is limited by the overheads associated with the need for frequent communication between processors. However, current research in parallel computing tends to reduce these overheads, hence facilitating increasingly fine grain tasks.

Acknowledgements

I wish to thank Professor John Gurd of the Department of Computer Science at Manchester University for his kind assistance and comments during the preparation of this paper.
The author has a grant from the Greek State Scholarships Foundation.

References

[1] A.V.Aho, J.E.Hopcroft, J.D.Ullman, *The Design and Analysis of Computer Algorithms*, Addison-Wesley, 1976.

[2] T.Beth, M.Frisch, G.J.Simmons (eds.), *Public-Key Cryptography*, LNCS **578**, Springer-Verlag, 1992.

[3] M.Gardner, "A new kind of cipher that would take millions of years to break", *Scientific American*, **237** (Aug. 1977), pp. 120–124.

[4] R.W.Hockney, C.R.Jesshope, *Parallel Computers 2*, Adam Hilger, 1988.

[5] D.E.Knuth, *The Art of Computer Programming, vol. 2 : Seminumerical Algorithms*, Addison-Wesley, 2nd edition, 1981.

[6] R.S.Lehman, "Factoring Large Integers", *Mathematics of Computation*, **28** (1974), pp. 637–646.

[7] H.W.Lenstra Jr., "Factoring integers with elliptic curves", *Annals of Mathematics*, **126** (1987), pp. 649–673.

[8] A.K.Lenstra, M.S.Manasse, "Factoring with two large primes", *Advances in Cryptology, Proc. of EUROCRYPT '90*, I.B.Damgard (ed.), LNCS **473**, Springer-Verlag, 1990, pp. 72–82.

[9] M.A.Morrison, J.Brillhart, "A Method of Factoring and the Factorization of F_7", *Mathematics of Computation*, **29** (1975), pp. 183–205.

[10] J.M.Pollard, "Theorems on Factorization and Primality Testing", *Proc. Cambridge Philosophical Society*, **76** (1974), pp. 521–528.

[11] J.M.Pollard, "A Monte Carlo method for factorization", *BIT*, **15** (1975), pp. 331–334.

[12] C.Pomerance, "The Quadratic Sieve Factoring Algorithm", *Advances in Cryptology, Proc. of EUROCRYPT '84*, T.Beth et al. (eds.), LNCS **209**, Springer-Verlag, 1985, pp. 169–182.

[13] H.Riesel, *Prime Numbers and Computer Methods for Factorization*, Birkhäuser, 2nd printing 1987, chapter 5, pp. 146–222.

[14] R.L.Rivest, A.Shamir, L.Adleman, "A Method for Obtaining Digital Signature and Public Key Cryptosystems", *Communications of the ACM*, **21** (1978), pp. 120–126.

[15] R.Sakellariou, MSc Thesis, Department of Computer Science, University of Manchester (in preparation).

[16] R.D.Silverman, "The Multiple Polynomial Quadratic Sieve" *Mathematics of Computation*, **48** (1987), pp. 329–339.

[17] S.S.Wagstaff Jr., J.W.Smith, "Methods of Factoring Large Integers", *Lecture Notes in Mathematics*, **1240**, Springer-Verlag, 1987, pp. 281–303.

[18] H.C.Williams, M.C.Wunderlich, "On the Parallel Generation of the Residues for the Continued Fraction Factoring Algorithm", *Mathematics of Computation*, **48** (1987), pp. 405–423.

[19] M.C.Wunderlich, "Implementing the Continued Fraction Factoring Algorithm on Parallel Machines", *Mathematics of Computation*, **44** (1985), pp. 251–260.

FROM PARALLEL TO SEQUENTIAL: KEEPING OPTIMALITY IN ALGORITHMS

Panagiotis Metaxas* Michael Mytilinaios[†]
Wellesley College[‡] Athens University of Economics[§]

Abstract

In this paper we first give a survey that describes the reductions between several fundamental parallel graph problems. Then, we examine the possibility of designing simple and practical sequential algorithms from parallel ones. Interestingly enough, we show how to design fast (in fact *optimal*) sequential algorithms from the optimal parallel algorithms for a graph theoretic problem that arizes often in many areas including Economics and Operations Research In particular, we give two optimal algorithms that solve the *vertex updating for a minimum spanning tree* problem. Finally, we show how these algorithms can be used to obtain incremental algorithms for computing the minimum spanning tree (MST) of a graph. The MST algorithms are optimal for dense graphs.

1 Introduction

Parallelizing Sequential Algorithms. Ever since the first attempt of building parallel machines, researchers have looked into the existing sequential algorithms in an effort to parallelize them efficiently. Unfortunately, most of the time, this proved to be a difficult task, because some of the basic sequential algorithms turned out to be inefficient in their parallel versions. Breadth-first search of a graph, one of the basic graph searching techniques, has a fast parallel version which runs in $O(\log^2 n)$ time, where n is the number of graph vertices. However, this algorithm uses a large number (almost n^3) of processors, making the solution impractical. Others, like the well known depth-first search algorithm, do not seem to have a corresponding fast

*Email address: takis@bambam.wellesley.edu
[†]Email address: mmit@isosun.ARIADNE-t.gr
[‡]Department of Computer Science, 106 Central Street, Wellesley, MA 02181, USA.
[§]Department of Informatics, 76 Patission Street, Athens, Greece 104 34.

parallel version at all. (By "fast" we mean an algorithm running in polylogarithmic parallel time using a polynomial number of processors.)

Reductions. These observations suggest that the reductions between parallel graph algorithms do not resemble those of sequential algorithms. The reduction graph of Figure 1 outlines a survey that depicts this situation. In this graph, an arc from problem X to problem Y implies that Y uses as subroutine an algorithm that solves X. A label on the graph refers to the paper in the bibliography that establishes the reduction. A preliminary version of this survey appeared in [Met91].

Optimality and Efficiency. Before we proceed, we need a few definitions. We say that a sequential algorithm for some problem of size n is *optimal* if it runs in time that matches a lower bound for the problem to within a constant factor. Let $t(n)$ denote the parallel running time for some parallel algorithm, and $p(n)$ denote the number of processors employed by the algorithm. Then, $w(n) = t(n) \times p(n)$ denotes the *work* performed by the algorithm. A parallel algorithm for some problem is said to be *optimal* if it has polylogarithmic parallel running time and the work $w(n)$ performed by the algorithm is $O(T(n))$, where $T(n)$ is the running time of the best sequential algorithm for the same problem. If, instead, $w(n)$ is within a polylogarithmic factor of the optimal speedup, the algorithm is called *efficient*.

It is well known that a parallel algorithm that performs work $w(n)$ can be simulated on a sequential machine in time $O(w(n))$. Such simulations, however, have usually a large overhead because the uni-processor machine has to switch context very often in order to simulate a single step on each processor. Thus, these simulations are considered impractical and complicated. In our approach, we will be looking for algorithms for which we can separate the *computing* and *scheduling* parts. We can think of the computing part as the "heart" of the algorithm, i.e. the part of the code that the solution is computed. On the other hand, the scheduling part is the code that "directs" the solution. In the algorithms we present, the computing part is identical in both the parallel and the sequential algorithms.

In the remaining of this paper we examine the possibility of designing simple and practical sequential algorithms from parallel ones. Interestingly enough, we show how to design fast (in fact optimal) sequential algorithms from the optimal parallel algorithms for a graph theoretic problem that arizes often in many areas including Economics and Operations Research. In particular, we give two optimal algorithms that solve the vertex updating for a MST problem. Finally, we show how these algorithms can be used to obtain incremental algorithms for computing the minimum spanning tree (MST) of a graph. The MST algorithms are optimal for dense graphs.

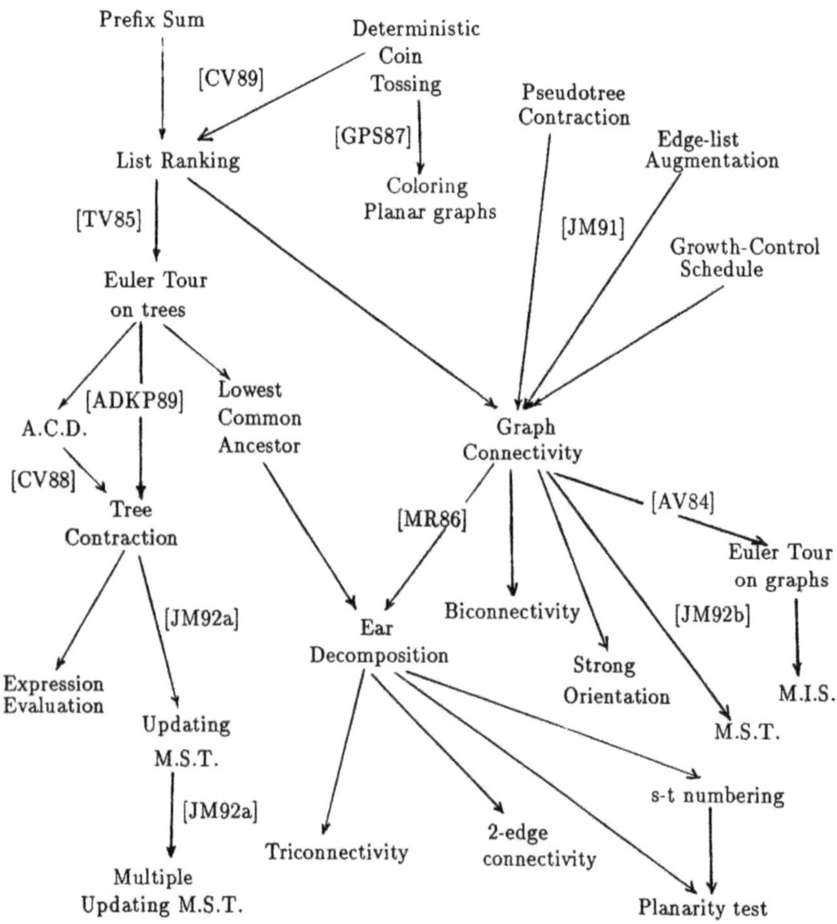

Figure 1: The reduction graph of parallel algorithms for some basic graph problems.

1.1 Definition of the Problem.

The vertex updating problem for a minimum spanning tree is defined as follows: We are given a weighted graph $G = (V, E_G)$ with $n = |V|$ vertices and $m = |E_G|$ edges, along with a MST $T = (V, E)$. The graph is augmented by a new vertex z and n weighted edges connecting z to every vertex in V. We want to compute a new MST $T' = (V \cup \{z\}, E')$.

The updating MST problem arizes often in practice in very diverse areas that vary from VLSI design and Artificial Intelligence to Operations Research and Economics. The problem has been addressed in the past in both the parallel [PR86, VD86, JM88, JM92a] and sequential settings. Optimal sequential algorithms have been given in [SP75, CH78]. In [JM92a] a set of *rules* is given which is used in conjunction with *parallel tree contraction* methods to develop optimal parallel algorithms for the EREW PRAM (exclusive-read exclusive-write parallel random access machine) model, the weakest of the PRAM models. These algorithms run in $t(n) = O(\lg n)$ time (where $\lg n$ denotes $\log_2 n$) using $p(n) = n/\lg n$ processors, thus achieving work optimality of $w(n) = t(n) \times p(n) = O(n)$. Note that the vertex updating problem has a lower bound of $\Omega(n)$ sequential time. To see that, consider the time it takes to find the maximum weighted edge on a tree composed of a single path of length $n - 1$, and whose ends are connected to the newly introduced vertex z.

Two Useful Observations We will use the rules of [JM92a] to develop simple and practical optimal sequential algorithms for the updating MST problem. These rules try to break *small* cycles (i.e. cycles of length 3 or 4) and are based on the following two observations:

1. The edge with minimum weight incident to some vertex will always be included into the MST. In fact, many sequential and parallel algorithms are based on this observation (Prim's algorithm and [SP75, CLC82] actually use this fact). In our algorithms, edge *inclusion* makes use of this observation.

2. Whenever some edge is found to correspond to the maximum weighted edge (MaxWE) of some cycle it can be removed from the tree without affecting the computation of the remaining graph. (Kruskal's MST algorithm makes use of this fact.) In the algorithms we describe, edge *exclusion* is based on this observation.

2 The Algorithms

Representation. Upon introducing the new vertex z along with n weighted edges,

$\binom{n}{2}$ cycles are created. [1] If we break these cycles by deleting the maximum weight edge (MaxWE) that appears in each cycle, the resulting tree will be the new MST T'. It is easy to see that at most n of these $2n-1$ weighted edges will be included into T'. No non-tree edge of the old graph can be included because all of them are already MaxWE on some existing cycle in the original graph, so we need not consider any such edge. For this reason we may take the input to be a tree T with $n-1$ weighted edges (corresponding to the given MST) and n weighted nodes (corresponding to weights of the newly introduced edges to z). We will call this object a *weighted tree*.

Without loss of generality we may assume that the weighted tree is binary. If this is not the case, there is an easy transformation which, given a weighted tree having vertices of unbounded degree, transforms it into an equivalent binary weighted tree $T = (V_T, E_T)$. This transformation is as follows: Each node $v = u_0$ with $k > 2$ children is augmented with $k-2$ fake nodes u_i to form a right path. Each of the children v_j of node v is attached as a left child of u_{j-1}, while u_{k-2} has a right child as well. The edges connecting the u_i's have weights $-\infty$ which makes them unremovable by the algorithm. The fake nodes do not have weights. They are introduced only to fix the order of processing of v's children. At the end of the algorithm the right path is always included in the MST of the binarized problem, giving a unique obvious solution to the general problem.

We assume that for each node of the binary tree the following pointers are defined: `par(v)` and `ch(v)` denote the parent and the child of node v. Note that whenever we refer to a child of some node v, there will be no ambiguity because v will only have one child. Also, we have a function *weight* : $V_T \cup E_T \rightarrow R$. `weight(v,w)` represents the weight of edge $(v, w) \in G$, while `weight(edge(v))` is the weight of weighted node v. By that we mean the weight of the newly introduced edge (z, v) or, during the execution of the algorithm, the weight of the MaxWE on the path from z to v via examined nodes.

The rules, presented in the following two subroutines, apply on tree nodes by "pruning" nodes with degree 1 (leaves of the tree or roots with one child) and "short-cutting" nodes of degree 2 (roots or nodes with only one child).

```
procedure prune(v)
   a := weight(v, par(v));
   b := weight(edge(v));
   c := weight(par(v));
   if a = max{a,b,c} then
      exclude(v, par(v)); include(edge(v));
   else if b = max{a,b,c} then
```

[1] If, in an instance of the problem, z is not connected to some vertex x, we can assume an edge (z, x) having maximum weight.

```
      exclude(edge(v)); include(v, par(v));
   else if c = max{a,b,c} then
      exclude(edge(par(v))); include(min{(v, par(v)), edge(v)});
      edge(par(v)) := max{(v, par(v)), edge(v)};

procedure shortcut(v)
   a := weight(edge(v));
   b:= weight(v, ch(v));
   c:= weight(edge(ch(v)));
   d:= weight(v, par(v));
   e := weight(par(v));
   if a = max{a,b,c} or a = max{a,d,e} then
      exclude(edge(v)); include(min{(v, ch(v)), (v, par(v))});
      (par(v), ch(v)) := max{(v, ch(v)), (v, par(v))};
   else if b := max{a,b,c} then
      exclude(v, ch(v)); prune(v);
   else if d = max{a,d,e} then
      exclude(v, par(v)); prune(v);
   else if c = max{a,b,c} and e = max{a,d,e} then
      exclude(edge(par(v))); exclude(edge(ch(v)));
      include(min{edge(v), (v, ch(v)), (v, par(v))});
      edge(par(v)) := max{edge(v), (v, par(v))};
      edge(ch(v)):= max{edge(v), (v, ch(v))};
```

The algorithms are based on the existence of a valid tree-contraction schedule. A *valid tree-contraction schedule* is one which schedules the nodes of the binary tree for pruning and shortcutting in such a way that (i) when a node is operated upon it has degree one or two, and (ii) neighboring nodes are not operated upon simultaneously. For the sequential algorithms only the first condition needs to be satisfied.

We now present two optimal sequential algorithms for the updating MST problem.

Postorder. The simpler algorithm to implement is the following: Visit the nodes of the weighted tree in postorder. A node is processed after all its children have been processed, therefore only the pruning rules are needed. Since each node is processed at most once, we have an $O(n)$ sequential algorithm.

Remove on the fly. Use depth-first-search to visit the nodes of the tree. Every time a node of degree 1 or 2 is encountered, process it using a pruning or a shortcutting rule, respectively. Each node will be visited at most twice (one on the way down the tree and one on the way up the tree), so its running time is $O(n)$.

Let us remark here that more algorithms can be derived by using other valid tree-contraction schedules like the ones reported in [ADKP89, CV88].

3 Incremental MST Algorithms

Using the previously described updating algorithms we can obtain incremental algorithms that compute the minimum spanning tree of a graph G in time $\sum_{1\leq i\leq n} i = O(n^2)$. These algorithms are optimal for dense graphs, i.e. for graphs having $\Theta(n^2)$ edges.

In the remaining part of this section we give the MST algorithm. In brief, the algorithm starts from the empty graph G_0 and computes the MST of the given graph $G_n = G$ in n steps by augmenting in each step G_{i-1} to G_i. Each augmentation introduces a new vertex v_i along with the edges that connect it to vertices already in G_{i-1}, and then calls the vertex updating routine.

Algorithm Incr-MST

Let $v_1, \ldots, v_n$ be a numbering of the vertices.
$G_0 = (V_0, E_0) \leftarrow (\emptyset, \emptyset)$;
$MST(G_0) \leftarrow \emptyset$;
for $i \leftarrow 1$ to n **do**
    Compute $MST(G_i)$ where $G_i = (V_{i-1} \cup v_i, E_{i-1} \cup (v_i, x))$
        and $x \in V_{i-1}$

4 Conclusions

In this paper we have presented an efficient way of designing sequential algorithms from existing parallel ones. Our approach requires that the scheduling and computing parts of the parallel algorithm are well defined and can be separated. This is the case of algorithms that use a valid tree-contraction schedule. We have shown that in this case, optimal, simple algorithms can be derived.

Our work shows that the design of sequential algorithms can be enriched form the research in parallel algorithms. Usually the research has followed the opposite direction.

References

[ADKP89] K. Abrahamson, N. Dadoun, D.G. Kirkpatrick, and T. Przytycka. A simple parallel tree contraction algorithm. *Journal of Algorithms*, 10:287–302, 1989.

[AV84] M. Atallah and U. Vishkin. Finding Euler tours in parallel. *Journal of Computer and System Sciences*, 29:330–337, 1984.

[CH78] F. Chin and D. Houck. Algorithms for updating minimal spanning trees. *Journal of Computer and System Sciences*, 16(12):333–344, 1978.

[CLC82] F.Y. Chin, J. Lam, and I-N. Chen. Efficient parallel algorithms for some graph problems. *Communications of ACM*, 25(9):659–665, September 1982.

[CV88] R. Cole and U. Vishkin. The accelerated centroid decomposition technique for optimal parallel tree evaluation in logarithmic time. *Algor.*, 3:329–346, 1988.

[CV89] R. Cole and U. Vishkin. Faster optimal parallel prefix sums and list ranking. *Information and Computation*, 81:334–352, 1989.

[GPS87] A. Goldberg, S. Plotkin, and G. Shannon. Parallel symmetry-breaking in sparce graphs. In *Proc. 19th Annual ACM Symp. on Th. of Comp.*, pages 315–324, 1987.

[JM88] H. Jung and K. Mehlhorn. Parallel algorithms for computing maximal independent sets in trees and for updating minimum spanning trees. *Information Processing Letters*, 27(5):227–236, April, 28 1988.

[JM91] D.B. Johnson and P. Metaxas. Connected components in $O(\log^{3/2} n)$ parallel time for the CREW PRAM. In *Proc. of 32nd IEEE Symposium on the Foundations of Computer Science (FOCS'91)*, October 1991.

[JM92a] D.B. Johnson and P. Metaxas. Optimal algorithms for the vertex updating problem of a minimum spanning tree. In *Proc. of the 6th Intl Parallel Proccessing Symposium (IPPS '92)*, pages 306–314, March 1992.

[JM92b] D.B. Johnson and P. Metaxas. A parallel algorithm for computing minimum spanning trees. In *Proc. of the 4th Annual ACM Symposium on Parallel Algorithms and Architectures (SPAA'92)*, June 1992.

[Met91] P. Metaxas. *Parallel Algorithms for Graph Problems.* PhD thesis, Dept. of Math. and Computer Science, Dartmouth College, Hanover, NH, July 1991.

[MR86] G.L. Miller and V. Ramachandran. Efficient parallel ear decomposition with applications. Technical report, MSRI, Berkeley, CA, 1986.

[PR86] S. Pawagi and I.V. Ramakrishnan. An $O(\log n)$ algorithm for parallel update of minimum spanning trees. *Infor. Proc. Lett.*, 22(5):223–229, April, 28 1986.

[SP75] P.M. Spira and A. Pan. On finding and updating spanning trees and shortest paths. *SIAM Journal on Computing*, 4(3):375–380, September 1975.

[TV85] R.E. Tarjan and U. Vishkin. An efficient parallel biconnectivity algorithm. *SIAM Journal of Computing*, 14(4):862–874, 1985.

[VD86] P. Varman and K. Doshi. A parallel vertex insertion algorithm for minimum spanning trees. In *13th ICALP, Lect. Not.*, volume 226, pages 424–433, 1986.

VISUALIZING MATHEMATICS AND COMPUTATION

David P. Dobkin+
Department of Computer Science
Princeton University
Princeton, NJ 08544

ABSTRACT

As computers become more powerful, they become useful tools to aid in research tasks. The availability of high speed high resolution graphics devices has had a significant impact on tasks which are visual in nature. We describe applications of visualization to the modeling of topological surfaces and the animation of geometric algorithms.

1. INTRODUCTION

During the last half of this century, the computer has become essential to the business world for data processing, spreadsheets and a variety of other tasks. Similarly, word processing and desktop publishing have become essential tools to anyone who writes and communicates the printed word. Number crunchers were the initial motivating application for high speed digital computers and they continue to drive the development of supercomputers of unprecedented computational power. More recently, the technology has allowed for powerful graphics displays to be accessible to a wider class of users allowing the user to interact visually with data within applications. The development of visualization techniques promises to have a significant impact on the nature of science and engineering providing new techniques for the development and exploration of ideas. Our focus here is on visualizing geometry and topology. We use visualization techniques to explore complex mathematical shapes and to view their transformations. And, we develop animations of geometric algorithms using the "moving

+This research supported in part by the National Science Foundation under Grant Number CCR90-02352.

frame'' to help explain their function.

Topology is the branch of mathematics devoted to the study of shapes and their transformations. Topologists consider problems asking whether one shape can be transformed into another while preserving aspects of its structure. They also define invariants describing individual shapes so that equivalence classes of shapes can be created having particular properties. Traditionally, topologists have worked, as have all mathematicians with fertile minds and pad and pen to prove properties of objects. In addition, topologists often kept cord or rope in their offices to enable them to create knots whose properties they wished to study. The computer can be an invaluable tool for doing the work which was done by these ''visual aids''. In the section that follows, we describe work aimed at providing the topologist with better visual aids. This work considers a class of topological structures and describes the techniques used to model them and explore them.

Over the past two decades, computational geometry has developed as a branch of theoretical computer science concerned with the development of algorithms for problems of a geometric nature. In part this development has been motivated by problems of computer graphics concerning the display of images represented by complex polyhedral models. Many of the problems that arise in this study concerning the detection and computation of intersections among members of a collection of objects. Perhaps the most basic such problems ask us to determine which, if any, of a collection of n line segments in the plane have common points. In section 3 below, we will use this as a model problem to explore ideas which have developed in computational geometry for its solution. Recently, an optimal algorithm has been developed to solve this problem. While this algorithm is complicated and currently of limited practical import, it has been implemented. We build upon this implementation to develop a computer animation of the algorithm in action which aids in our understanding of the algorithm.

The ideas here are meant to give a flavor of the role that visualization can play in the development of ideas of a geometric flavor in mathematics and computer science. It is clear that we have only begun to develop tools for the future. This paper should be considered as a ''static'' view of a subject which can be viewed dynamically. In accompanying videos, many of the concepts presented here are brought to life through the addition of motion [TCD, D3]. We have included individual images in this text to give a flavor or the shapes shown in the videos.

2. The animation of topological shapes

Perhaps the simplest shape of interest to the topologist is the torus.* The torus occurs in daily life as a doughnut. It is a shape we are used to seeing. Yet, when confronted with the four views of the same torus given in Figure 1, it requires some time to decide that each represents the same object. When considering these pictures, there is also a desire to slightly move the head and to see the object move. The images of the torus in this paper and in the related video were created using standard techniques of computer graphics. The reader is referred to [FvD] for further details. We describe in this section modeling techniques which were used to build the models which could be rendered.

Geometrically, a torus is the locus of points at a fixed distance from a given circle. In other words, a torus is created by sweeping a circle around another circle. For the images shown here, this is approximated by sweeping a regular polygon around another regular polygon. While this is straightforward for the case of the torus, it is worth looking at the relevant equations to ease the transition to the more advanced shapes described below.

Suppose that we are sweeping a circle of radius r about a circle of radius R. Points on the initial circle can be defined parametrically as

$$P(s) = (R\sin(2\pi s), R\cos(2\pi s), 0.0)$$

These points then form the centers of the circles we sweep out. We derive equations for the circles which are swept out by introducing a coordinate system at each center P(s). The coordinate system has as its z-axis the tangent to the circle at that point. This vector normalizes to

$$P_z(s) = (\cos(2\pi s), -\sin(2\pi s), 0.0)$$

The $x-\text{axis}(P_x)$ can then be chosen as the vector $(0,0,1)$ making the $y-\text{axis}$

$$P_y(s) = (-\sin(2\pi s), -\cos(2\pi s), 0.0)$$

* Further details on the topological ideas discussed here can be found in any text on the subject. For example, the book by Rolfsen [Ro] is a good source.

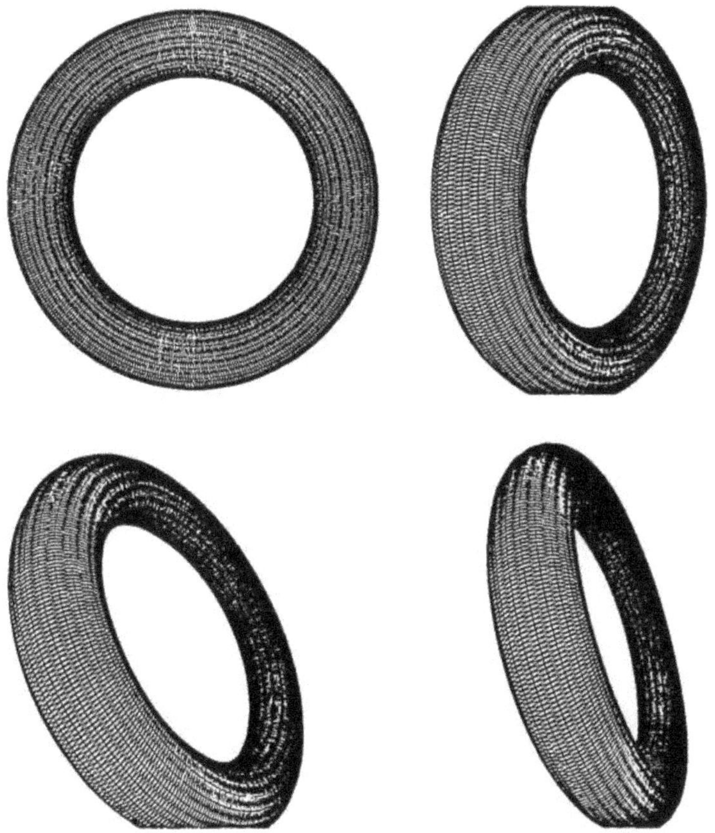

Figure 1: Four views of a tours

The circle of radius r centered at P(s) is parametrized by t as

$$C(s,t) = P(s) + r\,(P_x \sin(2\pi t) + P_y \cos(2\pi t))$$

This gives a description of the torus surface. We obtain approximations as polyhedra by discrete sampling of s and t which corresponds to sweeping a regular polygon about a regular polygon.

Our main interest in the torus is as a building block for more complex shapes. Two classes of such objects are of particular interest to us. The first class consists of objects built by combining many toruses to build shapes. The second class consists of shapes built by wrapping curves about toruses, the so-called torus knots. We describe each of these constructions below.

2.1. Building surfaces of high genus

The torus is the generic surface of genus greater than 0. The genus of a surface is an invariant measure of its number of holes. We use toruses to build surfaces of higher genus by various topological constructions. Connecting 2 toruses can be done to create a surface of genus 2. Similarly, these surfaces and their variations can be used to build surfaces of arbitrary genus. Computationally, the modeling of these surfaces is a topic of interest.

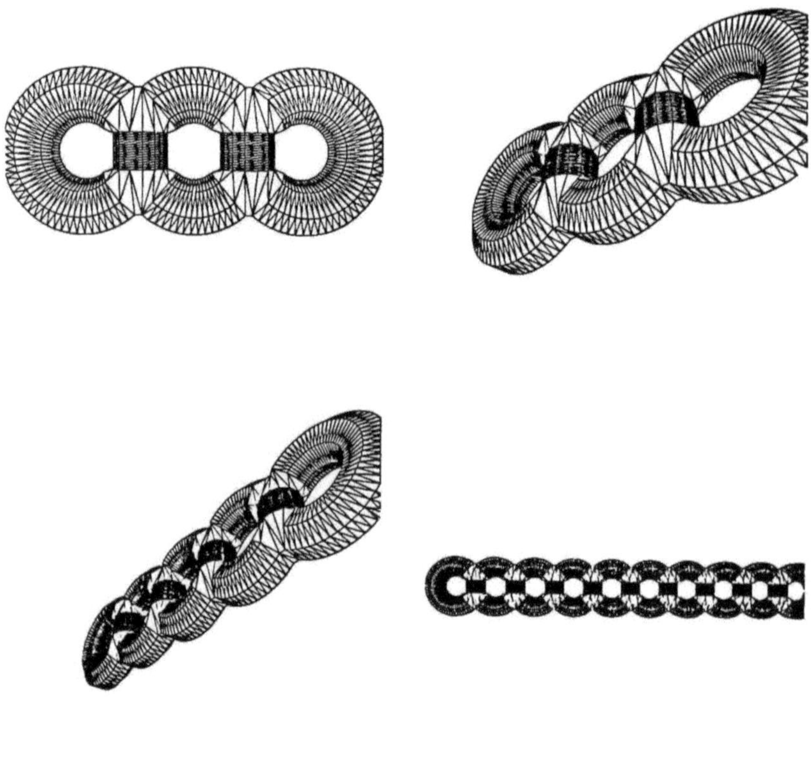

Figure 2: Surfaces of higher genus

Surfaces of genus higher than 0 correspond to spheres with handles or attached collections of toruses. Two views of a surface with genus 3 along with surfaces of higher genus are shown in figure 2. We build these surfaces by simple interconnections of basic building blocks.

The surfaces of Figure 2 are composed of the four parts shown in Figure 3. The first two of these parts are portions of a torus. The third is a cylinder and the fourth is a bridge to attach cross-sections. These are implemented by subroutines which can generate lists of vertices and faces which define the relevant polyhedron. The main program can then be very simple, it need only choose parts from the subroutine catalog, position them appropriately and attach them. The resulting structure is then a multi-holed torus as shown in Figure 2.

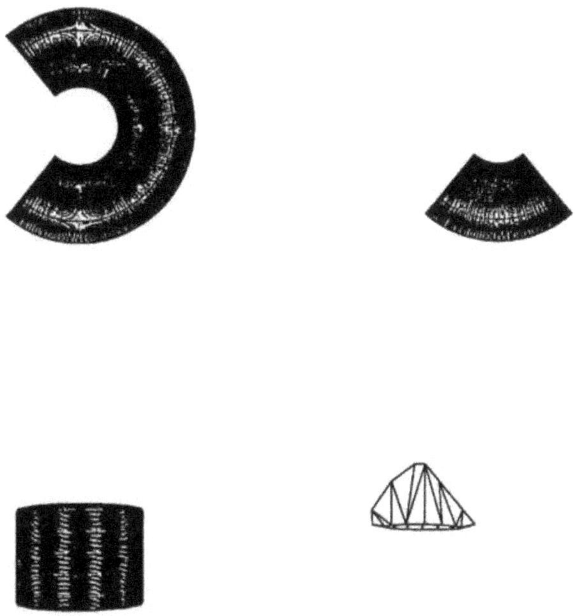

Figure 3: Pieces needed to build higher genus surfaces

The only complex part of the construction is the building of the bridges which join cylindrical cross-sections to torus cross-sections. These construction are often referred to as a "pair of pants" since they join a waist represented by a circular cross-section to two legs each of which is a circular cross-section. We realized a torus by joining together cross-sections which were polygons of equal number of vertices. This involves building ruled cylinders with equal number of vertices on top and bottom. To build a pair of pants, we must join a pair of cross-sections to a single cross-section, and this simple solution won't work. We solve this problem by letting the cylinder cross-sections have twice as many vertices as do the torus cross-sections. As a result, we can build the bridges by bifurcating a polygon of 2k vertices into 2 polygons each of k vertices.

It is interested to observe that the Euler relation among the numbers of vertices(V), edges(E) and faces(F) of a polyhedron states that

$$V - E + F = 2$$

This generalizes to the equation

$$V - E + F = 2 - 2*\text{genus}$$

for objects of positive genus. In our construction, all pieces are polyhedra of genus 0. Thus for each of the objects shown in Figure 3, we have V-E+F=2. However, when attaching pairs of pants, we identify 4k vertices, 4k edges and 3 faces of the pair of pants with the relevant features of the torus sections (or cylinder). In a n holed torus, we would use 2 pieces from the torus section in the upper left of Figure 3, $2n-4$ of the type in the upper right, $n-1$ cylindrical necks and $2(n-1)$ pairs of pants to do connections. This is a total of $5n-5$ pieces. For each piece, $V-E+F$ is 2. Totaling over all pieces, this results in an excess of $10*(n-1)$. We now observe, that each pair of pants removes 4k vertices and 4k edges (by identifying pairs that only occur on both the pants and the torus (or cylinder) section. Also removed are 6 faces, since the 3 faces of the pair of pants are identified with the 3 faces of the section to which it is joined and all 6 faces disappear. This results in a deficit of $12*(n-1)$ since there are $2*(n-1)$ pairs of pants. Combining these results gives the equation above.

2.2. Building torus knots

An interesting class of objects for topological study are those shapes formed by twisting a cord around a torus. This family of objects are known as torus knots. They are often used as building blocks for creating more complex mathematical structures. A basic shape in this class is the trefoil knot shown in multiple views in Figure 4. Careful study of the trefoil knot shows that it is created by twisting a cord around the spine of the torus (so as to make petals) 3 times. These 2 twists happen as we make 2 complete revolutions about the center of the torus (and so are separated by 240 degrees). We will describe below how to create the trefoil knot and other knots of this family.

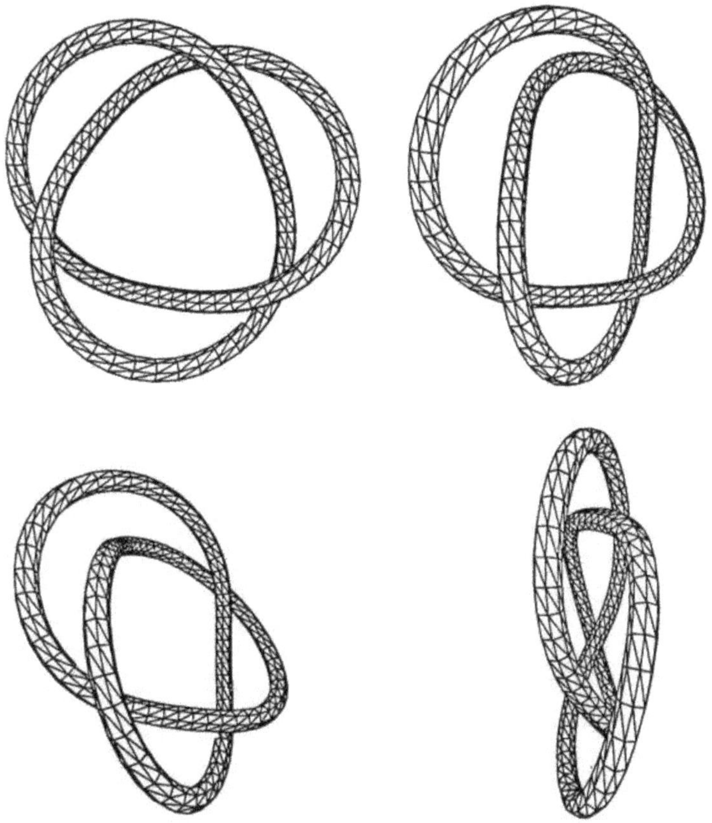

Figure 4: Four views of a trefoil knot

Our initial observation is that the trefoil knot can be described parametrically via the relations

$$x(\theta)=\sin(2\theta)(r+\cos(3\theta))$$

$$y(\theta)=\cos(2\theta)(r+\cos(3\theta))$$

$$z(\theta)=\sin(3\theta)$$

where r gives the the radius of the torus about which we are wrapping.

Clearly this is one of a general family of knots. We could create a torus knot by creating more petals in fewer revolutions about the center of the torus or by creating fewer petals in more revolutions. Examples of different knots viewed from the same point are shown in Figure 5. We observe that there are only 3 parameters which define an instance of a torusknot, the number of petals, the number of revolutions and the radius. Indeed, the general formula for the spine of such a surface is given as follows:

$$x(\theta)=\sin(p\theta)(r+\cos(q\theta))$$

$$y(\theta)=\cos(p\theta)(r+\cos(q\theta))$$

$$z(\theta)=\sin(q\theta)$$

Where p and q represent the number of petals and the number of revolutions. For example, the knots in Figure 5 have (p,q) pairs of ((3,5), (5,13), (7,19) and (3,53).

Having the equation above for the points on the torusknot, we now introduce a Frenet frame, a moving coordinate system which traces about the knot. The coordinate system takes as its z–axis the tangent to the curve. The x– and y– axes are in the plane perpendicular to this vector. Care must be taken to choose the y-axes in a consistent fashion between samples so that the circular (or polygonal) cross-sections will be joined in cylindrical (rather than twisted cylindrical) fashion.

3. Intersection computation

Computer graphics tasks can often be broken into two components -- modeling and rendering. In the previous section, we described a modeling task to give an idea of that part of the process. Rendering techniques vary depending on the quality of the image desired and the time available for rendering. A central aspect of any rendering process relates to hidden surface removal. The eye is very capable of looking at a scene and determining which objects lie in front of other objects. Designing an algorithm for the same task can be difficult. We devote this section to considering techniques and algorithms for one specific intersection problem which will provide an excellent overview of work in the area. We will trace the problem from

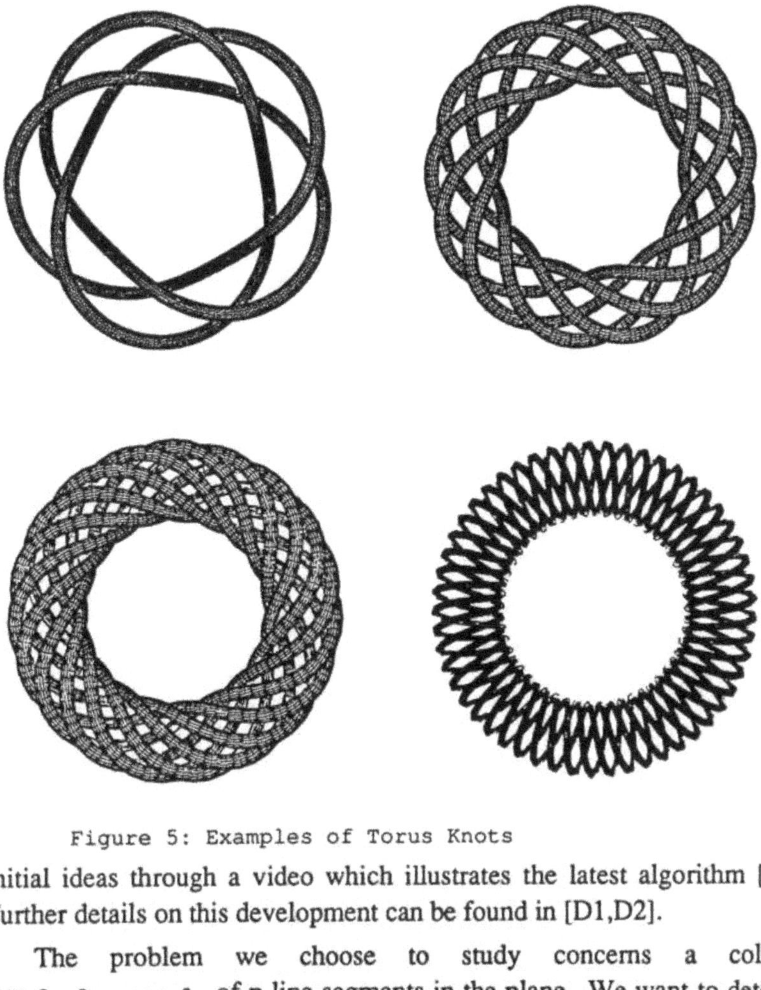

Figure 5: Examples of Torus Knots

initial ideas through a video which illustrates the latest algorithm [TDC]. Further details on this development can be found in [D1,D2].

The problem we choose to study concerns a collection $S = s_1, s_2, \cdots s_n$ of n line segments in the plane. We want to determine all pairs of segments which intersect. The analogy to computer graphics is fairly straightforward. The segments could be the projections of the boundaries of the triangles generated in our model of a torus knot for a given eyepoint and other viewing parameters. In this case, intersections would identify situations where triangles overlap in the final image. Using the information given by an intersection algorithm, we could determine which segments (and portions of segments) to draw.

We will change gears for a moment and ask for the assymptotically optimal algorithm for identifying all intersecting pairs of segments. We first observe that there are input sets which generate a quadratic number of intersections and others which generate few or no intersections. This means that estimating running time merely in terms of n may not be as good as a measure which also considers the actual number of intersection I. Our goal then is to find an algorithm of running time F(n,I) such that this is optimal.

We begin with the observation that no algorithm can run in time smaller than I since every intersection must be observed and reported. Also, it can be shown that any scheme for detecting whether a collection of n segments has any intersection must also be able to determine whether any two of a collection of n numbers are equal[SH]. This latter problem is known to be at least as hard as sorting[DL] and so O(n log n) is a lower bound on the assymptotic complexity of F(n,I). Combining these ideas shows that

$$F(n,I)=\Omega(n \log n + I)$$

Next, we turn to the problem of finding an algorithm which approaches this time bound. The standard algorithms for this problem build by having efficient means of going from one intersection to the next (or showing that no further intersections exist). The simplest algorithms work by sweeping vertically across the input segments in search of one (or many) intersections. Imagine that for each x value encountered on the sweep we have ordering the segments by y-coordinate. It is then observed that, if two segments intersect they must be adjacent in one of these ordered lists. Indeed, they are adjacent right before and right after their intersection. And their order switches at the point of intersection. This suggests an algorithm. All that remains is to determine those x-coordinates where a change in order could occur and to report the intersections observed at these points. In Figure 6, we give a set of line segments and mark the points at which their order changes. Note that it is not immediately obvious how to identify these points.

Shamos and Hoey[SH] were the first to propose a method for detecting intersections. The considered the problem of finding whether their was any intersection among the n line segments. They observe that it is sufficient to perform tests only at those x-coordinates where a segment is inserted or deleted. At the time of insertion, we determine where a segment fits in the

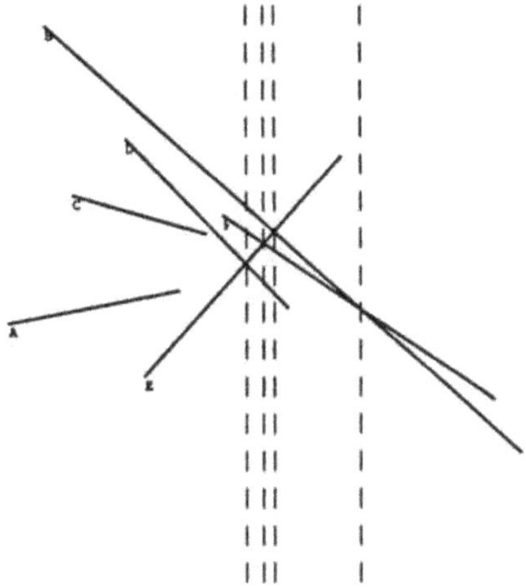

Figure 6: Line segment order changes at intersections (D,E), (E,F), (E,B), (F,B)

vertical order for that x-coordinate. We also determine whether it ever intersects either of its current neighbors in this ordering. At the time of deletion, we determine whether the neighbors of this segment intersect each other. They are able to prove that their algorithm correctly determines in O(n log n) operations whether any two line segments intersect.

Bentley and Ottman[BO] extended the Shamos-Hoey idea to report not just one but all line segment intersections. The algorithm given above processes events at x-coordinates where there is a segment inserted or deleted. This is extended to also consider events at intersection points. An event is created when it is noticed that an intersection will occur (note here that we always can predict an intersection *before* it happens). We continue processing with a vertical sweep line as above. However, at intersection points, we change the order of intersecting segments and proceed as though each was inserted at that time. That is, we test it against it's new neighbor for intersection. And, each time we detect an intersection, we create a new intersection event.

This is meant to be a brief discussion of the algorithm with further details available in the reference. The implementation of the algorithm involves keeping track of the current ordering on the vertical sweep and keeping an event list. The current ordering involves insertion new segments, deleting existing segments and switching the order of pairs of intersecting segments. The event list is created initially from the original line segment endpoints. Insertions are made to this list when intersections are detected. All of these data structure updating operations can be done in logarithmic time by maintaining balanced tree structures. This leads to an algorithm of running time $O(n \log n + I \log n)$. This running time is not optimal but it is the best possible using a seeping line algorithm of this type. We observe this since the intersection points are given sorted by x-coordinate.

For a decade, work proceeded trying to improve this result to the optimal. A partial result was given by Mairson and Stolfi[MS] who demonstrated that we could determine all I intersections among a collection of n red line segments and n green line segments in optimal time if there were no monochromatic intersections. Finally, Chazelle an Edelsbrunner[CE] were able to reach optimality in the general case. Their algorithm built off of the observation that the sweeping line could not be expected to keep things in sorted order. And, indeed, the sweeping line might itself want to be only piecewise linear. Rather than provide the details of their algorithm, the interested reader is referred to the video[TCD] where it is described in detail.

The discussion in this section is mainly a theoretical discussion of a practical problem. While many of the ideas here have contributed to methods used in practice, it is often the case that practical techniques can be adjusted to the problem at hand to enhance efficiency. For example, the rendering of the knots described in section 2 involved solving various hidden line removal problems in which all lines are very short and few pairs of lines intersect. Specializations of the algorithms mentioned here can be used in such cases. There are also hardware solutions in which simple brute force algorithms are easily built into hardware to give performance gains. Finally, we would be remiss if we did not mention that roundoff errors can sometimes be the dominant factor in deciding which algorithm is most appropriate to the task.

4. Conclusions

We have described two domains in which visualization adds to the nature of research in the mathematical sciences. There are many others which we do not mention here. As computers develop and hardware becomes more powerful and prices drop, these applications will be pervasive. Indeed, the next generation will learn computing and mathematics in a completely different manner because of the graphics display.

5. References

[BO] J. L. Bentley and T. A. Ottmann, ''Algorithms for reporting and counting geometric intersections'', *IEEE Trans. Comput.*, 1979, vol. C-28, pp. 643-647.

[CE] B. Chazelle and H. Edelsbrunner, ''An optimal algorithm for intersecting line segments in the plane'', *Proc. 29th Annu. IEEE Sympos. Found. Comput. Sci.*, 1988, pp. 590-600.

[D1] D. P. Dobkin, ''Computational Geometry -- Then and Now'', in *Theoretical Foundations of Computer Graphics and CAD*, edited by R. A. Earnshaw, Springer-Verlag, 71--110, 1988.

[D2] D. P. Dobkin, ''Computational geometry and computer graphics'', *Proc. of the IEEE*, to appear.

[D3] D. P. Dobkin, ''Torus knots in space'', unpublished video, Princeton U. 1989.

[DL] D. P. Dobkin and R. J. Lipton, ''On the complexity of computations under varying sets of primitives'', *J. Comput. Syst. Sci.*, 1979, vol. 18, pp. 86-91.

[FvD] J. D. Foley and A. Van Dam, *Fundamentals of Interactive Computer Graphics*, Addison-Wesley, Reading, MA, 1982.

[MS] H. Mairson and J. Stolfi, ''Reporting and counting intersections between two sets of line segments'', in *Theoretical Foundations of Computer Graphics and CAD*, edited by R. A. Earnshaw, Springer-Verlag, 1988.

[Ro] D. Rolfsen, ''Knots and Links'', Publish or Perish Press, Wilmington, DE, 1976.

[SH] Shamos, M.I. and Hoey, D., "Geometric intersection problems", *Proc. 17th Annu. IEEE Sympos. Found. Comput. Sci.*, 1976, pp. 208-215.

[TCD] A. Tal, B. Chazelle, and D. Dobkin, "The New Jersey line segment saw massacre", in *Animation and geometric algorithms: A video review*, edited by M. Brown and J. Hershberger, Technical Report 87b, DEC System Research Center, Palo Alto, CA.

NUMERICAL COMPUTATIONS FOR SOLIDIFICATION PROBLEMS WITH CHANGE OF VOLUME

T. Hanada (University of Ellectro-Communication)
H. Imai (University of Tsukuba)
H. Kawarada (Chiba University)
M. Natori (University of Tsukuba)

Abstract

A new formulation for free boundary problems in phase transition phenomena including volume changing is proposed. This is an extension of classical Stefan formulation to consider volume changing induced by the difference of density in each phase. With this formulation in three dimension, these phenomena are simulated in the axisymmetric cases reduced into two dimension by the fixed domain method, and in the flat cases reduced into one dimension by the enthalpy method.

1 Introduction

The phase transition problems have been mostly formulated as Stefan problems, to simulate or analyze [4][8][6]. In phase transition phenomena, material constants, such as heat conductivity, density, and specific heat, are different in each phase. The change of volume neglected in classical Stefan formulations is taken into consideration by J.Crank [2]. Extending his formulation, we will reformulate a phase transition problem including change of volume.

In our formulation for the solidification, there is another free boundary than the moving interface between liquid and solid. It is the free surface between liquid and the outside, that is moving with the interface. We further pay attention to the surface extension effects on the free surface and to the conservation of mass. Reduced into a case of one dimension, the whole mass is exactly conserved and the surface extension effects vanish. Nevertheless it may not conserve the mass in our formulation in a case of multi-dimension, it is expected to be satisfactorily approximated.

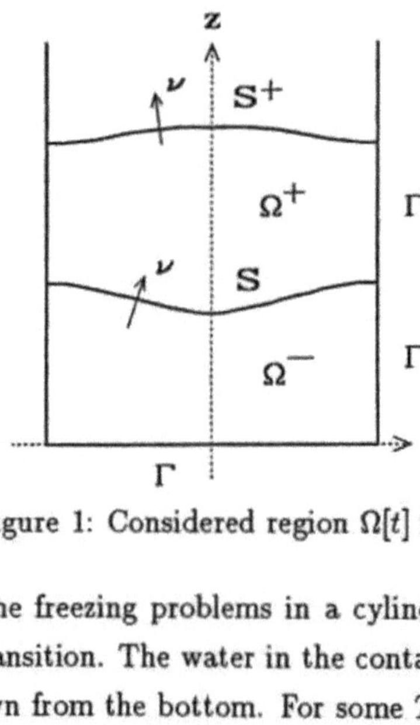

Figure 1: Considered region $\Omega[t]$ in G

We consider the freezing problems in a cylindrical container as a model of phase transition. The water in the container G is frozen into ice by cooling down from the bottom. For some $T \in \mathbb{R}^+$, we consider the problem in a domain $\Omega \subset G_T =]0,T[\times G$, where G is a domain in $\mathbb{R}^3$. It is the problem to find temperature θ, fluid velocity u, pressure p and free boundaries (their prescriptions Φ and Φ^+). Temperature is normalized so that the melting point to be zero.

It is difficult to find fully three dimensional solutions, then the lower dimensional solutions are searched. In § 3 flat solutions reduced into one dimension are searched, and axisymmetric solutions reduced into two dimension in § 4. In § 5 the equations in our formulation are discretized, and the numerical results got from that discretizations are shown in § 6.

2 Formulation

2.1 Basic equations

In the solid region Ω^- where temperature is negative, we consider only the heat diffusion equation

$$\rho c\,\dot{\theta} + k\nabla^*\nabla\theta = 0. \tag{1}$$

In this equation the specific heat c, heat conductivity k, and density ρ are assumed to be constants in each phase. Here

$$\dot{v} = \frac{\partial v}{\partial t}, \nabla v = \text{grad} v = (v_{,x}, v_{,y}, v_{,z})$$

and

$$\nabla^* v = -\text{div}\, v = -(v^x_{,x} + v^y_{,y} + v^z_{,z}).$$

In the liquid region Ω^+ of positive temperature, we are not able to neglect the motion of material due to the change of volume in phase transition. So we have the diffusion equation with convection

$$\rho c\dot{\theta} + k\nabla^*\nabla\theta - \rho c\nabla^*(u\theta) - (\nabla u)\sigma = 0, \tag{2}$$

where σ is the stress tensor. Thus the temperature θ is a function defined in the whole Ω. On the other hand, the fluid velocity u and the pressure p are defined only in Ω^+. Since the liquid is considered as an incompressible viscous fluid, we have the equation of continuity

$$\nabla^* u = 0 \tag{3}$$

and the Navier-Stokes equation

$$\rho\dot{u} + \rho(u \cdot \nabla)u = -\nabla p - \eta\nabla^*\nabla u, \tag{4}$$

where η is the viscosity of fluid [7].

And initial state for each quantity is assumed to be suitable as

$$\theta[0] = \theta_o, \tag{5}$$

$$u[0] = 0. \tag{6}$$

2.2 Boundary conditions

The conditions satisfied on the boundary of the regions Ω^- and Ω^+ are given as follows. Since we consider that the container G is cooled from the bottom Γ (Fig.1), we assume the Dirichlet condition

$$\theta = \theta_\Gamma (< 0) \text{ on } \Gamma. \tag{7}$$

We assume that the wall $\Gamma^+ \cup \Gamma^-$ consists of heat-insulated material, we have the adiabatic condition

$$\theta_{,\nu} = 0 \text{ on } \Gamma^+ \cup \Gamma^-. \tag{8}$$

Here ν shows the outer normal unit vector on $\partial\Omega^-$ or $\partial\Omega^+$ (Fig.1). It is moreover assumed that the fluid flows along Γ^+ in order that the periphery of the interface is able to move, then

$$\nu \cdot u = 0, \tag{9}$$

$$(u_\tau)_{,\nu} = 0 \text{ on } \Gamma^+ \tag{10}$$

are satisfied. Here u_τ is the tangential component of u. These boundaries Γ and $\Gamma^\pm$ stay on ∂G_T, but the instantaneous wall

$$\Gamma^\pm[t] = \{x \in \partial G | (t, x) \in \Gamma^\pm\}$$

at time t may vary.

On the interface $S = \partial\Omega^- \cap \partial\Omega^+$, the continuity of temperature

$$\theta^- = \theta^+ = 0 \tag{11}$$

and the Stefan condition

$$\rho^- \ell \dot{\Phi} = -(k^- \nabla\theta^- - k^+ \nabla\theta^+) \cdot \nabla\Phi \tag{12}$$

are satisfied. Here

$$\theta^\pm = \theta|_{\Omega^\pm},\ k^\pm = k|_{\Omega^\pm},\ \rho^- = \rho|_{\Omega^-}$$

and ℓ is the latent heat required to melt the solid matter. Φ is a C^1 function which prescribes the interface S such that

$$\Phi[t, x] \lesseqgtr 0 \Leftrightarrow \theta[t, x] \lesseqgtr 0.$$

On the interface S, moreover, the fluid is assumed to flow as

$$u = -(1-\alpha)\dot{\Phi}\nu, \tag{13}$$

where $\alpha = \frac{\rho^-}{\rho^+}$ is the coefficient of expansion. Since this condition is based on the equations (27) later in one-dimensional case (§3), the conservation of mass is expected to hold approximately.

Lastly, conditions on the free surface $S^+ = \partial\Omega^+ \setminus (\Gamma^+ \cup S)$ are given. For temperature we simply assume that the Neumann condition

$$\theta_{,\nu} = 0 \tag{14}$$

or the Dirichlet condition

$$\theta = \theta_{S^+}$$

holds. For the motion of the free surface, we assume that

$$\begin{aligned} \dot{\Phi}^+ &= -u \cdot \nabla\Phi^+ && (15) \\ \sigma\,\nu &= 0, && (16) \end{aligned}$$

from the conservation of the particle and the momentum on the free surface, and that

$$p = \kappa H \tag{17}$$

from the continuity of stress. Here κ is the surface extension, and $H = \frac{1}{R_1} + \frac{1}{R_2}$, where R_j is a radius of the principal curvature.

2.3 Reformulation (in non-dimensional forms)

As the above basic equations (1- 17), the solidification problem has been formulated physically. It is reformulated in non-dimensional forms to be intended for mathematical analysis and numerical computations by the transformations of variables

$$x = L\overline{x},\ u = U\overline{u},\ \theta = \Theta\overline{\theta}$$

in use of the standard length L, velocity U and temperature Θ. With their accompanying transformations

$$t = \frac{L}{U}\overline{t},\ p = \rho U^2\overline{p},\ \Phi = L\overline{\Phi},\ R_i = L\overline{R_i},\ \ell = U^2\overline{\ell},\ H = \frac{1}{L}\overline{H}$$

and

$$c = \frac{U^2}{\Theta}\overline{c},\ k = \frac{\rho L U^3}{\Theta}\overline{k},\ \sigma = \frac{\eta U}{L}\overline{\sigma},\ \kappa = \rho U^2 L\overline{\kappa},\ Re = \frac{\rho L U}{\eta},$$

the basic system is rewritten as follows (overlines are omitted):

$$c\dot{\theta} + k\nabla^*\nabla\theta = 0 \qquad \text{in } \Omega^-$$

$$\left.\begin{aligned} c\dot{\theta} + k\nabla^*\nabla\theta &= c\nabla^*(u\theta) + \tfrac{1}{Re}(\nabla u)\sigma \\ \nabla^* u &= 0 \\ \dot{u} + (u\cdot\nabla)u + \tfrac{1}{Re}\nabla^*\nabla u &= -\nabla p \end{aligned}\right\} \qquad \text{in } \Omega^+$$

$$\theta = \theta_\Gamma \qquad \text{on } \Gamma$$

$$\theta_{,\nu} = 0 \qquad \text{on } \Gamma^\pm$$

$$\left.\begin{aligned} \nu\cdot u &= 0 \\ (u_\tau)_{,\nu} &= 0 \end{aligned}\right\} \qquad \text{on } \Gamma^+$$

$$\left.\begin{aligned} \theta^- = \theta^+ &= 0 \\ \ell\,\dot{\Phi} &= (k^-\nabla\theta^- - \tfrac{1}{\alpha}k^+\nabla\theta^+)\cdot\nabla\Phi \\ u &= -(1-\alpha)\dot{\Phi}\nu \end{aligned}\right\} \qquad \text{on } S$$

$$\left.\begin{aligned} \theta_{,\nu} &= 0 \\ \dot{\Phi}^+ &= -u\cdot\nabla\Phi^+ \\ \sigma\nu &= 0 \\ p &= \kappa H \end{aligned}\right\} \qquad \text{on } S^+.$$

Note that this system is reduced to the classical Stefan formulation if the expansion α is equal to one.

3 One dimensional case

In order to analyze mathematical structure of solutions of the solidification problem, we search flat solutions dependent only on t and z at first. By elimination of dependence on x and y as

$$\begin{aligned} \theta[t,x,y,z] &= \theta[t,z], \\ u[t,x,y,z] &= (0,0,u[t,z]), \\ p[t,x,y,z] &= p[t,z], \end{aligned}$$

the problem is first reduced into one dimensional case. Since the equation (3) of continuity is rewritten as

$$u' = 0 \tag{18}$$

(where a z-derivative is denoted by a prime ($'$)), the fluid velocity u is shown to be dependent only on time. We assume that the moving interface S and the free surface S^+ are regular and represented as $z = s[t]$ and $z = s^+[t]$ by one-valued functions. Then we can obtain

$$\Phi[t,z] = z - s[t], \ \Phi^+[t,z] = z - s^+[t].$$

Thus the system of equations (1-17) is rewritten as follows:

$$c\dot{\theta} = k\theta'' \qquad \text{in } \Omega^-, \tag{19}$$

$$c\dot{\theta} = k\theta'' - cu\theta' \qquad \text{in } \Omega^+, \tag{20}$$

$$\dot{u} = -p' \qquad \text{in } \Omega^+, \tag{21}$$

$$\theta = \theta_\Gamma \qquad \text{on } \Gamma, \tag{22}$$

$$\theta^- = \theta^+ = 0 \qquad \text{on } S, \tag{23}$$

$$\ell\dot{s} = k^-(\theta^-)' - \tfrac{k^+}{\alpha}(\theta^+)' \quad \text{on } S, \tag{24}$$

$$p = 0 \quad \text{on } S^+, \tag{25}$$

$$\theta' = 0 \quad \text{on } S^+, \tag{26}$$

$$u = (1-\alpha)\dot{s} = \dot{s}^+. \tag{27}$$

From equations (21, 25) on pressure p,

$$p[t,z] = \dot{u}[t](s^+[t] - z),$$

then it is able to solve independent on other quantities. And the equation (27) is gained from the relation

$$\rho^- ds = \rho^+(ds - udt)$$

at the moving interface [2].

Secondly in order to vanish freedom of surface, we transform the coordinate system as

$$d\tilde{z} = \begin{cases} \alpha dz, & z < s[t], \\ dz, & s[t] < z < s^+[t], \end{cases}$$

and the variables correspondingly as

$$\tilde{\theta} = \begin{cases} \alpha^2 k\theta[z], \\ k\theta[z], \end{cases} \quad \tilde{c} = \begin{cases} c/\alpha^2 k, \\ c/k. \end{cases}$$

Since the free boundaries are transformed as

$$\tilde{s} = \alpha s, \ \widetilde{s^+} = \alpha s + (s^+ - s),$$

the free surface is invariant in time.

Thus the system is reduced to the following (tildes are omitted) :

$$c\dot{\theta} = \theta'' \quad \text{in } \Omega^-, \tag{28}$$

$$c\dot{\theta} = \theta'' - cu\theta' \quad \text{in } \Omega^+, \tag{29}$$

$$\theta = \theta_\Gamma \quad \text{on } \Gamma, \tag{30}$$

$$\theta = 0, \ \ell\dot{s} = (\theta^-)' - (\theta^+)' \quad \text{on } S, \tag{31}$$

$$\theta' = 0 \quad \text{on } S^+, \tag{32}$$

$$u = (\tfrac{1}{\alpha} - 1)\dot{s} \tag{33}$$

This system (28-33) is, lastly, reformulated in the weak sense in mathematics (the enthalpy formulation [3][1][10]), in use of the monotone discontinuous function γ such that

$$\gamma[\theta] = \begin{cases} c\theta - \ell, & \text{if } \theta < 0, \\ c\theta, & \text{if } \theta > 0, \end{cases}$$

and the characteristic function χ of the liquid region such that

$$\chi[\theta] = \begin{cases} 0, & \text{if } \theta < 0, \\ 1, & \text{if } \theta > 0. \end{cases}$$

Definition. *A function $\theta \in H^1_\cdot(\Omega)$ is called a weak solution of (28-33) in Ω if for any functions $\phi \in C^1(\Omega); \phi = 0$ on $\Gamma, \phi[T] = 0$,*

$$\theta = \theta_\Gamma \text{ on } \Gamma,$$
$$u = (\tfrac{1}{\alpha} - 1)\dot{s}, \ \theta[t, s[t]] = 0,$$
$$\textstyle\int_\Omega(-\gamma[\theta]\dot{\phi} + \theta'\phi' - \chi u\gamma[\theta]\phi') + \int_{S+} u\gamma[\theta]\phi = \int_{\Omega[0]} \gamma[\theta_\circ]\phi[0]. \quad (34)$$

In this formulation, $\gamma[\theta]$ is called the enthalpy. Using this formulation, we get numerical solutions being to be frozen everywhere for the Neumann condition, and those converged to steady states provided the Dirichlet condition is assumed in place of (32).

4 Axisymmetric case

In this section we search more complicated solutions than reduced in cases of one dimension. In order to simulate other solutions than flat ones the problems is considered in two dimension by assuming axisymmetry. Introducing the cylindrical coordinate system (r, ϑ, z), all variables are independent on ϑ by the assumption. So we take the region G as a two dimensional rectangular domain. Other domains and their boundaries are denoted by the same as before, the z-axis is considered newly as a boundary and it is denoted by Γ^a. Note that in the cylindrical coordinates the divergence in the equations in § 2.3 is replaced by $\overline{\nabla}$:

$$\overline{\nabla}(u^r, u^z) = u^r_{,r} + \frac{u^r}{r} + u^z_{,z}.$$

Numerical computations are discussed in § 5.2.

5 Discretization

Now we describe some numerical discretizations concerning the solidification problem in one dimensional and two dimensional cases.

5.1 Flat solutions

For the weak formulation

$$(\gamma[\theta])^{\cdot} + \chi u(\gamma[\theta])' - \theta'' = 0 \tag{35}$$

by (34) in §3, we will apply the Chernoff method [9]. It must be extended to be able to apply to our formulation. Approximating the above (35) with finite-differences in time, we have

$$e^{i+1} - e^i + \Delta t(\chi^i u^i (e^i)' - (\theta^{i+1})'') = 0. \tag{36}$$

here $\Delta t > 0$ is the time step, and $\theta^i = \theta[i\Delta t]$, $e^i = \gamma[\theta^i]$, $u^i = u[i\Delta t]$, $\chi^i = \chi[i\Delta t]$ are the approximating values of temperature, enthalpy, fluid velocity, and characteristics at time $i\Delta t$. In this scheme, it is redefined as

$$\Omega^- = \{(t, z) \in \Omega | \gamma[\theta[t, z]] < 0\}.$$

Hence, the complete scheme is constituted by the initial states

$$e^0 = \gamma[\theta^0], \quad u^0 = 0, \quad s^0 = \text{a zero point of } e^0$$

for the given data θ^0, and the iteration

$$\chi^i = \begin{cases} 0 & \text{if } e^i < -\ell, \\ e^i/\ell + 1 & \text{if } -\ell < e^i < 0, \\ 1 & \text{if } 0 < e^i, \end{cases} \tag{37}$$

$$E^{i+1} - \frac{\Delta t}{\mu}(E^{i+1})'' = \theta^i - \frac{\Delta t}{\mu}\chi^i u^i (e^i)', \tag{38}$$

$$e^{i+1} = e^i + \mu(E^{i+1} - \theta^i), \tag{39}$$

$$s^{i+1} = \text{a zero point of } e^{i+1}, \tag{40}$$

$$\theta^{i+1} = \gamma^{-1}[e^{i+1}], \tag{41}$$

$$u^{i+1} = \frac{s^{i+1} - s^i}{\Delta t}. \tag{42}$$

Here $\mu > 0$ is the relaxation parameter such that

$$\mu\gamma^{-1} \text{ is Lipschitz with constant 1,}$$

to simulate stabilized. This algorithm is an effective numerical scheme after discretizing in space; namely, e^i is approximated by continuous piecewise linear finite elements and the variables θ^i by piecewise constants.

5.2 Axisymmetric solutions

In an axisymmetric problem, the temperature and the fluid velocity are computed from the following [11]:

$$c\dot{\theta} = k\overline{\nabla}\nabla\theta \qquad \text{in } \Omega^-, \tag{43}$$

$$c\dot{\theta} = k\overline{\nabla}\nabla\theta - c(u\nabla)\theta \qquad \text{in } \Omega^+, \tag{44}$$

$$\dot{u^r} = -(u\nabla)u^r - p_{,r} + \tfrac{1}{Re}(\overline{\nabla}\nabla u^r - u^r/r^2) \quad \text{in } \Omega^+, \tag{45}$$

$$\dot{u^z} = -(u\nabla)u^z - p_{,z} + \tfrac{1}{Re}\overline{\nabla}\nabla u^z \qquad \text{in } \Omega^+, \tag{46}$$

$$\theta = \theta_\Gamma \qquad \text{on } \Gamma, \tag{47}$$

$$\theta_{,r} = u^r = u^z_{,z} = 0 \qquad \text{on } \Gamma^+ \cup \Gamma^- \cup \Gamma^a. \tag{48}$$

In order to simulate this problem, we apply the fixed domain method using the general coordinate system (τ, ξ, η) introduced in [5]. In this method, functions

$$\begin{aligned} \tau &= t \\ \xi &= \xi(t, r, z) \\ \zeta &= \zeta(t, r, z) \end{aligned} \tag{49}$$

are chosen to transform the domain $\Omega[t]$ in the physical space into a regular domain in the computational space. In the problem, we use different mapping functions ξ and ζ in each region $\Omega^-[t]$ and $\Omega^+[t]$. Then the free boundaries, those are the image of $\zeta =$ constant , can be treated precisely.

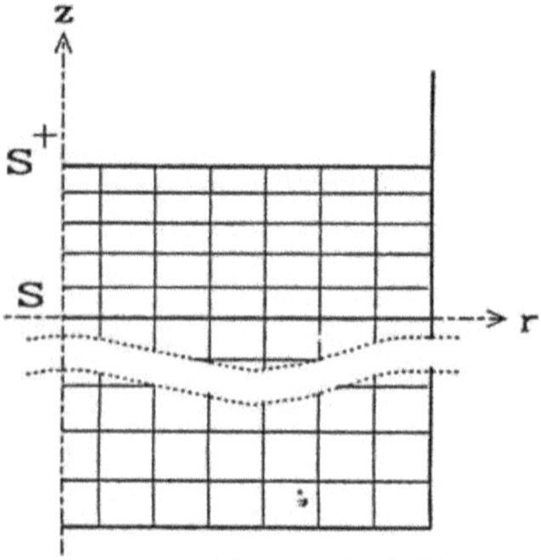

Figure 2: Grid at the initial

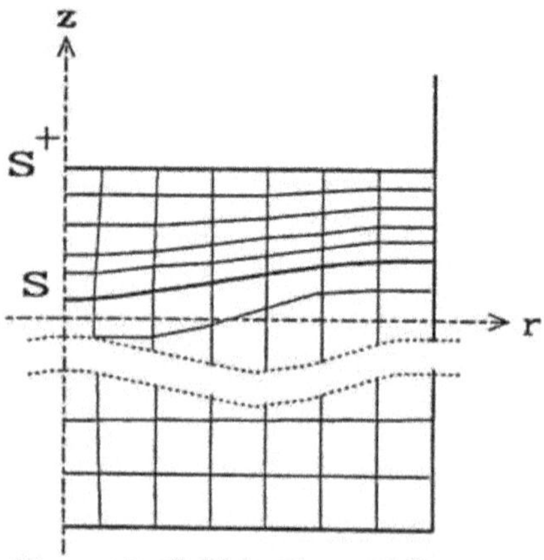

Figure 3: Grid in the middle

The basic system is now transformed using the following relations:

$$\begin{aligned} J &= r_{,\xi} z_{,\zeta} - r_{,\zeta} z_{,\xi}, \\ f_{,r} &= (f_{,\xi} z_{,\zeta} - f_{,\zeta} z_{,\xi})/J, \\ f_{,z} &= (r_{,\xi} f_{,\zeta} - r_{,\zeta} f_{,\xi})/J, \\ \dot{f} &= f_{,\tau} - f_{,r} r_{,\tau} - f_{,z} z_{,\tau}, \\ (u\nabla)f &= ((u^r z_{,\zeta} - u^z r_{,\zeta}) f_{,\xi} + (u^z r_{,\xi} - u^r z_{,\xi}) f_{,\zeta})/J, \end{aligned}$$

$$\overline{\nabla}\nabla f \;=\; -\Delta f + \Delta z f_{,r} + \Delta r f_{,z} - \frac{f_{,r}}{r}.$$

where

$$\Delta f = ((r_{,\zeta}^2 + z_{,\zeta}^2) f_{,\xi\xi} + (r_{,\xi}^2 + z_{,\xi}^2) f_{,\zeta\zeta} - 2(r_{,\xi} r_{,\zeta} + z_{,\xi} z_{,\zeta}) f_{,\xi\zeta})/J^2.$$

Thus we can describe the algorithm of simulation. At first in each iteration we compute the pressure distribution as follows [11]:

$$\overline{\nabla}\nabla p = -\dot{D} - ((u\nabla)u^r)_{,r} - ((u\nabla)u^z)_{,z} - \tfrac{1}{r}(u\nabla)u^r \quad \text{in } \Omega^+, \tag{50}$$

$$p_{,r} = 0 \quad \text{on } \Gamma^a, \tag{51}$$

$$p_{,r} = \tfrac{1}{Re}(u^r_{,rr} + \tfrac{1}{r}u^r) \quad \text{on } \Gamma^+, \tag{52}$$

$$p = \kappa H \quad \text{on } S^+. \tag{53}$$

Here boundary value of p on S is given by extrapolation.

We use a special mesh to avoid the singularity on Γ^a [5] and the central difference in space for the simplicity. As for time derivatives, $r_{,\tau}$ and $z_{,\tau}$ are computed explicitly, $u^r_{,\tau}$ and $u^z_{,\tau}$ are computed implicitly.

6 Numerical result

6.1 In one dimensional cases

In this section, we adopt constants of virtual material, then

$$\widetilde{c^-} = 1, \quad \widetilde{\rho^-} = 0.75, \quad \bar{\ell} = 3,$$
$$\widetilde{c^+} = 2, \quad \widetilde{\rho^+} = 1.00.$$

The boundary values are taken as

$$\widetilde{\theta_\Gamma} = -12, \; \widetilde{\theta_{S^+}} = 6.$$

From these and

$$\bar{s}[0] = 0.25, \; \widetilde{s^+} = 1.0,$$

initial status are taken to be linear in each region.

To consider at first the problem whose stationary state is known, we take the Dirichlet condition

$$\theta = \theta_{S^+} \text{ on } S^+$$

in place of the condition (32). As the values on the boundaries are assumed to be constant, we have the unique stationary state

$$\tilde{s}_\infty = \frac{-\theta_\Gamma}{\theta_{S^+} - \theta_\Gamma}, \quad \tilde{\theta}_\infty[\tilde{z}] = (\theta_{S^+} - \theta_\Gamma)\tilde{z} + \theta_\Gamma,$$

therefore

$$s_\infty = \frac{-\theta_\Gamma}{\alpha(\theta_{S^+} - \theta_\Gamma)}, \quad s^+_\infty = \frac{\alpha\theta_{S^+} + \theta_\Gamma}{\alpha(\theta_{S^+} - \theta_\Gamma)}.$$

Shown in Figure 4, the computed free boundaries for $n = 512$ (where n is the number of partition in space and time) converge to the stationary state $s_\infty = \frac{8}{9}$ and $s^+_\infty = \frac{11}{9}$, as $t \to \infty$. Table 1 shows the positions of the computed free boundaries at fixed time. They seem to converge to some values near the stationary state as $n \to \infty$.

Secondly, we simulate the Neumann problem with the same constants. In Figure 5 the time-variation of the computed free boundaries for $n = 512$. In Table 2 there are shown the positions of the free boundaries at fixed time, and the time when the liquid region disappears in simulation.

6.2 Axisymmetric solutions

In this simulation, we will consider more realistic case of freezing of water. Constants are taken as

$$\rho^- = 917, \quad \overline{k^-} = 24, \quad \overline{c^-} = 1.2, \quad \bar{\ell} = 100,$$
$$\rho^+ = 999, \quad \overline{k^+} = 5.6, \quad \overline{c^+} = 5.6, \quad \bar{\kappa} = 100, \quad Re = 20$$

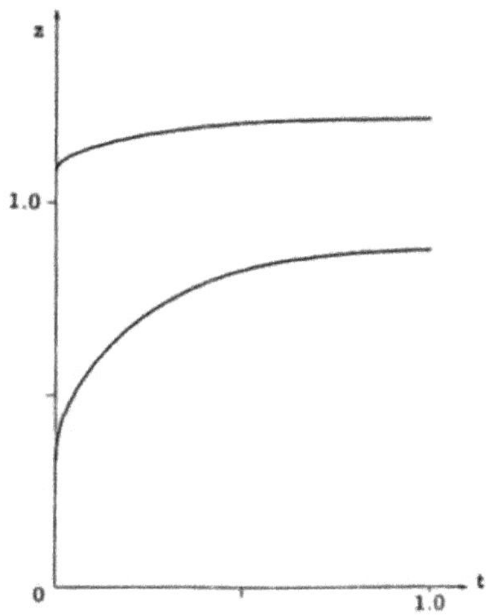

Figure 4: Dirichlet case.

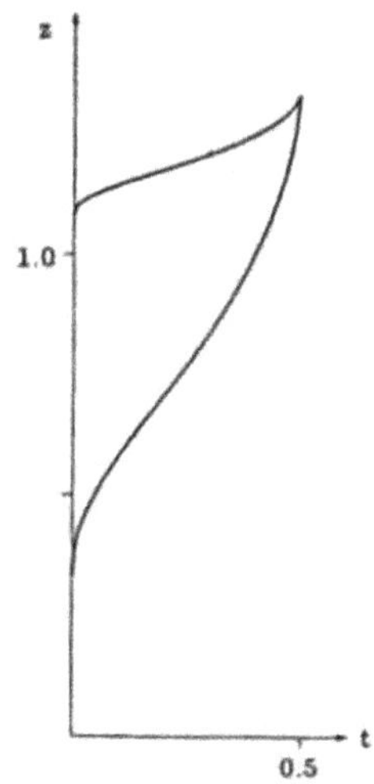

Figure 5: Neumann case.

n	(at $t = 1.00$) interface	surface
16	.809	1.202
32	.846	1.211
64	.867	1.217
128	.875	1.219
256	.880	1.220
512	.881	1.220

Table 1: Dirichlet problem.

n	interface (at $t = 0.25$)	surface	freezing time
16	0.712	1.178	.625
32	0.748	1.187	.562
64	0.763	1.190	.500
128	0.771	1.193	.492
256	0.776	1.194	.484
512	0.779	1.195	.486

Table 2: Neumann problem.

provided that $U = 10^{-3}, L = 10^{-1}, \Theta = 10^{-6}$. We have taken the value of ℓ smaller than the right value. As initial conditions are

$$\Omega^-[0] =]0,1[\times]-1,0[, \quad \theta_\Gamma = -2,$$
$$\Omega^+[0] =]0,1[\times]0,0.2[, \quad \theta_{S^+} = 0.1$$

and regions $\Omega^-[0]$ and $\Omega^+[0]$ are partitioned into 40×40 pieces. Figure 6–8 show the computed free boundaries. Table 3 shows the variation of the normalized mass of both regions and of the whole region.

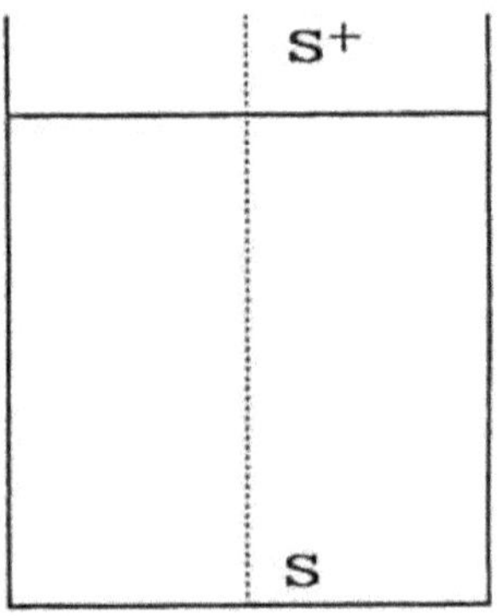

t=0. 000

Figure 6: At the initial.

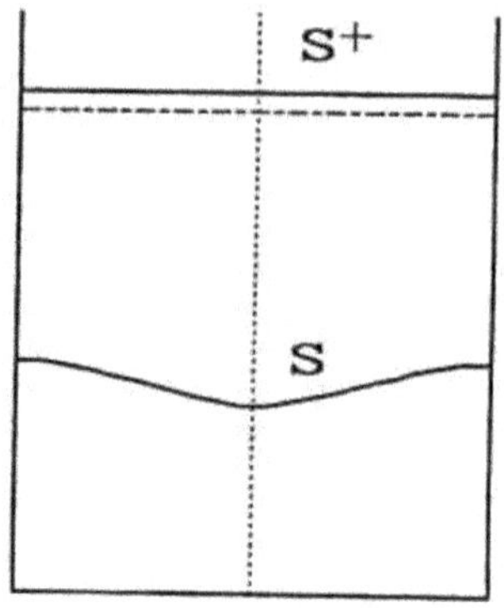

t=0. 152

Figure 7: Halfway.

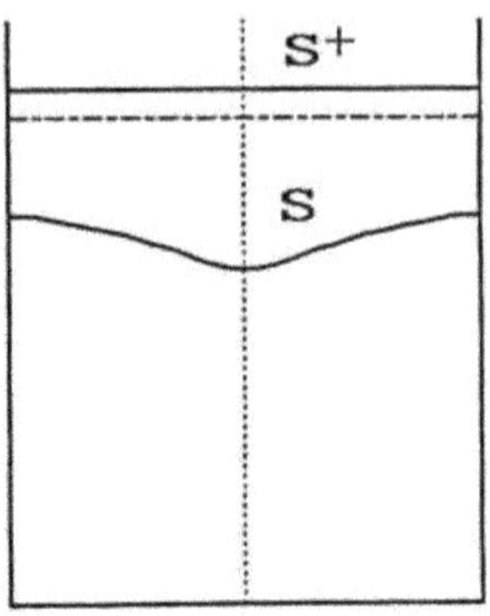

t=0. 244

Figure 8: At the last.

time	ice	water	total
0	0.9171	0.1998	1.1170
0.08	0.9574	0.1595	1.1170
0.16	1.0059	0.1111	1.1170
0.24	1.0564	0.0605	1.1170

Table 3: Variation of mass (flat case).

Another simulation is carried out. Figure 9–11 show the computed free boundaries. Table 4 shows the variation of normalized mass.

time	ice	water	total
0	0.9324	0.1832	1.1156
0.08	0.9375	0.1781	1.1156
0.16	0.9430	0.1726	1.1156
0.20	0.9479	0.1677	1.1156

Table 4: Variation of mass (non-flat case).

Figure 12,13 show the time dependence of the positions of the free boundaries at the symmetric axis and on the wall.

7 Conclusions

In the paper, we proposed a new formulation by which we can simulate the moving interface and the free surface in the solidification with change of volume. The aims of this formulation are to conserve the whole mass and to obtain the shape of the free surface with the effect

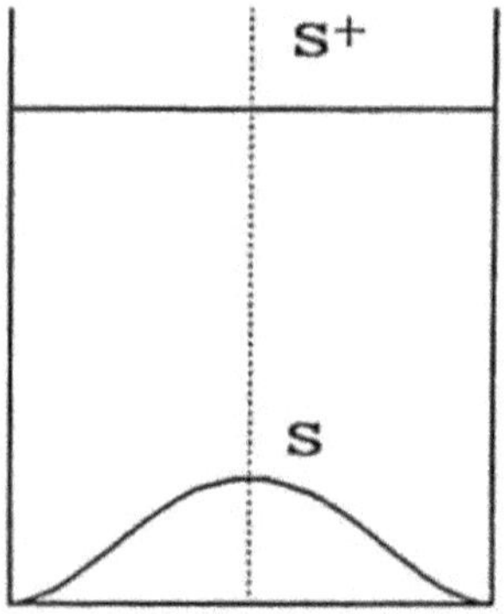

Figure 9: At the initial.

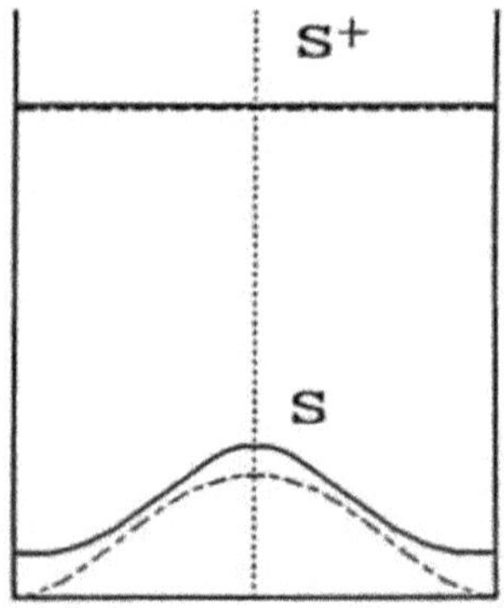

Figure 10: Halfway.

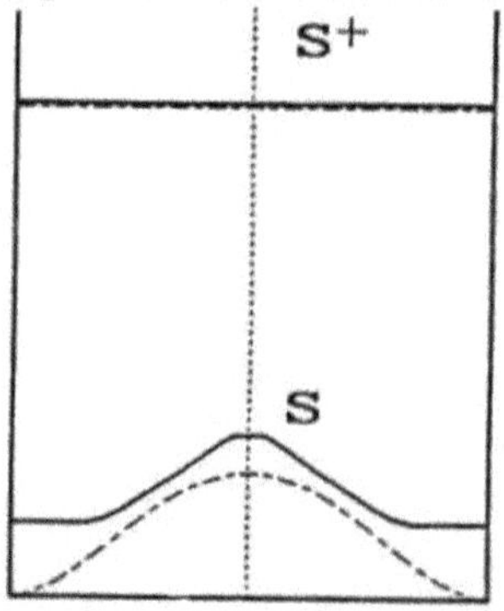

Figure 11: At the last.

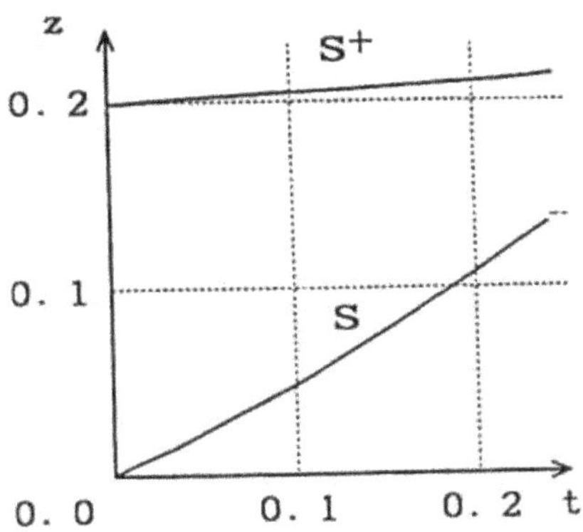

Figure 12: At the axis.

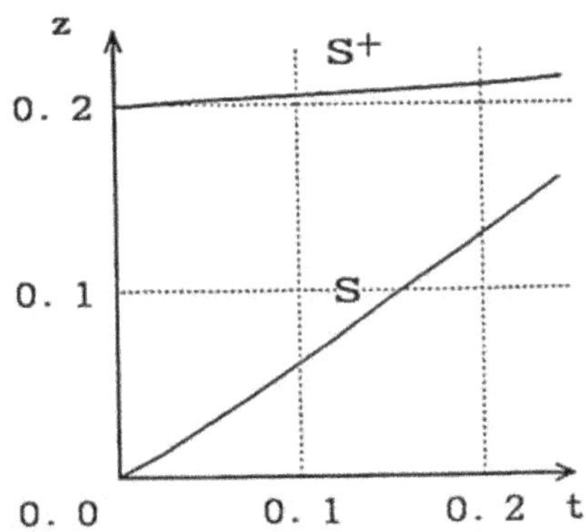

Figure 13: On the wall.

of surface tension. On the conservation of mass it is satisfactorily approximated. And we can see the effect of the surface tension from the flatness of the free surface in spite of the curving of the moving interface. For the solidification problem, we have shown numerically there are flat and axisymmetric solutions, but it is unknown whether fully three dimensional solutions exist. From the numerical results, it seems to us that the flat solutions are unstable in two dimensional structure. Hence the stability of the axisymmetric solutions in three dimensional structure is an open problem. There are moreover left problems on the asymptotic behavior of the free boundaries when the whole is solidified, and on the effect of the boundary conditions on the wall or on the free surface. In numerical solution, our scheme using mapping functions (§ 5.2) may be applied to multi-phase free boundary problems effectively.

Numerical computations were carried out on HITAC M-682H and S820 at the Computer Centre of the University of Tokyo.

References

[1] J.R.Cannon, E.DiBenedetto: *On the existence of weak-solutions to an n-dimensional Stefan problem with nonlinear boundary conditions*, SIAM J. Math. Anal., 11, 1980, pp.632-645

[2] J.Crank: *Free and moving boundary problems*, Clarendon Press, Oxford, 1984

[3] A.Fasano, M.Primicerio, S.Kamin: *Regularity of weak solutions of one-dimensional two-phase Stefan problem*, Ann. Mat. Pura Appl.(4), 115, 1977

[4] A.Friedman: *The Stefan problem in several variables*, Trans. Amer. Math. Soc. 133, 1968, pp.51-87

[5] Y.Katano, T.Kawamura, H.Takami: *Numerical study of drop formation from a capillary jet using a general coordinate system*, Theor. Appl. Mech. 34, 1986, pp.3-14

[6] H.Kawarada: *Free Boundary Problems (in Japanese)*, Univ. of Tokyo Press, 1989

[7] K.Murata, M.Natori, Y.Karaki: *Large Scale Numerical Simulation (in Japanese)*, Iwanami

[8] M.Natori, H.Kawarada: *Numerical analysis of free boundary problems (in Japanese)*, Japan. Phy. 31, 1976, pp.547-551

[9] R.H.Nochetto, C.Verdi: *An efficient linear scheme to approximate parabolic free boundary problems: error estimates and implementation*, Math.Comp. 51, 1988, pp.27-53

[10] R.H.Nochetto, C.Verdi: *Approximation of degenerate parabolic problems using numerical integration*, SIAM J. Numer. Anal., 25, 1988, pp.784-814

[11] D.Takahashi, Y.Takeda, H.Takami: *Numerical simulation of collision of liquid droplets*, Theor. Appl. Mech. 36, 1988, pp.3-15

[12] Joe F.Thompson, Z.U.A.Warsi, C.Wayne Mastin: *Numerical grid generation*, North-Holland, 1987

Takao Hanada:
The University of Ellectro-Communications,
Faculty of Ellectro-Communications,
Chofugaoka, Chofu-shi, Tokyo 182, Japan

Hitoshi Imai, Makoto Natori:
University of Tsukuba,
Institute of Information Sciences and Electronics,
Tsukuba City, Ibaraki 305, Japan

Hideo Kawarada:
Chiba University,
Faculty of Engineering,
Chiba 260, Japan

ON A STABLE QUADRATIC PROGRAMMING ALGORITHM FOR l_2 DATA APPROXIMATION BY NON-NEGATIVE DIVIDED DIFFERENCES

I.C. DEMETRIOU and E.A. LIPITAKIS

Department of Informatics
Athens University of Economics
76 Patission Street
104 34 Athens, Greece

Abstract For given univariate function values contaminated by random errors, the least squares smoothing subject to non-negative divided differences of order m may be required. We overview a stable, two phase, approach. First calculate an initial approximation that comes close to the optimum. Then use this approximation as the starting point of a quadratic programming algorithm that completes the calculation of the optimum in the dual space. Its implementation utilizes duality, QR factorization and updating procedures that preserve the banded matrices that occur. A computer program has been written and some numerical results illustrate the method.

1980 Mathematics Subject Classification (1985 Revision): 49D40, 65D10
Key words and phrases: least squares, data smoothing, divided differences, quadratic programming, duality, QR factorization

1. Introduction

A stable implementation [5] of a data smoothing method [1] is overviewed where the required smoothing accuracy goes much beyond the data at an order determined by higher divided differences.

Let n be a positive integer and let $\{\phi_i: i=1,2,\ldots,n\}$ be measurements on the data points $x_1<x_2<\ldots<x_n$, where each ϕ_i has been altered by random error. We regard the measurements as a vector $\boldsymbol{\phi}$ in $\mathbb{R}^n$

For a given positive integer m much smaller than n, we consider the strictly convex quadratic programming calculation that minimizes the objective function

$$\Phi(\mathbf{y}) \quad \|\boldsymbol{\phi}-\mathbf{y}\|_2^2 \tag{1.1}$$

subject to the constraints

$$\mathbf{a}_i^T \mathbf{y} = y[x_i, x_{i+1}, \ldots, x_{i+m}] \geq 0, \quad i=1,2,\ldots,n-m \tag{1.2}$$

where, after excluding zero factors in the denominator, (see, for example, Powell [6, p,46])

$$y[x_i,x_{i+1},\dots,x_{i+m}] = \frac{y(x_i)}{(x_i-x_{i+1})\cdots(x_i-x_{i+m})} + \frac{y(x_{i+1})}{(x_{i+1}-x_i)\cdots(x_{i+1}-x_{i+m})}$$
$$+ \dots + \frac{y(x_{i+m})}{(x_{i+m}-x_i)\cdots(x_{i+m}-x_{i+m-1})} \qquad (1.3)$$

denotes the value of the i-th divided difference of **y** of order m. The constraints (1.2) on **y** are linear with linearly independent normals $\mathbf{a}_i$, where each normal consists of just m+1 consecutive non-zero components that give a structure useful to our numerical implementation.

Demetriou & Powell [3] developed a fast and efficient algorithm for the case when m=2 that makes good use of the structure of the problem. Cullinan [1] let m be any positive integer and presented a feasible point quadratic programming procedure that starts from the point where all constraints are imposed to be active.

We also address this problem but due to minimizing the dual function

$$L(\lambda) = \| \sum_{i=1}^{n-m} \lambda_i \mathbf{a}_i + \phi \|_2^2, \; \lambda \in \mathbb{R}^n \qquad (1.4)$$

subject to the constraints

$$\lambda_i \geq 0, \; i=1,2,\dots,n-m. \qquad (1.5)$$

This provides us with computational advantages not only because the constraints are simpler, but also because (1.4) depends immediately upon the structure of the divided differences. We describe the method of Lipitakis & Demetriou [5] for this calculation, which consists of two phases stated in Sections 2 and 3 respectively. In Section 2 we present an algorithm that generates a starting approximation for the quadratic programming calculation in only $O(m^2n)$ computer operations due to QR factorization and duality. The quadratic programming algorithm for the dual problem is briefly described in Section 3.

Section 4 describes techniques for updating the QR factorization, initially provided by the starting procedure, as the active set changes by inserting and deleting columns, one at a time, according to the quadratic programming calculation.

Finally, Section 5 reports some preliminary numerical results in order to demonstrate the performance of the algorithms and the reliability of the smoothing method in certain data sets.

We only outline here the algorithms. A detailed description of stable procedures for updating matrix factorizations is given in [5]. Justifications of the complexity of the algorithms and extensive numerical results are also included in [5].

2. The initialization procedure

We let $y(A)$ denote the vector $y \in \mathbb{R}^n$ that minimizes the objective function (1.1) subject to the equality constraints (active constraints)

$$y[x_i,x_{i+1},\dots,x_{i+m}]=0, \quad i \in A, \tag{2.1}$$

where A is any subset of the constraint indices $\{1,2,\dots,n-m\}$.

The initialization procedure begins with $A=A^0=\emptyset$, so $y(A)=\phi$, and continues generating a sequence of sets $\{A^i:\ i=1,2,\dots,n-m\}$ subject to the inclusion relation

$$A^{i-1} \subseteq A^i \quad \text{for} \quad i=1,2,\dots,n-m,$$

where A^i is an instance of the set A. Then for $i=1,2,\dots,n-m$, the least squares approximation $y(A^i)$ to ϕ is calculated from the approximation $y(A^{i-1})$ which was chosen in relation to A^{i-1}, but $y(A^i)$ need not be feasible. To obtain $y(A^i)$, the constraint $y[x_i,x_{i+1},\dots,x_{i+m}]\ge 0$ is introduced into the calculation. Then, the procedure defines $y(A^i)=y(A^{i-1})$ unless $y(A^{i-1})$ violates this constraint, in which case it sets $A^i=A^{i-1}\cup\{i\}$ and considers the vector $y(A^i)$. So A^{n-m} denotes the active set at the end of this process. If the data satisfies the constraints, then just the indices of any zero divided differences of the data are accumulated in A^{n-m}. Otherwise A^{n-m} includes the index of the first infeasible constraint and probably some indices of later infeasible constraints too. Therefore this procedure identifies either the solution or whether changes need to be made to the original data.

The procedure requires the sign of $y[x_i,x_{i+1},\dots,x_{i+1}]$ rather, than the calculation of each component of $y=y(A^i)$ as the i-th constraint is inserted into the process. Therefore it suffices to know the $m+1$ components $\{y_j:\ j=i,i+1,\dots,i+m\}$ of y. Since $y(A)$, for $A=A^i$, is exactly the residual in the dual problem of the equality constrained minimization problem stated in the beginning of this section, we are based on the following calculation. Any time the matrix A is augmented by the addition of a new constraint index, we update the orthogonal part, say Q, of the QR factorization of the matrix $[a_i:\ i\in A]$ and derive the mentioned $m+1$ components of the residual in at most $O(m^2)$ operations by considering only the necessary part of the product $Q^T\begin{bmatrix}0\\d\end{bmatrix}$, where $Q\phi=\begin{bmatrix}c\\d\end{bmatrix}$. Note that the vectors c and d have $|A|$ and $n-|A|$ components respectively, where $|.|$ denotes set cardinality. Therefore, the computational procedure terminates in $O(m^2n-m^3)$ computer operations after inserting all the constraints into the calculation. So far the Lagrange multipliers are ignored in this process.

3. The quadratic programming algorithm

Having found the initial approximation, several decisions need to be made to complete the calculation of the optimum. We delete constraints from A^{n-m} where negative Lagrange multipliers occur untill all multipliers are non-negative. Then the quadratic programming algorithm of Demetriou & Powell [3] which is a variant of the dual approach of Goldfarb & Idnani [4], is employed to derive the numerical results in Section 5. However we follow the QR implementation that is referred to in Section 2 instead of that in [4].

The quadratic programming algorithm generates a sequence of sets of indices of equality constraints that are subsets of $\{1,2,...,n\text{-}m\}$. For each member, say A, of the sequence, the unique vector $\lambda=\lambda(A)$ in $\mathbb{R}^{n-m}$ is derived by the minimization calculation of (1.4) subject to the constraints $\{\lambda_i=0: i\notin A\}$. Thus the remaining components $\{\lambda_i: i\in A\}$ are found by solving the upper triangular system $R\mu=c$, with respect to the $|A|$-vector μ, given the factorization $QR=[a_i: i\in A]$. Accordingly, the vector $y(A)$ is derived as it has already been described in Section 2. If A is not the final set of the sequence, then $y(A)$ is infeasible.

4. Modifying the QR factorization

Due to the banded form of the constraints (1.2), the constraint normals are packed into an $m\times(n\text{-}m)$ working array in our implementation. Similarly for the matrices Q and R. In order to update Q and R as the active set changes, we apply suitable Givens rotations that preserve the existing sparsity structure and avoid fill-in. The effectiveness of these operations depends heavily on expedient coding.

Given the QR factorization of $[a_i: i\in A]$, dropping the k-th column is an operation that in only $O(m|A|)$ computer operations updates R and obtains $\lambda(A)$.

Consider revising the QR factorization of $[a_i: i\in A]$ after inserting the constraint normal a_j into $[a_i: i\in A]$. We replace A by $A\cup\{j\}$ so that A remains ordered and let $q=\min\{k,j\}$, where k is the index of the column that has just previously been deleted from the active set. We introduce the notation $Q=[Q_1|Q_2]$ and $R=[R_1|R_2]$, where Q_1 and R_1 consist of the first q-1 columns of Q and R respectively. We leave Q_1 and R_1 unchanged and complete the QR factorization by inserting sequentially into the calculation all remaining columns that have their indices in A. Further, due to the choice of q, matrix Q recovers its desirable form as a product of Givens rotations that had been distorted after deletion. The main disadvantage of this approach is that if q is too small compared to

$|\mathcal{A}|$ then $O(m^2|\mathcal{A}|)$ operations will be needed to complete the QR factorization of $[a_i: i\in\mathcal{A}]$. It is advantageous, however, that since we preserve the constraint structure, fill-in in the employed data structures is avoided and, Q and R are just m-banded arrays. Finally, after reconstructing Q and R between the columns a_j and a_k, we might preserve major parts of these matrices, by taking into account possible gaps of more than m consecutive constraints that usually (but not necesserily) occur in the active set.

5. Numerical results and discussion

This section presents preliminary results that illustrate the performance of the smoothing method on some ill conditioned problems. The numbers

$$\varrho = \max_{k\in\mathcal{A}} |a_k^T \cdot y|/\|a_k^T\| \quad \text{and} \quad \sigma = \max_{1\le i\le n} |2(y_i-\phi_i) \sum_{k\in\mathcal{A}} \lambda_k(\mathcal{A})\, a_{k(i)}|$$

provide a measure of the accuracy of the computer program that implements the procedures of this paper, where $a_{k(i)}$ is the i-th component of the k-th constraint normal and $\mathcal{A}$ is the optimal active set. Both, ϱ and σ, should be zero in exact arithmetic. Further, we introduce the function $I(q)=-log_{10}|q|$ which measures the number of decimal places to which a small real quantity q is zero. The program includes several extensions to the implementation that is outlined here.

We consider certain underlying functions f(x) (see Tables 1-6) and the data $\{\phi_i: i=1,2,\dots,n\}$ is created by adding to $f(x_i)$ a random number from the uniform distribution on [-r,r]. Each table can be explained as follows. The actual numbers of r and n are given in table columns 1 and 2. The third column displays the number $|\mathcal{A}^{n-m}|$; the fourth column displays the number of active constraints that are common to both $\mathcal{A}^{n-m}$ and to the optimal active set (denoted by Cm); the fifth column shows the number of active constraints after some deletions from $\mathcal{A}^{n-m}$ in order to achieve the first feasible Lagrangian vector. The number of constraint additions and deletions of the quadratic programming procedure is displayed in columns 6 and 7. Columns 8 and 9 present the cpu times to perform the calculations, in single precision arithmetic on a common INTEL 8087 mathematical coprocessor operating in MSDOS. The remaining columns are self explanatory. Certain chosen figures illustrate the method.

Data sets in Table 1. The algorithm terminated with small errors within n active set changes. The values of ϱ and $I(\varrho)$ show that the calculation of the smoothed values is very accurate. The accuracy for the Kuhn-Tucker conditions is slightly reduced as the values of σ and $I(\sigma)$ show.

The results for $(m=4, r=0.25)$ are displayed in Figure 1.

Data sets in Tables 2 and 3. When $m=3$, the algorithm terminated with small errors within $n/2$ active set changes. When $m=5$, the active sets remained large throughout the calculation and roughly $3n/2$ active set changes were required for termination. The values of ϱ, σ, $I(\varrho)$ and $I(\sigma)$ show that the smoothing process is very accurate. The results for $(m=5, r=0.025)$ are displayed in Figure 2.

Data sets in Tables 4 and 5. The underlying function is a Fejer kernel. It has a sharp peak in the mean of its range and undulates twice symmetrically in a long tail. It is impossible to approximate by polynomials. Even when using splines, one has to be particularly careful about where to place the knots. The choice $m=4$ was not sufficient for a good model to the data despite the termination of the program. The values of ϱ, σ, $I(\varrho)$ and $I(\sigma)$ show that the calculation is less accurate for the Lagrange multipliers. When $m=10$, the data modelling is notably good, although the calculation of the Lagrange multipliers is not so. Large active sets occured, while $3n/2$ active set changes were roughly required for termination. Figure 3 illustrates the case $(m=10, r=0.01, n=100)$. Figure 4 illustrates the cubic smoothing spline approximation with tension parameter equal to 2 to the data used in Figure 3.

Data sets in Table 6. The initialization procedure identified immediately the optimal active set.

The method we presented provides the optimal piecewise polynomial approximation to the data, as it is implied by the constraints (1.2). It is a global method that determines automatically where the polynomial pieces fit together by relaxing certain divided differences. The smoothing accuracy is determined by the chosen order of the divided differences. In this respect it is more advantageous than the method of the smoothing splines.

It should be noted that the choice of m is qualitative. As m increases the divided differences (1.2) become very sensitive to the data errors. Therefore, the numerical results indicated that in some ill conditioned cases, the accuracy of the Lagrange multipliers went sloppy as m became large. In contrast, the calculation of the smoothed values remained reliable.

If the data undulates, it seems reasonable to increase m but due to the underlying polynomial feature of the constraints (1.2) instability may occur in the smoothing calculation similar, in some extend, to those in polynomial calculations (cf. Powell [6, p. 212]).

Sometimes, the criterion (1.2) alone is not sufficient for data modelling, even if certain derivatives of the function underlying the data satisfy positiveness. The remarks suggest that some knowledge about the nature of the data is likely to improve the required fit.

The tables show that the initialization procedure of Section 2 is able to identify a large subset of the final active constraints. However, in view of the many active set changes that are implied by columns 6 and 7, it is worth investigating strategies so as to make a better use of the initial closeness to the final active set. Further, recent methods that allow more than one constraint at a time to correct infeasible constraints (Demetriou [2]) are expected to speed the termination when applied.

Finally, we are going to publish some results of Lipitakis & Demetriou [5] as well as results of more elaborated algorithms that are currently being developed.

Acknowledgements

This work was supported by the Research Center of Athens University of Economics.

REFERENCES

1. Cullinan, M.P., *Data smoothing using non-negative divided differences and l_2 approximation.* IMA J. Numer. Anal. 10(1990), pp 583-608.
2. Demetriou, I.C., *On a quadratic programming algorithm for l_2 discrete convex approximation.* Report INFOS/ΠΥΜ-90-32, Athens University of Economics, Greece. (Revision, Math. Computation 1991).
3. Demetriou, I.C. and M.J.D. Powell, *The minimum sum of squares change to univariate data that gives convexity.* IMA J. of Numer. Anal., 11(1991), pp 433-448.
4. Goldfarb, D. and G. Idnani, *A numerically stable dual method for solving strictly convex quadratic programs,* Math. Programming, 27(1983), pp 1-33.
5. Lipitakis, E.A. and I.C. Demetriou, *Least squares data smoothing by non-negative divided differences Report under research grand Nr E130-1/07-10-1991 by the Research Center of Athens University of Economics.* Report INFOS/ΠΥΜ-92-56, Athens University of Economics, Greece, 1992.
6. Powell, M.J.D., **Approximation Theory and Methods**, Cambridge University Press, Cambridge, 1981.

Tables of number of active constraints, CPU times and program accuracy

TABLE 1 (m=4) $f(x)=x^4-3x^3/4+3x^2-5x/2,\ x\in[0,1]$

		Initial active sets			QP active set changes		CPU times min:sec.100ths		Accuracy				Approximation squared error
r	n	A^{n-m}	Cm	Dual feas	Addit	Delet	Initial procedure	QP	ρ	l(ρ)	σ	l(σ)	
$\frac{4}{n^2}$	50	27	25	41	36	15	0.38	21.31	1.11E-7	6.9	1.78E-7	6.7	3.28E-5
	100	60	55	45	88	46	0.82	1:59.02	1.45E-7	6.8	2.65E-7	6.5	3.76E-6
$\frac{100}{n^2}$	50	27	25	21	44	22	0.32	28.78	8.76E-8	7.0	3.31E-6	5.5	2.16E-2
	100	60	57	45	127	80	0.82	3:19.10	1.25E-7	6.9	1.06E-6	5.9	2.46E-3

TABLE 2 (m=3) $f(x)=xe^{-x},\ x\in[-1,5]$

$\frac{10}{n^2}$	50	24	22	23	18	7	0.21	8.95	4.42E-8	7.3	1.07E-7	6.9	1.48E-4
	100	51	47	45	63	36	0.49	1:01.02	1.16E-7	6.9	3.14E-7	6.5	5.94E-5
$\frac{100}{n^2}$	50	27	24	23	21	7	0.21	10.27	6.97E-8	7.1	3.01E-7	6.5	1.76E-2
	100	56	51	48	70	35	0.60	1:12.39	1.85E-7	6.7	5.22E-7	6.3	2.23E-3

TABLE 3 (m=5) $f(x)=xe^{-x},\ x\in[-1,5]$

$\frac{10}{n^2}$	50	30	26	20	33	15	0.49	24.71	3.14E-8	7.5	1.57E-7	6.8	1.81E-4
	100	62	57	40	107	62	1.04	3:02.57	1.44E-7	6.8	9.77E-3	2.0	2.19E-5
$\frac{100}{n^2}$	50	31	28	20	45	25	0.54	39.71	5.59E-8	7.2	3.14E-6	5.5	1.91E-2
	100	62	58	40	131	84	1.04	4:04.19	9.29E-8	7.0	5.30E-6	5.2	2.30E-3

TABLE 4 (m=4) $f(x)=\sin^2(3x)/(12\pi\sin^2(x/2)),\ x\in[\pi,\pi]$

$\frac{4}{n^2}$	50	36	35	28	43	27	0.49	38.39	5.58E-8	7.2	2.05E-5	4.7	1.48E+0
	100	79	77	64	129	99	1.15	4:16.94	5.49E-8	7.3	5.31E-4	3.3	3.01E+0
$\frac{50}{n^2}$	50	40	39	28	62	46	0.60	52.61	3.69E-8	7.4	2.36E-5	4.6	1.48E+0
	100	61	60	48	202	156	0.82	5:59.48	2.54E-8	7.6	2.94E-4	3.5	3.01E+0

TABLE 5 (m=10) $f(x)=\sin^2(3x)/(12\pi\sin^2(x/2)),\ x\in[\pi,\pi]$

$\frac{4}{n^2}$	50	26	26	9	120	91	1.04	4:04.73	3.31E-8	7.5	7.31E-2	1.1	3.21E-1
	100	60	45	22	125	85	2.41	6:46.66	7.94E-8	7.1	1.34E-1	3.8	1.09E-5
$\frac{50}{n^2}$	50	24	23	9	119	90	0.87	4:04.25	6.54E-8	7.2	8.72E-2	1.1	3.25E-1
	100	55	50	24	279	224	2.14	22:12.05	4.55E-8	7.3	7.55E-1	1.2	8.37E+0

TABLE 6 (m=4) $f(x)=(1-x^2)^{1/2},\ x\in[-1,1]$

$\frac{4}{n^2}$	50	46	46	46	0	0	0.60	2.03	9.51E-8	7.0	9.30E-5	4.0	1.17E-1
	100	96	96	96	0	0	1.31	2.36	9.81E-8	7.0	1.13E-3	2.9	2.06E-1
$\frac{50}{n^2}$	50	46	46	46	0	0	0.60	2.63	9.58E-8	7.0	8.06E-5	4.1	1.27E-1
	100	96	96	96	0	0	1.26	2.47	1.03E-7	7.0	1.15E-3	2.9	2.12E-1

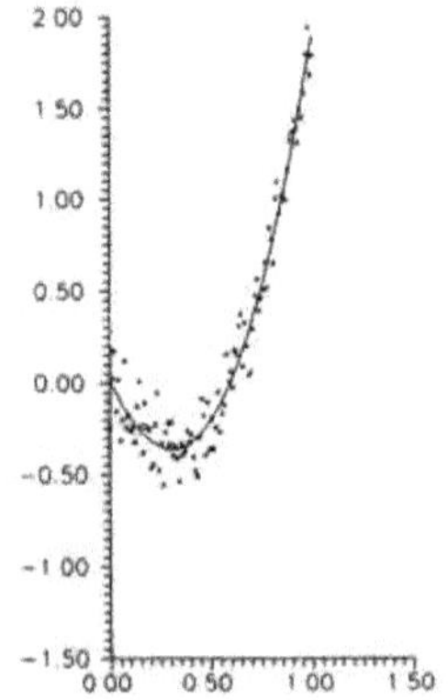

Figure 1: Best approximation to $n=100$ data points when $f(x)=5x/2-3x^2+3x^3/2-x^4$, $x\in[0,1]$ and $(m=4, r=0.25)$

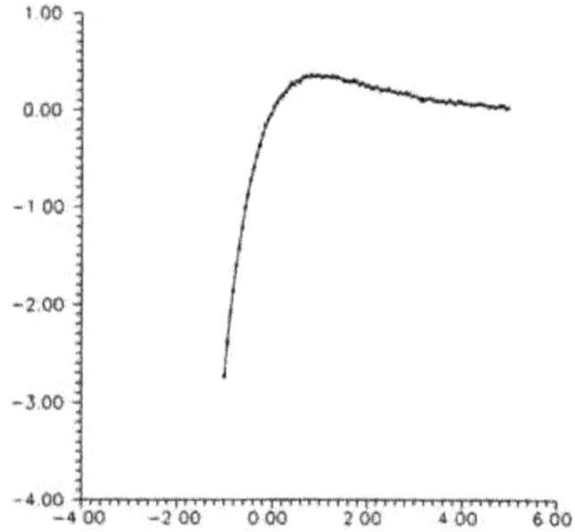

Figure 2: Best approximation to $n=100$ data points when $f(x)=xe^{-x}$, $x\in[-1,5]$ and $(m=5, r=0.025)$

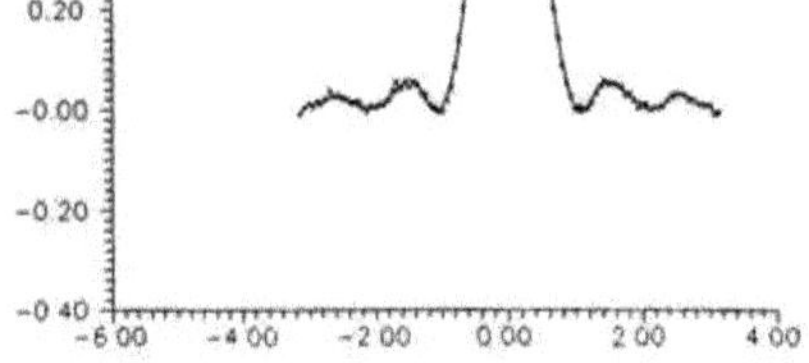

Figure 3: Best approximation to $n=100$ data points when $f(x)=\sin^2(3x)/(12\pi\sin^2(x/2))$, $x\in[-\pi,\pi]$ and $(m=10, r=0.01)$

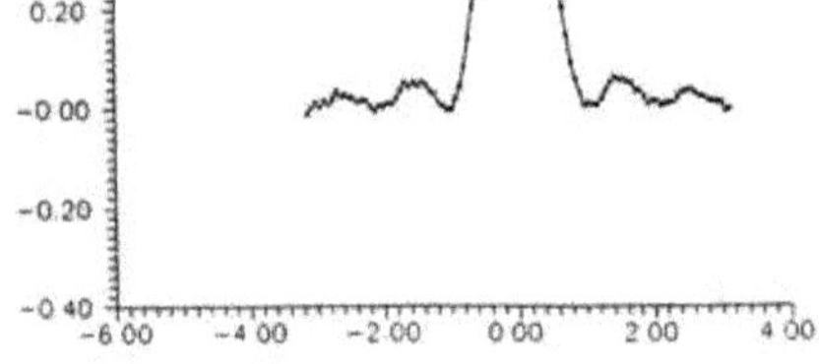

Figure 4: Cubic spline approximation with the tension parameter equal to 2 to the data points of Figure 3

SOLUTION METHODS TO LINEAR-QUADRATIC TWO-STAGE DYNAMIC GAMES: AN APPLICATION TO THE STRATEGIC EXPORT SUBSIDY PROBLEM[1]

Konstantine Gatsios[2], Tryphon Kollintzas[2], and Costas Lagopoulos[3]

ABSTRACT

This paper is part of a research project aiming at a general methodology for solving stochastic dynamic two-stage linear-quadratic games. The well-known problem in the strategic trade literature, whereby the domestic government seeks an optimal subsidy for a domestic firm that along with a foreign firm produces a homogeneous product for export to a third country, is used as a vehicle to illustrate the solution methodology. In particular, it is shown that the matrix polynomial associated with the characteristic equations of the Euler conditions that describe the equilibrium of the two-stage game can have a stable factorization. This factorization makes it possible to obtain an analytic representation of the equilibrium. Although the results of this paper are derived only with explicit reference to the time-inconsistent equilibrium of the game, most of them apply to the time-consistent equilibrium as well.

[1] An earlier version of this paper was presented at the First Hellenic Conference on Mathematics and Informatics, Athens, September 25 & 26, 1992.

[2] Athens University of Economics and Business (ASOEE); IMOΠ; and CEPR.

[3] Athens University of Economics and Business (ASOEE).

1. INTRODUCTION

In many problems in economics, the outcome of the interaction between agents takes place sequentially and repeatedly in a stochastic environment. Despite the commonality of these problems, however, it has been proven difficult to characterize the outcome of these interactions analytically. One of the main reasons for this, is the fact that the decision rules of the agents involved have not an outright recursive structure, due to the nature of the constraints facing them. To be more concrete, consider the well known example in the strategic trade literature of a domestic and a foreign firm producing a homogeneous product for export to a third country. In this context it is known that governments have an incentive to intervene and subsidize the production of their domestic firm. Clearly, the decisions made by governments and firms take place in an environment characterized by strategic interdependence. Moreover, this environment is both dynamic, in the sense that current decisions have to be made as part of a plan about future decisions as well, and stochastic, meaning that firms are uncertain about the future demand and cost conditions. Yet, the studies of the optimal export subsidy ignore these dynamic and stochastic features.[1] See, for instance, Brander and Spencer (1985), Eaton and Grossman (1986), and Dixit (1984).

In particular, these features are ignored because in their presence the decision rules of firms that should enter as constraints in the optimal export subsidy problem solved by governments, depend on the entire future profile of these

---

[1] Brander and Spencer (1983) and Gatsios (1989) have attempted to introduce models with dynamic features into this literature, but in a deterministic environment.

subsidies. This dependence makes the latter problem non recursive. In turn, this non-recursiveness gives rise to the well known problem of time inconsistency, as governments may not be able to commit credibly to an announced policy (Kydland and Prescott (1977)).

This paper is part of a research agenda that aims to deal with the above-mentioned gap in the literature. A related attempt to deal with these problems is Hansen, Epple, and Roberds (1985). They obtained time consistent and time inconsistent solutions of one-stage dynamic stochastic games in a linear-quadratic set up. However, they did so by exploiting special features of their model to get around the matrix factorization problem referred to below. Their work presents a starting point in obtaining solutions to general multi-stage dynamic stochastic games.

The purpose of this paper, then, is to state and solve a linear-quadratic two-stage dynamic game of optimal subsidies in an international context. The solution method developed herein takes care of the time inconsistent case. However, it can also serve as a guide to the solution of the time consistent problem as well. We are primarily concerned with the technical aspects of the problem. Most notably, we are concerned with factoring the matrix polynomial associated with the characteristic equation of the Euler conditions that describe the equilibrium of the two-stage game.

The plan of the paper is as follows. Section 2 states the problem and discusses its applicability. Section 3 solves the time inconsistent case. Section 4 presents some concluding remarks and discusses the difficulties associated with the solution to the time consistent case.

2. THE MODEL

The model we employ as a vehicle to deliver our ideas, is the well-known in the strategic trade literature paradigm of two firms, a domestic and a foreign one, producing a homogeneous product for export to a third country. We assume that this product is not consumed domestically by the exporting countries. This is a fascilitating assumption; it can be relaxed without altering the main insights of the problem. In such a context governments have an incentive to intervene and subsidize the output of their domestic firm. The reason for this is simple: subsidies imply that the actual production cost faced by firms is reduced. As a result, their best response functions shift outwards in the output space, meaning that for each output level produced by its opponent, each firm has now an incentive to increase its output. In game theoretic jargon, government's intervention alters the Nash equilibrium in the post-subsidy game played by the two firms. In fact, Brander and Spencer (1985) have shown that each government in choosing its subsidy, attempts to make its own firm a Stackelberg leader in the output game, increasing thereof its output and market share as well as both its gross and net of subsidy profits. The latter clearly explains why governments intervene in the first place; the increase in net profits implies an increase in the country's welfare. The underlying idea, of all this is credibility. A government by announcing its subsidy policy can "help" its domestic firm to commit to an output level to which, it could not credibly commit on its own.

As stated in the Introduction our aim is to introduce both dynamic and stochastic features in this model as a way to study the solution to two-stage dynamic stochastic games in a linear-quadratic set-up exhibiting a non-recursive structure. In what follows we will assume that only one government, say that of Country 1, is active in commercial policy[2]. The rules of the game in hand are as follows: In each period the shocks affecting demand and costs take place and become known to the players (firms and governments) so that period "t"'s shocks are in their information set. Then, the government of country 1 sets the subsidy level to the production of its domestic firm. Finally, the two firms (knowing the subsidy level) play Nash against each other in quantities. There is complete information - all players know everything about the structure of the model.

The objective of firm i (i=1,2), is to choose an output contingency plan $\{q_{it}\}_{t=0}^{\infty}$ so as to maximize its expected discounted future stream of profits:

$$E_0\sum_{t=0}^{\infty} \beta^t \left[(p_t + \tau_{it-1})q_{it} - a_{it}\, q_{it} - \frac{1}{2}b_i\, q_{it}^2 - c_i\, q_{it}\, r_{it-1} \right] \tag{1}$$

$$s.t. \qquad p_t = d_t - e(q_{1t}+q_{2t}) \tag{2}$$

[2] It is straightforward, however, to extend the model so that both governments set subsidies. What will be in our case a subsidy level, will be in the latter case a best response function in the subsidy game, and, subsidy levels will be determined by the joint solution of the two best response functions in the usual way.

$$r_{it} = r_{it-1} + q_{it} \tag{3}$$

$$q_{it} \geq 0 \tag{4}$$

$$r_{i,-1} \quad given \tag{5}$$

where $\beta \in (0,1)$ is a common constant discount factor in all periods; p_t is the price of output in period t; τ_{it-1} is the subsidy per unit of output of the ith country in period t[3]; r_{it} is the accumulated amount of output that the firm of the ith country has produced up to the end of period t, a_{it} is a time-dependent cost parameter that incorporates the influence of a random cost shock in period t; d_t a time-dependent parameter that incorporates a random demand shock in period t; and finally, b_i , c_i and e (e>0) are constant parameters. The first term in (1) represents the revenues of firm i in period t. The structure of these costs is taken from Hansen et al. (1985), with the difference that we do not require b_i and c_i to be positive. This allows us to consider both the case that the good is exhaustible (i.e., $c_i>0$), and that the production of this good is subject to learning by doing (i.e., $c_i<0$), Clarke et al. (1982). Note here that when c_i is equal to zero the cost structure, and hence the problem of the firm, becomes static. This is because there is no term (like $c_i q_{it} r_{it-1}$) that links current period's output decision with all

[3]This implies that the subsidy payments in period t are expressed as $[\tau_{1t-1} q_{1t}]$. At this point a remark concerning the subscript of the subsidy τ_{1t-1} is in order. The reason why τ_{1t-1} appears in period t's problem of both the firm and the government is to indicate that the subsidy in period t is announced and known one period earlier altering thus the set of credible actions that the domestic firm has available to choose from when it moves in period t. From the point of view of the government though, τ_{1t-1} represents notationally the subsidy rate that the government chooses optimally and applies on output of period t.

the past output decisions. Furthermore, it will allow for the possibility of decreasing (i.e., $b_i>0$), constant (i.e., $b_i=0$), and increasing (i.e., $b_i<0$) returns to scale in temporal production. Nevertheless, some restrictions on these parameters will be imposed later on as a part of the solution. Finally, (2) represents the aggregate (inverse) world demand in period t. For the purposes of this paper we do not need to be explicit about the probability laws of motion of the exogenous variables of the model, i.e. the demand and supply shocks. It should be understood however, that the conditional expectations on information available up to the beginning of period t, $E_t(\cdot)$, is with respect to the probability law of motion of these shocks.

The linear structure of the world demand curve and the quadratic structure of the cost function are necessary to exploit the Linear Rational Expectations Equilibrium (LREE) methodology developed by Hansen and Sargent (1980,1981), Hansen et al. (1985), Kollintzas (1985,1986), and Cassing and Kollintzas (1991). The LREE models are stochastic dynamic equilibrium models where: objective functions are quadratic, transition equations of exogenous variables are linear, subjective discount factors are constant, expectations of exogenous state variables are replaced by their MMSE counterparts, and subjective expectations of endogenous states variables are generated by the equilibrium of the model. Examples of LREE models can be found in Aoki (1989), Lucas and Sargent (1981), Sargent (1989), and Kollintzas (1989). LREE models are a special class of what is commonly referred to as Linear-Quadratic Stochastic Control models.

The objective of the government of country 1 is to choose,

prior to the output decisions made by firms, a subsidy contingency plan so as to maximize the expected discounted future stream of profits made by its domestic firm, net of subsidy payments. Hence government 1's objective function is given by

$$E_0 \sum_{t=0}^{\infty} \beta^t \left[p_t \, q_{1t} - a_{1t} \, q_{1t} - \frac{1}{2} b_1 \, q_{1t}^2 - c_1 \, q_{1t} \, r_{1t-1} \right] \tag{6}$$

3. SOLUTION - THE TIME INCONSISTENT CASE

Starting in the usual way from the last stage of the game we observe from (1)-(5) that if $\{q_{it}\}_{t=0}^{\infty}$ is to be a solution to firm i's problem, then $\{r_{it}\}$ must be a solution to the following[4] :

$$\begin{aligned} \max_{\{r_{it} \geq r_{it-1}\}} & \left[\left\{ \left([d_t - e(q_{1t} + q_{2t}] + \tau_{it-1} \right) q_{it} - a_{it} q_{it} - \frac{1}{2} b_i \, q_{it}^2 - c_i \, q_{it} \, r_{it-1} \right\} + \right. \\ & \left. + E_t \, \beta \left\{ (p_{t+1} + \tau_{it}) q_{it+1} - a_{it+1} q_{it+1} - \frac{1}{2} b_i \, q_{it+1}^2 - c_i \, q_{it+1} \, r_{it} \right\} \right], \quad \forall \, t \geq 0 \end{aligned} \tag{7}$$

Hence, the following conditions must hold for any interior (i.e., $r_{it} > r_{it-1}$) solution to (7):

$$\begin{aligned} & E_t \, [\, (b_i - c_i + 2e) r_{it+1} + e r_{jt+1} - [(\beta^{-1} + 1)(b_i - c_i + 2e) + (\beta^{-1} - 1) c_i] r_{it} - (\beta^{-1} + 1) e r_{jt} + \\ & + \beta^{-1} (b_i - c_i + 2e) r_{it-1} + \beta^{-1} e r_{jt-1} - (\xi_{it+1} + \tau_{it}) + \beta^{-1} (\xi_{it} + \tau_{it-1}) \,] = 0 \, ; \quad \forall t \geq 0; \; i,j = 1,2; \; i \neq j \end{aligned} \tag{8}$$

---

[4] See Stokey and Lucas (1989, pp. 280-283)

$$where \qquad \xi_{it} = d_t + a_{it} \tag{9}$$

$$\left[(\beta^{-1}+1)(b_i-c_i+2e) + (\beta^{-1}-1)c_i\right] \geq 0 \; ; \quad i=1,2 \tag{10}$$

Equations (8) are the Euler conditions and inequalities (10) are the Legendre conditions. According to the former, the ith firm produces up to the point where the net benefit accruing from its last unit in period t equals the expected discounted net benefit from producing that unit in period t+1. And the latter condition requires that the difference between these two net benefits is non-increasing at the margin.

Notice that the two Euler equations (8) can be written in matrix form as:

$$E_t\left(Ar_{t+1} - Br_t + \beta^{-1}Ar_{t-1} - D\,[\,\xi_{t+1} + \beta^{-1}\xi_t\,]-G\tau_{t-1}\right) = 0 \tag{11}$$

where $r_t=[r_{1t}\ r_{2t}]'$, $\xi_t=[d_t\ a_{1t}\ a_{2t}]'$, $\tau_t=[\tau_{1t}\ \tau_{2t}]'$, and the matrix coefficients of this equation are defined as follows :

$$A = \begin{bmatrix} b_1-c_1+2e & e \\ e & b_2-c_2+2e \end{bmatrix},$$

$$B = \begin{bmatrix} (\beta^{-1}+1)(b_1-c_1+2e)+(\beta^{-1}-1)c_1 & (\beta^{-1}+1)e \\ (\beta^{-1}+1)e & (\beta^{-1}+1)(b_2-c_2+2e)+(\beta^{-1}-1)c_2 \end{bmatrix}, \tag{12}$$

$$D = \begin{bmatrix} 1 & -1 & 0 \\ 1 & 0 & -1 \end{bmatrix}, \quad G=\begin{bmatrix} 1 & 0 \\ 0 & 0 \end{bmatrix}$$

At this point two remarks are in order. First, equation (6) represents in implicit form the best response functions of the two firms in the second stage of the game. The Euler conditions take this form with the third matrix being β^{-1} times the first matrix in the left hand side of (11). Alternatively, A is always a symmetric matrix. Second, matrix B can be expressed in terms of A and $C=\begin{bmatrix} c_1 & 0 \\ 0 & c_2 \end{bmatrix}$ as follows:

$$B = (\beta^{-1}+1)\,A + (\beta^{-1}-1)\,C \tag{13}$$

Clearly then the Legendre conditions require that the diagonal elements of that matrix be nonnegative. Now, we assume that the $\{\xi_t : t\in N\}$ process takes values in $\mathcal{H}$. That is, the space of sequences $\chi=(\chi_0, \chi_1, \chi_2, \ldots)$, $y=(y_0, y_1, y_2, \ldots)$ and so on, such that:

$$\sum_{t=0}^{\infty} \beta^t E_t\, \chi_t' y_t < \infty, \quad \chi_t, y_t \in \mathbf{R}, \quad \forall\, t\in \mathrm{N}. \tag{14}$$

Effectivelly, $\mathcal{H}$ is the space of sequences of all real valued stochastic vectors, the expected value of which cannot grow more than $\beta^{-1/2}$. In what follows we shall consider solving the equilibria of the two stages of the game in this space.

Moving now to the first stage of the game, the government chooses its subsidy plan to maximize (6), given the solution to the second stage game represented implicitly by (8). To solve this problem we first follow the methods used by Hansen et al. (1985) and transform (8), (which are now treated as constraints in the government's problem), so as to capture the dependence of the two firms' decisions on the anticipated decisions of the government.

$\mathcal{F}$ denotes the forward shift operator, defined by $\mathcal{F}^k E_t r_{it} = E_t r_{it+k}$, $\forall$ k≥0. Then the two Euler conditions in (8) can be written as:

$$E_t\,[\theta_i(\mathcal{F})r_{it-1}+\zeta(\mathcal{F})r_{jt-1}+\eta(\mathcal{F})(\xi_{it}+\tau_{it-1})] = 0 \quad \forall t\geq 0\ ; \quad i,j=1,2;\ i\neq j \tag{15}$$

where:

$$\theta_i(\mathcal{F})=k_i\mathcal{F}^2-((\beta^{-1}+1)k_i+(\beta^{-1}-1)c_i)\mathcal{F}+\beta^{-1}k_i$$

$$\zeta(\mathcal{F}) = e[\mathcal{F}^2-(\beta^{-1}+1)\mathcal{F}+\beta^{-1}] \quad -\beta^{-1}e(1-\mathcal{F})(1-\beta\mathcal{F})$$

$$\eta(\mathcal{F})=-(\mathcal{F}-\beta^{-1})$$

$$k_i=b_i-c_i+2e$$

Clearly, given the regularity condition:

[R1] $k_i\neq 0$ $(i=1,2)$

$\theta_i(\mathcal{F})$ can be factored as follows:

$$\theta_i(\mathcal{F}) \quad -\beta\sigma_i k_i(1-\beta\sigma_i\mathcal{F})(\mathcal{F}-\sigma_i) \tag{16}$$

where

$$\sigma_i+(\beta\sigma_i)^{-1}=(\beta^{-1}+1)+(\beta^{-1}-1)c_i k_i^{-1} \tag{17}$$

Since the norm of $\mathcal{S}$ in $\mathcal{H}$ is $\beta^{-1/2}$ (Cassing and Kollintzas (1991)), it follows that the necessary condition for the invertibility of the operator $(1-\beta\sigma_i\mathcal{S})$ in $\mathcal{H}$ is that $|\sigma_i|<\beta^{-1/2}$. A condition which is equivalent to:

[R2] $\qquad (\beta^{-1}-1)c_i > -(\beta^{-1/2}-1)^2 k_i$

Clearly, since $\beta\in(0,1)$, this condition is automatically satisfied when c_i and (b_i-c_i+2e) are positive. However, it can also be satisfied for negative c_i as long as they are not "too" negative.[5] Collecting results, we have:

<u>PROPOSITION 1</u>: Given [R1] and [R2] it follows that the Nash equilibrium rule of the second stage of the game is given by:

$$E_t q_{it}=\sigma_i q_{it-1}+\frac{\zeta(\mathcal{S})}{\beta\sigma_i k_i(1-\beta\sigma_i\mathcal{S})}E_t q_{jt-1}+\frac{\eta(\mathcal{S})}{\beta\sigma_i k_i(1-\beta\sigma_i\mathcal{S})}E_t(\xi_{it}+\tau_{it-1}) \qquad (i=1,2) \qquad (18)$$

Since $(1-\beta\sigma_i\mathcal{S})^{-1}=\Sigma_{t=0}^{\infty}(\beta\sigma_i)^t\mathcal{S}^t$ in $\mathcal{H}$, it follows that the current output decision of firms depend on the entire expected future path of subsidies. It is this dependence that makes the problem of the active government in the first stage non-recursive.

Having thus transformed the constraints we can now state the problem of the active government as choosing a contingency plan $\{q_{1t}, q_{2t}, \tau_{1t-1}\}_{t=0}^{\infty}$ so as to maximize (6) subject to (8), (2)-(5). Clearly, if $\{q_{1t}, q_{2t}, \tau_{1t-1}\}_{t=0}^{\infty}$ is a solution to the government's

---

[5]See Cassing and Kollintzas (1991) for further interpretations.

problem[6], then $\{r_{1t}, r_{2t}, \tau_{1t-1}\}_{t=0}^{\infty}$ must be a solution to the following:

$$
\begin{aligned}
\max_{r_{1t}, r_{2t}, \tau_{1t-1}} E_t \Bigg\{ & \Bigg[\Big[d_t - e\,(q_{1t}+q_{2t}) - a_{1t} - \frac{b_1}{2}\, q_{1t} - c_1 r_{1t-1}\Big]\, q_{1t} \Bigg] + \\
& + \beta \Bigg[\Big[d_{t+1} - e(q_{1t+1}+q_{2t+1}) - a_{1t+1} - \frac{b_1}{2}\bar{q}_{1t+1} - c_1 r_{1t}\Big]\, q_{1t+1} \Bigg] + \\
& + m_{1t} \Bigg[-(\mathcal{F}-\sigma_1) E_t r_{1t-1} + \frac{\zeta(\mathcal{F})}{k_1(1-\beta\sigma_1\mathcal{F})} E_t r_{2t-1} + \frac{\eta(\mathcal{F})}{k_1(1-\beta\sigma_1\mathcal{F})} E_t(\xi_{1t}+\tau_{1t-1}) \Bigg] + \\
& + m_{2t} \Bigg[-(\mathcal{F}-\sigma_2) E_t r_{2t-1} + \frac{\zeta(\mathcal{F})}{k_2(1-\beta\sigma_2\mathcal{F})} E_t r_{1t-1} + \frac{\eta(\mathcal{F})}{k_2(1-\beta\sigma_2\mathcal{F})} E_t(\xi_{2t}+\tau_{2t-1}) \Bigg] + \\
& + \beta m_{1t+1} \Bigg[-(\mathcal{F}-\sigma_1) E_t r_{1t} + \frac{\zeta(\mathcal{F})}{k_1(1-\beta\sigma_1\mathcal{F})} E_t r_{2t} + \frac{\eta(\mathcal{F})}{k_1(1-\beta\sigma_1\mathcal{F})} E_t(\xi_{1t+1}+\tau_{1t}) \Bigg] + \\
& + \beta m_{2t+1} \Bigg[-(\mathcal{F}-\sigma_2) E_t r_{2t} + \frac{\zeta(\mathcal{F})}{k_2(1-\beta\sigma_2\mathcal{F})} E_t r_{2t} + \frac{\eta(\mathcal{F})}{k_2(1-\beta\sigma_2\mathcal{F})} E_t(\xi_{2t+1}+\tau_{2t}) \Bigg] \Bigg\}
\end{aligned}
\tag{19}
$$

Differentiating this expression with respect to r_{1t}, r_{2t}, and τ_{1t-1} gives the Euler conditions[7]:

$$
\begin{aligned}
E_t \Big[& (b_1 - c_1 + 2e) r_{1t+1} \quad [(\beta^{-1}+1)(b_1 - c_1 + 2e) + (\beta^{-1}-1)c_1] r_{1t} + \beta^{-1}(b_1 - c_1 + 2e) r_{1t-1} \Big] + \\
& + e r_{2t+1} \quad (\beta^{-1}+1) e r_{2t} + \beta^{-1} e r_{2t-1} + (\beta^{-1}\xi_{1t} \quad \xi_{1t+1}) + \\
& + (-\beta^{-1} m_{1t} + (\beta\sigma_1)^{-1} m_{1t+1}) - [\beta^{-1} e (1-\mathcal{F})(1-\beta\mathcal{F})]\, m_{2t-1}^{E} = 0
\end{aligned}
\tag{20}
$$

$$
\textit{where} \quad m_{2t-1}^{E} \quad \frac{m_{2t-1}}{(1-\sigma_2\mathcal{F}^{-1})}
$$

[6]It should be clear that government 1 chooses only subsidies. The fact that r's and m's are in our problem is due to the fact that we have obtained the solution of the second stage game only in its implicit and not its explicit form.

[7]Note that since τ appears as τ_{1t-1} in the objective function, equation (22) holds for $t^{*} = t-1$ as well, i.e. $m_{1t-1} = E_t m_t$ for $t = 0, 1, 2 \ldots$

$$E_t \,[\![\, er_{1t+1} - e(\beta^{-1}+1)r_{1t} + \beta^{-1}er_{1t-1} + e(k_1)^{-1}[m_{1t}(-(\beta^{-1}+1)+\sigma_1^-)+m_{1t+1}] + \\ + [-(1-\beta\sigma_2\mathscr{F})(\mathscr{F}-\sigma_2)]\, m^E_{2t-1} \,]\!] = 0 \tag{21}$$

$$m_{1t} - E_t m_{1t+1} \tag{22}$$

where all equations hold for t=0,1,2..., and m_t=0 for negative values of t as in Hansen et al. The second order (Legendre) condition is:

$$B \geq 0; \quad i.e., \; B \text{ is positive semidefinite} \tag{23}$$

The ensuing five first order necessary conditions make up a system of five equations which have to be solved simultaneously in order to give the solution to the five unknowns (i.e. r_{1t}, r_{2t}, τ_{1t}, m_{1t}, m^E_{2t}). It turns out, however, that we only have to solve a three equation system, because two of the variables can be solved for independently[8]. This is done in three steps: In step

---

[8]The reason that enables us to solve in this fashion, i.e. first for m_{1t} then for r_{1t}, r_{2t}, and, m^E_{2t}, and subsequently for the subsidy, is simple: If we arrange the five equations into a matrix format we get the following :

$$\begin{bmatrix} \theta_1(\mathscr{F})+\zeta(\mathscr{F}) & \zeta(\mathscr{F}) & \zeta(\mathscr{F}) & 0 & k_1(-1+\beta\sigma_1^-\mathscr{F}) \\ \zeta(\mathscr{F}) & \theta_2(\mathscr{F}) & 0 & 0 & 0 \\ \zeta(\mathscr{F}) & 0 & \theta_2(\mathscr{F}) & 0 & e[-(\beta^{-1}+1)\sigma_1^-+\mathscr{F}] \\ \theta_1(\mathscr{F}) & \zeta(\mathscr{F}) & 0 & \beta^{-1}-\mathscr{F} & 0 \\ 0 & 0 & 0 & 0 & 1-\mathscr{F} \end{bmatrix} \begin{bmatrix} r_{1t-1} \\ r_{2t-1} \\ m^E_{2t-1} \\ \tau_{1t-1} \\ m_{1t-1} \end{bmatrix} - E_t \begin{bmatrix} -\beta^{-1}(1-\beta\mathscr{F})\xi_{1t} \\ -\beta^{-1}(1-\beta\mathscr{F})\xi_{2t} \\ 0 \\ -\beta^{-1}(1-\beta\mathscr{F})\xi_{1t} \\ 0 \end{bmatrix}$$

The coefficient matrix in the lhs of this equation is what we call "block recursive". This property depends on the zero entries of the coefficient matrix. For example, we said above that equation (22) gives by itself, the solution to m_{1t}. This is exactly what the zeroes in the fifth row imply as well. Also, the zeroes in the fourth and fifth columns, of the first three rows imply that we can solve for r_{1t},r_{2t},m^E_{2t}, independently of the tariff.

one, equation (22) gives the solution of m_{1t} which is substituted in all the other four equations, thus reducing the problem to four equations in four unknowns. In step two, we solve the best response function of firm 1 for the subsidy variable, and call this the "subsidy equation"

$$\tau_{1t-1}=-\beta(1-\beta\mathcal{F})^{-1}\,\theta_1(\mathcal{F})\,E_t r_{1t-1}-\beta(1-\beta\mathcal{F})^{-1}\zeta(\mathcal{F})\,E_t r_{2t-1}-E_t\xi_{1t} \tag{24}$$

which gives the subsidy as a function of r_{1t}, r_{2t} and m_{2t}^E. In step 3, we solve the remaining three equations, i.e. (20),(21) and the reaction function of firm 2, simultaneously to get the optimal paths for r_{1t}, r_{2t}, and m_{2t}^E. Substituting these optimal values in the subsidy equation gives the optimal subsidy plan. The "reduced" system of the three equations thus obtained, can be arranged in matrix format as follows:

$$\begin{bmatrix}\theta_1(\mathcal{F})+\zeta(\mathcal{F}) & \zeta(\mathcal{F}) & \zeta(\mathcal{F})\\ \zeta(\mathcal{F}) & \theta_2(\mathcal{F}) & 0\\ \zeta(\mathcal{F}) & 0 & \theta_2(\mathcal{F})\end{bmatrix}\begin{bmatrix}r_{1t-1}\\ r_{2t-1}\\ m_{2t-1}^E\end{bmatrix} - E_t\begin{bmatrix}-k_1(-m_{1t}+\beta\sigma_1^- m_{1t+1})-\beta^{-1}\xi_{1t}+\xi_{1t+1}\\ -\beta^{-1}\xi_{2t}+\xi_{2t+1}\\ -em_{1t}[-(\beta^{-1}+1)+\sigma_1^-]-em_{1t+1}\end{bmatrix} \tag{25}$$

The above system, if expanded so that terms are grouped by period, gives an equation similar to (11) :

$$E_t\,[\,Ar_{t+1}-Br_t+\beta^{-1}A_1r_{t-1}-Dw_{t+1}-\beta^{-1}Gw_t\,]=0 \tag{26}$$

where $r_t=[r_{1t}\ \ r_{2t}\ \ m_{2t}^E]'$, $w_t=[\xi_{1t}\ \ \xi_{2t}\ \ m_{1t}]'$, and the coefficient matrices are now defined as :

$$A = \begin{bmatrix} (b_1-c_1+2e) & e & e \\ e & (b_2-c_2+e) & 0 \\ e & 0 & (b_2-c_2+e) \end{bmatrix},$$

$$B = \begin{bmatrix} (\beta^{-1}+1)(b_1-c_1+2e)+(\beta^{-1}-1)c_1 & (\beta^{-1}+1)e & (\beta^{-1}+1)e \\ (\beta^{-1}+1)e & (\beta^{-1}+1)(b_2-c_2+e)+(\beta^{-1}-1)c_2 & 0 \\ (\beta^{-1}+1)e & 0 & (\beta^{-1}+1)(b_2-c_2+e)+(\beta^{-1}-1)c_2 \end{bmatrix},$$

$$G = \begin{bmatrix} -1 & 0 & \beta k_1 \\ 0 & -1 & 0 \\ 0 & 0 & -e(-(1+\beta)+\beta\sigma_1^-) \end{bmatrix}, \qquad D = \begin{bmatrix} 1 & 0 & -\beta k_1\sigma_1^- \\ 0 & 1 & 0 \\ 0 & 0 & -e \end{bmatrix} \tag{27}$$

Before we proceed to the solution of (26), it is crucial that we examine and discuss the rhs of (25). First, because of the initial condition that $m_{1,-1}=0$, we have that all the terms including m_1 become zero when the expectation operator passes inside the rhs vector expression (since due to linearity, we have $E_t\,[-k_1(-1+\beta\sigma_1\mathcal{F})m_{1t}] = k_1\,E_t m_{1t}+\beta\sigma_1\,E_t m_{1t+1} = k_1 0+\beta\sigma_1 0 = 0$). So, the rhs vector of equation (25) becomes $E_t[\,\{-\beta^{-1}(1-\beta\mathcal{F})\xi_{1t}\},\{-\beta^{-1}(1-\beta\mathcal{F})\xi_{2t}\},\ 0]'$. We solve (25) following the method described in Kollintzas (1985), since the third matrix in the lhs of (26) is β^{-1} times the first matrix in the same side. Thus we have the following:

<u>Proposition 2.</u> <u>CASE I</u> $c_i>0$ (exhaustible resources)

Given,

[R3] $k_1,\ k_2-e\ ,\ k_1(k_2-e)-e^2,\ k_1(k_2-e)-2e^2\neq 0$

and

[R4] $[(\beta^{-1}+1)k_1+(\beta^{-1}-1)c_1][(\beta^{-1}+1)(k_2-e)+(\beta^{-1}-1)c_2]-2(\beta^{-1}-1)e^2>0$

there exists a nonsingular matrix $T=[t_1\ t_2\ t_3]$ such that:

$T'AT= \mathrm{diag}\{h_1, h_2, h_3\}^{-1}=$

$$diag\left\{(k_2-e)((\beta^{-1}-1)c_2)^{-1},\ \frac{\frac{k_1}{c_1}+\frac{k_2-e}{c_2}+\left[\left(\frac{k_1}{c_1}+\frac{k_2-e}{c_2}\right)^2-\frac{4(k_1(k_2-e)-2e^2)}{c_1c_2}\right]^{1/2}}{2(\beta^{-1}-1)},\right.$$

$$\left.\frac{\frac{k_1}{c_1}+\frac{k_2-e}{c_2}-\left[\left(\frac{k_1}{c_1}+\frac{k_2-e}{c_2}\right)^2-\frac{4(k_1(k_2-e)-2e^2)}{c_1c_2}\right]^{1/2}}{2(\beta^{-1}-1)}\right\}$$

and $T'CT=I$.

Further given,

[R5] $h_i \in [-2\beta^{-1/2}, 2\beta^{-1/2}]$ $(i=1,2,3)$,

there exists a unique solution to (26) in $\mathcal{H}$, given by:

$$x_t = Hx_{t-1} + M(\mathcal{F})E_t w_t \tag{28}$$

where:

$H= T^{-1}\lambda T$;

$\lambda= \mathrm{diag}\{\lambda_1, \lambda_2, \lambda_3,\}$, λ_i is the smallest root of

$$\lambda^2-[(\beta^{-1}+1)+h_i]\lambda+\beta^{-1}=0 \quad (i=1,2,3) \tag{29}$$

and therefore are real and less than $\beta^{-1/2}$ in absolute value; and

$$M(\mathcal{T}) = \sum_{i=1}^{3} [\beta^{-1}+1+h_i]\, t_i t_i' \sum_{j=0}^{\infty} (\beta\lambda_i \mathcal{T})^j (D\mathcal{T}+\beta^{-1}G) \tag{30}$$

CASE II $c_1<0$ (learning by doing)

In this case we use the transformation $T'(-C)T=I$ instead of $T'CT=I$, so that $T'AT=\text{diag}\,\{h^*_1, h^*_2, h^*_3\}^{-1}$ becomes

$$\text{diag}\left\{(k_2-e)(-(\beta^{-1}-1)c_2)^{-1},\ \frac{-\frac{k_1}{c_1}-\frac{k_2-e}{c_2}+\left[\left(-\frac{k_1}{c_1}-\frac{k_2-e}{c_2}\right)^2-\frac{4(k_1(k_2-e)-2e^2)}{c_1c_2}\right]^{1/2}}{2(\beta^{-1}-1)},\right.$$

$$\left.\frac{-\frac{k_1}{c_1}-\frac{k_2-e}{c_2}-\left[\left(-\frac{k_1}{c_1}-\frac{k_2-e}{c_2}\right)^2-\frac{4(k_1(k_2-e)-2e^2)}{c_1c_2}\right]^{1/2}}{2(\beta^{-1}-1)}\right\}$$

The solution in this case is similar to that of CASE I, except for the fact that we now use $-h^*_i$ instead of h_i in equations (28), (29) and (30).

Proof: Given [R3], A is a nonsingular symmetric matrix, and given [R4], B is symmetric and positive definite. Further, given [R5], it follows that the eigenvalues of $A^{-1}B=[(\beta^{-1}+1)+A^{-1}C]$

have the stated properties. The rest follows from Kollintzas (1985). ■

Equation (28) gives the optimal paths for r_{1t}, r_{2t}, and m^E_{2t} which we call r^*_{1t}, r^*_{2t}, and m^{E*}_{2t}. When we substitute them into the subsidy equation (24) we get the optimal subsidy plan τ^*_{1t} as:

$$\tau^*_{1t-1} = -e(1-\mathscr{F})E_t r^*_{1t-1} - e(1-\mathscr{F})E_t(m^E_{2t-1})^* \qquad (31)$$

This equation can be easily computed and used to simulate the optimal subsidies.

PROPOSITION 3 . The optimal export subsidy, τ_{1t}, moves the equilibrium in the output space, in each and every period, to what would be, in the absence of a subsidy, a Stackelberg equilibrium with firm 1 being the leader.

Proof. Making firm 1 the Stackelberg leader, the game is reduced to a one-stage game (this stage corresponding to stage-two of our game) with no government present and hence no tariffs. The first order conditions for this game are:

$$E_t \begin{bmatrix} \theta_1(\mathscr{F})+\zeta(\mathscr{F}) & \zeta(\mathscr{F}) & \zeta(\mathscr{F}) \\ \zeta(\mathscr{F}) & \theta_2(\mathscr{F}) & 0 \\ \zeta(\mathscr{F}) & 0 & \theta_2(\mathscr{F}) \end{bmatrix} \begin{bmatrix} r_{1t-1} \\ r_{2t-1} \\ m^E_{2t-1} \end{bmatrix} = E_t \begin{bmatrix} -\beta^{-1}(1-\beta\mathscr{F})\xi_{1t} \\ -\beta^{-1}(1-\beta\mathscr{F})\xi_{2t} \\ 0 \end{bmatrix} \qquad (32)$$

But note now that this is exactly the system we get with the government setting the subsidy (see equation (16)). Since both games produce the same set of equilibrium conditions they obviously have the same set of solutions. ■

4. CONCLUSION

This paper is a step towards developing a general methodology for solving stochastic dynamic two-stage linear-quadratic games. The model we employ as a vehicle to deliver our ideas, is a stochastic dynamic extension of the the well-known problem of the strategic trade literature, whereby the domestic government seeks an optimal subsidy for a domestic firm that along with a foreign firm produce a homogeneous product for export to a third country (Brander and Spencer (1985)). This extension is achieved by introducing random shocks in the aggregate demand and production cost functions and by allowing each firm's production cost in any given period to depend on that firm's cumulative output up to the beginning of this period. When this dependence is negative, it is a manifestation of learning-by-doing technology (Clarke, Darrough, and Heineke (1982)). And, when this dependence is positive, it is a manifestation of the exhaustible nature of the underlying product (Hansen, Epple, and Roberds (1985)). The game has two stages. In the second stage of the game, the firms observe the realizations of the random aggregate demand and production cost shocks as well as the subsidy offered to one firm by its government and choose a contingency plan for their output in the current and every future period so as to maximize their expected discounted future stream of profits, taking each other's plan as given. In the first stage of the game the active government observes the realizations of the random aggregate demand and production cost shocks and chooses a contingency plan for the subsidy so as to maximize the expected discounted future stream of the domestic firm's profits minus the expected discounted future stream of the subsidy costs, subject to the rules that characterize the firms' behavior in the second stage of the game.

We follow the Linear Rational Expectations Equilibrium methodology (Hansen and Sargent (1980, 1981), Kollintzas (1985)) to show that, under a sufficient set of conditions on the parameters of the model, the underlying Nash equilibrium of the second stage of the game exists and depends on a weighted future stream of all expected future subsidies. This, of course, makes the problem of the active government in the first stage of the game non-recursive. Thus, in principle, there are two possible solutions to the problem of the active government. In the first case, the active government takes into account the effect of its announcement about its current and future subsidies on the current and future output decisions of firms. This is the time inconsistent case, whereby the active government has no incentive to adhere to its announcement (Kydland and Prescott (1976)), unless there is a commitment technology that forces it to do so. In the second case, the active government takes into account only the effect of its current subsidy on the current and future output decisions of firms. This is the time consistent case, whereby the active government follows a subsidy policy that does not intend to renege. In either case, the above mentioned Nash equilibrium laws of motion enter the problem of the active government as constraints.

Then, it was shown that, under no additional restrictions on the parameters of the model, the characteristic matrix polynomial associated with the characteristic equations of the Euler conditions that describe the time-inconsistent solution to the active government's problem have a stable factorization, making it possible to obtain an analytic representation of that solution. This solution has several noteworthy features.

First, the optimal (stochastic dynamic) subsidy policy has a similar effect as that in Brander and Spencer (1985), in the sense that it changes the behavior (best response function) of the domestic firm in what this response would have been if the domestic firm was not taking the foreign firm's output as given, but instead it was choosing its own output subject to the foreign firm's rule of behavior in the second stage of the game. That is, convert the domestic firm into been a Stackelberg leader in the market in question.

Second, the optimal subsidy for any given period is recursive, in the sense that it depends on the current state of the system. In particular, it depends on the current value of the multiplier associated with the rule of behavior of the foreign firm. This is a manifestation of the time-inconsistent nature of the solution.

Third, the evolution of the system can be computed numerically. Hence, the solution can serve to simulate the optimal subsidy policy.

Finally, although the results of this paper are derived only with explicit reference to the time-inconsistent equilibrium of the game, most of them apply to the time-consistent equilibrium as well. For, it can easily be shown that in this case the solution of the active government's problem in the first stage is given by a system such as (16). The difficulty of solving (17) is that it cannot be factored by a straightforward method as in Proposition 1.

REFERENCES

Aoki, M., 1989, Optimization of stochastic systems. San Diego, Academic Press.

Cassing, S. and Kollintzas, T., 1991, "Recursive dynamic interrelations and endogenous cycling", International Economic Review, 32, 417-440.

Clarke, F. H., Darrough M. N., and Heineke, J. M., 1982, "Optimal pricing policy in the presence of experience effects", Journal of Business 53, 51-67.

Dixit, A., 1984, "International trade policy in oligopolistic industries", Economic Journal 94 (Supplement), 1-16.

Eaton, J., and Grossman G. M., 1986, "Optimal trade and industrial policy under oligopoly", Quarterly Journal of Economics, 101, 383-406.

Gatsios, K., 1989, "Imperfect competition and International trade", in F. H. Hahn (ed.): The Economics of Missing Markets, Information and Games. Oxford University Press.

Kollintzas, T., 1985, "The symmetric linear rational expectations model", Econometrica, 53, 963-976.

Kollintzas, T., 1986, "A nonrecursive solution for the linear rational expectations model", Journal of Economic Dynamics and Control, 10, 327-332.

Kollintzas, T., (Ed.) 1989, The Rational Expectations Equilibrium Inventory Model, New York, Springer-Verlag.

Kydland, F. and Prescott, E. C., 1977, "Rules rather than discretion: The inconsistency of optimal plans", Journal of Political Economy, 85, 619-637.

Lucas, R. E. Jr., and Sargent, T. J. (Eds), 1981, Rational Expectations and Econometric Practice, Minneapolis, The University of Minnesota Press.

Sargent, T. J., (Ed.), 1985, Energy, Foresight & Strategy, Washington, Resources for the Future, Johns Hopkins University Press.

Sargent, T. J., 1989, Macroeconomic Theory, New York, Academic Press.

Spencer, B. J., and Brander, J. A., 1983, "International R&D rivalry and industrial strategy", Review of Economic Studies, 50, 707-722.

Spencer, B. J., and Brander J. A., 1985, "Export subsidies and International market share rivalry", Journal of International Economics, 18, 83-100.

Stokey, N. L., and Lucas, R. E. Jr., 1989, Recursive Methods in Economic Dynamics, Cambridge, Harvard University Press.

www.ingramcontent.com/pod-product-compliance
Lightning Source LLC
LaVergne TN
LVHW060543180926
848329LV00007B/43

* 9 7 8 9 8 1 0 2 1 2 9 2 6 *